Greenhouse Technology and Management, 2nd Edition

My GH (Hot Climate)

→ Monitors
- RH
- CO_2
- Radiation (transmitivity)
- Soil pH
- Temperature (TC under shade + air)
- Soil temp.

→ White mulch

→ Along wind direction

→ Geothermal for temp maintenance
- N-S orientation (if wind permits)

→ transfer 30-60 GH volume/hr to get CO_2 inside consumed by plants during day Pg 40

→ Place near bangalore (outskirts)

To Elena, Nicolás and Carmen for their support.

Doubting is the principle of sapience.
Aristotle

Nothing would ever be discovered if we were satisfied with the discovered things.
Seneca

Greenhouse Technology and Management, 2nd Edition

Nicolás Castilla, PhD

Research Coordinator, Department of Horticulture
IFAPA (Institute for Agricultural Research and Training), Granada, Spain
E-mail: nicolas.castilla@juntadeandalucia.es

Based on the second edition of the book
Invernaderos de Plástico: Tecnología y Manejo
by Nicolás Castilla, PhD
Published by Ediciones Mundi-Prensa, Madrid (Spain) and Mexico

Translated by

Esteban J. Baeza, PhD

Agricultural Engineer
IFAPA, Spain

Reviewed by

A.P. Papadopoulos, PhD

Senior Research Scientist, Greenhouse Crops
Agriculture and Agri-Food Canada, Harrow, Ontario, Canada
Adjunct Professor (Laval and Guelph Universities, Canada)
E-mail: papadopoulost@agr.gc.ca

CABI is a trading name of CAB International

CABI
Nosworthy Way
Wallingford
Oxfordshire OX10 8DE
UK

Tel: +44 (0)1491 832111
Fax: +44 (0)1491 833508
E-mail: info@cabi.org
Website: www.cabi.org

CABI
38 Chauncey Street
Suite 1002
Boston, MA 02111
USA

Tel: +1 800 552 3083 (toll free)
Tel: +1 (0)617 395 4051
E-mail: cabi-nao@cabi.org

A catalogue record for this book is available from the British Library, London, UK.

Library of Congress Cataloging-in-Publication Data

Castilla, Nicolás.
[Invernaderos de plástico. English]
Greenhouse technology and management by / Nicolás Castilla; translated by Esteban J. Baeza; reviewed by A.P. Papadopoulos.
p. cm.
Translation of the second ed.: Invernaderos de plástico: tecnología y manejo.
Includes bibliographical references and index.
ISBN 978-1-78064-103-4 (alk. paper)
1. Greenhouses. 2. Plastics in agriculture. I. Title.

SB415C37813 2012
635.98′23--dc23

2012013384

ISBN: 978 1 78064 103 4

Commissioning editor: Sarah Hulbert
Editorial assistant: Alexandra Lainsbury
Production editor: Tracy Head

Typeset by SPi, Pondicherry, India.
Printed and bound by CPI Group (UK) Ltd, Croydon, CR0 4YY.

Contents

The colour plates can be found following p. 178.

Foreword

The introduction of polyethylene (PE) films for agriculture use in the 1960s produced a tremendous change in out-of-season production of vegetables and flowers in southern Europe, especially in the frost-free coastal regions around the Mediterranean Sea.

PE films allowed the construction of very simple and cheap greenhouse structures well adapted to the so-called mild-winter climate and to the incipient technology locally available. Growers' ingenuity turned the new mild-winter greenhouses into a very efficient tool for the production of vegetables and flowers during the winter season. This was the start of a 'plastic boom', which was fuelled by produce exports and increasing home consumption.

The environmental control of mild-winter greenhouses had many specificities and strong limitations. Dr Nicolás Castilla was one of the first scientists to realize that this new type of structure should be optimized in order to improve crop productivity and produce quality. With this in mind he soon initiated a successful research and development programme at the Experimental Station 'Las Palmerillas', Almeria, Spain. This book incorporates the long experience of the author in mild-winter greenhouse production systems where the most successful solutions are those that integrate greenhouse design and environmental control with crop husbandry.

The content is focused on plastic greenhouses for vegetable growing but includes cross references to glasshouses, high-tech environmental control methods and other cultivation systems when these are needed to fill some gaps and to justify the basic concepts. The major attention is on the principles and techniques of greenhouse technology and management. However, plastic mild-winter greenhouses cannot be dissociated from local socio-economic conditions and crop characteristics. This explains why this book is not restricted to greenhouse design and environmental control but also deals with integrated production, crop physiology, soilless cultivation, irrigation and fertilization, plant protection, postharvest, economics, marketing and production strategies.

Due to the attention devoted to the principles and the diversity of technological solutions the content of this book is also applicable to many other regions of the world where plastic greenhouses are currently used. It is well written and profusely illustrated, and combines theoretical, technical and practical information.

After the success of the Spanish editions the English version is expected to attract a wider audience. It is an excellent tool for specialists on greenhouse cultivation, an original textbook for students and an occasional resource for experienced growers.

António A. Monteiro
President of the International Society for Horticultural Science

Preface to the First Edition (Spanish)

Greenhouse cultivation has expanded during the last few decades around the world to different areas of 'mild winter climate', an expression coined by Portuguese colleagues Carlos Portas and António Monteiro in December 1985, on the occasion of an ISHS (International Society for Horticultural Sciences) Symposium, held in Faro (Portugal). These mild winter climate conditions are characteristic of the Mediterranean region but are not exclusive to the countries of the Mediterranean Basin.

This book is mainly focused on plastic greenhouses, the majority in Spain, and chiefly oriented towards the cultivation of vegetable crops. The reader will often find references to works carried out in the north of Europe and other areas with a tradition of growing in sophisticated greenhouses, when it is necessary to fill some gaps in information.

The text is aimed at both qualified horticulturists and field technicians, as well as at students and specialists in greenhouse cultivation.

It is not a general treatise on protected horticulture, enumerating and describing the growing techniques, providing recipes on fertilization, plant protection or other similar aspects. It deals with the current technology and the management of plastic greenhouses, describing the principles on which they are based, with a pragmatic and quantitative approach, when possible, and also dealing with other interesting aspects within the integrated approach of protected cultivation (economic analysis, marketing, production strategies). I trust that it may contribute to improving the protected cultivation agricultural systems in plastic greenhouses of very diverse areas.

The encouragement received from José María Hernández, director of Mundi-Prensa, during the last few years and from Joaquín Hernández during the slow writing process has been very important.

Among the numerous people who have provided suggestions and data contributing to enrich the text, I would like to highlight Juan Ignacio Montero, Carmen Giménez, Marisa Gallardo, Rodney Thompson, Javier Calatrava, Enrique Espí, Tomás Cabello, Julio Gómez, Isabel Cuadrado, Jan van der Blom and the scientists from Caja Rural of Granada (Ignacio Escobar) and from Cajamar (Jerónimo Pérez Parra and his team: Juan Carlos López, Juan José Magán, Esteban Baeza, Juan Carlos Gázquez and Guillermo Zaragoza). The expertise of Laura García Quesada, with the collaboration of Teresa Soriano, allowed the figures to be made more intelligible. The contributions of some figures from Juan Ignacio Montero, Jerónimo Pérez Parra and Joaquín Hernández and of several photographs from Jan van der

Blom have contributed to improve the comprehension of the text. A special mention must be dedicated to Maribel Morales, for the tedious task of transcription and compendium, and to Joaquin Hernandez, for reviewing the whole text. To all of them, I express my most sincere gratitude.

Finally, a special reference is dedicated to Carmen, my wife, and to Nicolás and Elena, my son and daughter, who suffered, with great understanding and support for several years, my partial absence from family life.

The sponsorship received, co-editing this work, from Caja Rural of Granada and Cajamar, highlights the excellent work with which both cooperative banks are contributing to the technological development of the agricultural sector.

Nicolás Castilla
Granada (Spain), June 2004

Preface to the Second Edition (Spanish)

The favourable reception, among readers interested in greenhouses, of the first edition of this book suggests maintaining its structure in this second edition.

Detected misprints have been corrected, some photographs have been substituted, new pieces of text have been incorporated and some data has been updated, which may contribute to maintaining the validity and accuracy of the text.

Among the people who have provided suggestions and information for this second edition I would like to highlight Pilar Lorenzo, Enrique Espí, Juan Ignacio Montero, Ignacio Escobar, Mari-Cruz Sánchez-Guerrero, Juan Carlos López, Jean Claude Garnaud and Kwen Woo Park. In addition, Carmen Cid, Tomás Cabello, Julio Gómez, Richard Vollebregt and Jan van de Blom kindly provided some photographs. To all of them I express my gratitude.

Moreover, I thank everyone who has honoured us by reading the pages of the first edition of this work and, very specially, those who favoured us with their comments in Spanish or foreign magazines and other publications.

Nicolás Castilla
Granada (Spain), June 2007

Acknowledgements

The author thanks very much the following people, who provided slides and photographs: Carmen Cid (Photo 4.1), Richard Vollebrecht (Photos 4.6 and 8.3), Jan van der Bloom (Photo 13.1, Plates 25 and 26) Tomas Cabello (Photos 13.2 and 13.3) and Julio Gomez (Photo 13.4). All other slides come from the author's archives.

Several figures and tables have been adapted, as indicated in each one of them, from different sources with copyright. Specific permissions were granted from the following publishers: FAO (Food and Agriculture Organization of the United Nations), INRA (Institut National de la Recherche Agronomique, France), Elsevier Science BV (Pergamon Press, Agricultural Meteorology and Elsevier), World Meteorological Organization (Switzerland), Wageningen Pers. (The Netherlands), Technique et Documentation (France), CTIFL (Centre Technique Interprofessionnel des Fruits et Légumes, France), Ediciones Omega (Spain), *Netherlands Journal of Agricultural Science* (The Netherlands), Cooperative Extension of NRAES (Natural Resource, Agriculture, and Engineering Service, Ithaca, New York) and Matias Garcia-Lozano (Spain). The author is grateful to all of them.

1

Protected Cultivation

1.1 Introduction

Protected cultivation is a specialized agricultural system in which a certain control of the soil–climate ecosystem is exercised modifying its conditions (soil, temperature, solar radiation, wind, humidity and air composition). Plants are cultivated by means of these techniques modifying their natural environment to prolong the harvesting period, alter the conventional cropping cycles, increase yields, improve product quality, stabilize production and provide products when open field cultivation is limited (Wittwer and Castilla, 1995).

The main goal of protected cultivation is to obtain high value products (vegetables, fruits, flowers, ornamentals and seedlings). The most relevant determining factor of horticultural production activity is the climate. Among the most important limitations for horticultural production are the low solar radiation conditions, the unfavourable temperature and humidity conditions, unfavourable water and nutrient levels, presence of weeds, excessive wind and an inadequate concentration of carbon dioxide (CO_2) in the air. The majority of the above-mentioned limitations are climatic factors or factors directly related to the climate, which may be altered by means of protected cultivation.

The lack of water is the most important limitation for all agricultural activity. Losses caused by drought are about equal to those induced by all other climatic factors together, including excess of water, floods, cold, hail and wind (Boyer, 1982). Irrigation is, without doubt, the most ancient method to protect crops (from drought) and has permitted agricultural activity in arid and desert regions that, without irrigation, would not be productive. Nowadays protected cultivation goes far beyond providing irrigation, to include several plant protection techniques, and has reached an enormous importance during the last century.

1.2 Types of Protection

All vegetable species have an optimal range for each environmental parameter. Placing a screen near the plant modifies the environmental conditions affecting the whole or part of the plant.

The position of the screen or other similar protection, in relation to the plant, determines the type of protection (CPA, 1992). When the screen is placed over the soil and under the aerial parts of the plant, we call it mulch (Photo 1.1). Lateral screens or forms of protection are referred to as windbreaks (Photo 1.2). When the screens

Photo 1.1. Black polyethylene mulch in a strawberry crop.

Photo 1.2. Semi-porous windbreak.

are located over the plants, as a cover, we have a third type of protection: greenhouses, tunnels and floating covers.

In the case of floating covers, also known as direct covers and floating mulches ('baches' in the French literature), the protection directly rests over the plants, without any kind of a supporting structure (Photo 1.3).

Low tunnels are tunnels normally up to 1 m in height (Photo 1.4). High tunnels or macro-tunnels are high enough to allow workers to walk inside them, and for the cultivation of species of a similar height

Photo 1.3. Mechanized placement of a textile cover.

Photo 1.4. Low tunnels.

(Photo 1.5). Greenhouses differ from the other forms of protection in being more solid and high and wide enough for cultivation of tall plants, even fruit trees (Photo 1.6). The distinction between high tunnels and greenhouses is not always well made and it is not uncommon to see high tunnels and greenhouses being referred to indiscriminately.

1.3 Objectives of Protected Cultivation

The general objective of protected cultivation is to modify the natural environment, through different techniques, to reach the optimal productivity of the crops, increase yields, improve product

Photo 1.5. High tunnels.

Photo 1.6. Plastic film multi-span greenhouse.

quality, extend the harvest period and expand the areas of production (Wittwer and Castilla, 1995). In some regions, the reduction of solar radiation (shading), or the protection against wind, hail or rain are also objectives of protected cultivation. It is also intended to make a more efficient use of soil, water, energy, nutrients and space, as well as climatic resources such as solar radiation, temperature, humidity and CO_2 in the air (Wittwer and Castilla, 1995).

Besides protection of the crops against drought by irrigation, as outlined by Wittwer and Castilla (1995), other objectives of protected cultivation are to:

- *Reduce water use.* The use of different types of mulching (organic wastes, gravel, sand, plastic film, etc.) allows for a decrease in water losses by evaporation and prevents growth of weeds (which compete for the available water in the soil). The use of greenhouses, tunnels and other forms of protection, which limit solar radiation, allow for a reduction in the water requirement of plants and for a more efficient use of irrigation water.
- *Protect crops from low temperatures.* The use of individual forms of protection for each plant, or plant rows or whole plots, by means of individual 'caps', tunnels, direct covers or greenhouses, are typical examples. It is worth mentioning other complementary techniques, with this same objective, such as the use of wind machines (to mix the air layers and prevent the stratification of cold air close to the earth surface where plants are grown), anti-frost sprinkle irrigation, smoke generators or the use of burners to heat open field orchards.
- *Decrease wind velocity.* The use of windbreaks, both hedges and specific structures made with natural (dried cane, bamboo) or artificial materials, is not the only technique to reduce wind velocity, because other forms of protection may also provide a very important windbreak effect (i.e. tunnels and greenhouses).
- *Limit the impact of arid and desert climates.* In greenhouses located in arid or desert regions, the insulation from the outside environment allows for the generation of a proper microclimate for horticultural production. This is the case in many areas of the Mediterranean Basin, the Middle East and Africa, Australia and America (Mexico, the USA).
- *Decrease damage caused by pests, diseases, nematodes, weeds, birds and other predators.* In an isolated environment, such as a greenhouse, it is easier to provide plant protection (biological control, for instance). It is possible to fumigate the soil and the atmosphere, substitute the soil or use artificial substrates (e.g. growing plants in bags, rockwool and other sorts of hydroponics), as a means to fight against soil-borne diseases, pests, nematodes and weeds. The use of nets or screens to prevent damage caused by pests is quite efficient, as is the use of direct covers with textile or non-woven materials. Using a mulch for solarization (soil disinfection with solar heat) is an efficient technique without environmental impact, but it is only feasible in regions that have high solar radiation.
- *Extend areas of production and growing cycles.* The use of greenhouses of varying levels of sophistication, tunnels and mulches have increased yields all over the world enabling horticultural production in new areas and extending availability of many products outside their traditional periods of consumption.
- *Increase yields, improve product quality and preserve resources.* In addition to the increase in yields achieved with protected cultivation, the use of resources (soil, water, solar radiation, energy and atmospheric CO_2) is more efficient than with conventional cultivation. Besides, with protection against the wind, rain, hail, cold and the attack of insects and other pests protected cultivation also results in a better quality harvest.
- *Climate control allows for maximization of yields and optimization of product quality.* In greenhouses, the management of temperature and ambient humidity, as well as atmospheric CO_2 and light, allow for significant improvements in the yield and quality of horticultural products.
- *Stabilize the supply of high quality products to horticultural markets.* Protected cultivation avoids many of the risks of conventional horticulture and facilitates a regular supply to the markets, extending the marketing calendars for many species.

In order to achieve these objectives a higher investment than in conventional cultivation is normally required, as well as a higher use of inputs, which may imply a higher environmental impact if it is not properly managed. Modern greenhouse crop production is justifiably characterized as a high investment, high technology, high risk business.

1.4 History

The first documented attempts at protected cultivation, as recorded by the historian Columella, date back to the Roman Empire, during the reign of Emperor Tiberius Caesar, when small mobile structures were used for the cultivation of cucumber plants, which were taken in to the open air if the weather was good or kept under cover when the weather was inclement (Wittwer and Castilla, 1995). Sheets of mica and alabaster were used as enclosure materials. The philosopher Seneca considered these practices unnatural and condemned their use. These growing methods disappeared with the decline of the Roman Empire (Dalrymple, 1973) until the Renaissance (from the 16th to the 17th century) when the first precursors of greenhouses appeared, initially in England, The Netherlands, France, Japan and China (Enoch and Enoch, 1999). They were very rudimentary structures made with wood or bamboo, covered with glass or oiled paper panes, or glass bells to cover hot beds (Wittwer and Castilla, 1995). Later, in the northern hemisphere the first lean-to type greenhouses were built facing south, using a brick wall on the north side. The first attempts to use heating took place in such structures. During the night the plants were protected with straw and/or reed blankets, as insulators. Their use was very limited, for instance in botanical gardens (Photo 1.7). During the 19th century, the first gable-frame greenhouses appeared and the cultivation of grapes, melons, peaches and strawberries became common; by the end of that century tomatoes were introduced (a fruit vegetable that, years before, was considered poisonous).

Soon, the use of greenhouses expanded from Europe to America and Asia, appearing in areas neighbouring great cities (Enoch and Enoch, 1999). In the 20th century, economic development, especially after the Second World War, boosted the construction of glasshouses. By the middle of the century there were more than 5000 ha of glasshouses in The Netherlands, mostly

Photo 1.7. Traditional greenhouse (Brussels Botanic Garden).

devoted to tomato cultivation (Wittwer and Castilla, 1995).

However, it was the arrival of plastic films that facilitated an enormous expansion of the greenhouse industries in Asia (mainly Japan, Korea and China) and in the countries around the Mediterranean (with Spain and Italy leading in terms of total area). In Europe, the energy crisis and the introduction of plastics contributed to the partial shifting of greenhouse vegetable production from northern countries (mainly The Netherlands) to the Mediterranean Basin, where low-cost plastic greenhouses allowed for low-cost production of out-of-season vegetables (Castilla, 1994). Improvements in logistics facilitated the distribution of the products in the national and European markets, where demand was increased by economic development.

In parallel, a progressive change occurred in the greenhouse industries of Northern Europe where the cultivation of cut flower and ornamental plants, increased to the detriment of vegetable cultivation.

There are two basic greenhouse concepts (Enoch, 1986). The first one (typical of Northern Europe) aims at achieving maximum climate control to maximize productivity, requiring the use of sophisticated greenhouses. The second concept pursues minimum climate control using low technology greenhouses, making production possible under modified, but non-optimal, conditions at a low cost, and it is typical of Mediterranean-type greenhouses. Obviously, there are different gradations between these two extreme concepts. The type of greenhouse selected mainly depends on: (i) these two concepts (maximum or minimum climate control); (ii) the type of species to be cultivated; (iii) the locality; and (iv) the prevailing socio-economic conditions.

The supply of fresh fruits, vegetables and flowers which consumers are demanding, may be achieved in three ways (Enoch and Enoch, 1999): (i) growing in greenhouses which are near to centres of consumption; (ii) storing the products after their harvest, to sell them later; and (iii) transporting the products from other climatic regions, where they are naturally produced (in open air), to the consumption centres.

Nowadays, these three procedures not only coexist, but a hybrid method of production has become predominant in which produce is grown in greenhouses in mild climate areas, such as the Mediterranean, and transported to the big European consumption centres.

1.5 Importance

Windbreaks were the first type of protection used in agriculture and, although there are no precise statistics about their use, they are still very important all over the world.

The availability of plastic films has permitted the widespread use of mulch on many crops in some Mediterranean countries and, especially, in East Asia (China, Japan and Korea) (Table 1.1).

Table 1.1. Estimated areas of protected cultivation in the world in 2010 (adapted from Castilla and Hernández, 1995, 2005, 2007; Ito, 1999; Castilla *et al.*, 2001, 2004; Castilla, 2002; Jouet, 2004; Espi *et al.*, 2006; Park, K.W., 2006, personal communication; Zhang, 2006; Schnitzler *et al.*, 2007; Kan *et al.*, 2012).

	Geographical area (thousands of ha)					
Protection	Asia	Mediterranean	Rest of Europe[a]	America	Others	Total
Mulching	9,870	402	65	265	15	10,617
Direct cover	22	16	39	13	15	105
Low tunnel	1,505	133	9	20	5	1,672
Greenhouse[b]	1,630	201	45	25	4	1,905

[a]Excluding Mediterranean countries.
[b]Includes high tunnels.

Low tunnels allow for a temporary protection of crops and have also developed mainly in the Mediterranean area and in Eastern Asia (Tables 1.1 and 1.2).

Direct covers, in the absence of a structure to support them, are a simple semi-protection technique that is inexpensive and effective. The area protected by direct cover is relatively minor being limited to low height crops, but is increasing (Table 1.1).

High tunnels, which comprise those structures in which all crop-related work is done inside them, are included in the greenhouse group, because in fact they are a simplified variant of greenhouses.

The greenhouse industry in the Mediterranean area (which includes all the Mediterranean coastal countries and Portugal) covered 65,000 ha in 1987 (Nisen *et al.*, 1988), leading the world, while in 2006 it had exceeded 200,000 ha (Table 1.2) mainly in greenhouses covered with plastics (Photo 1.8). Plastic greenhouses in countries like Spain represent nearly 99% of the total greenhouse area (Photo 1.9), estimated at 53,843 ha in 2005 (Table 1.2), which was double the area that existed a decade earlier (Castilla, 1991).

Table 1.2. Global distribution of greenhouses and low tunnels in the Mediterranean area (2006) (adapted from Castilla, 2002; Jouet, 2004; Castilla and Hernández, 2005; Schnitzler *et al.*, 2007).

	Greenhouses (ha)	Low tunnels (ha)
Spain	53,843	13,055
Italy	42,800	30,000
Turkey	30,669	17,055
France	11,500	15,000
Morocco	11,310	3,770
Egypt	9,437	25,000
Israel	6,650	15,000
Algeria	6,000	200
Former Yugoslavia	5,040	–
Greece	5,000	4,500
Syria	4,372	50
Lebanon	4,000	700
Libia	3,000	–
Portugal	2,700	100
Jordan	1,989	718
Tunisia	1,579	7,316
Albania	415	–
Cyprus	280	280
Malta	55	102
Total	**200,639**	**132,846**

Photo 1.8. The Mediterranean area; the distribution of greenhouses has increased in this area during the last few decades.

Photo 1.9. Plastic greenhouses in the Poniente area in Almeria; plastic greenhouses in Spain represent nearly 99% of the total greenhouse area.

In Japan, the leading country in greenhouse production in the past, glasshouses represented only 5% of the total area (Ito, 1999). Similarly, in China the majority of greenhouses are those with plastic covers; there has been a spectacular growth in these since 1980 – their surface area in 2010 stood at 1,496,000 ha, of which over half belonged to the 'lean-to' type (see Fig. 7.5 in Chapter 7) and the rest to high (2.5–3.0 m ridge height) and middle (1.8–2.5 m ridge height) tunnels, while the glasshouse area was minimal (Zhibin, 1999; Zhang, 2006; Kan *et al.*, 2012). The figures for tunnels can vary as they can be erected or dismantled based on production needs from year to year (Kan *et al.*, 2012).

Glasshouses were in majority only in some areas of North America and in Northern Europe, where they represented 90% of the surface in Germany and up to 98% in The Netherlands, for a total of less than 25,000 ha within the European Union (Von Elsner *et al.*, 2000a,b).

1.6 Plastic Materials

Rather than the result of specific scientific research work, the birth of 'plasticulture' (the use of plastic materials in agriculture) was a consequence of the development of new plastic materials that coincided with a series of circumstances in agriculture (the need to decrease investment costs and to secure the harvests, and the scarcity of resources such as water).

Since the first plastic greenhouses, a simple wooden structure covered with cellophane built in 1948, the appearance of polyethylene plastic film (a material previously used only in military applications) in the 1950s in the US market and the assembly in 1962 (in Israel) of the first drip irrigation installation of a significant dimension (10 ha), the expansion of plastic applications has been enormous, especially in protected horticulture.

Among the advantages provided by plastics worth noting are: (i) its lightness, because of its low density (so for instance, 1 m^2 of polyethylene film of 25 μm thickness covering a greenhouse weighs 100 times less than 1 m^2 of horticultural glass of 4 mm thickness); (ii) its good mechanical resistance (e.g. to hail) as compared with glass; (iii) its durability (it resists corrosion by chemical agents such as fertilizers and biological agents such as bacteria and fungi); (iv) its safety for plants and animals; (v) its impermeability to water and gases; and (vi) its transparency to light.

Its low cost, in particular, has permitted the displacement of traditional materials in some applications (in greenhouses or mulches) and the generation of new uses that previously did not exist, such as tunnels, direct covers or drip irrigation, among others.

A good knowledge of the limitations of the use of plastics (see Chapter 4) will allow for better use of them: for instance, under extreme conditions of temperature (the thermal stability of some plastics at low or high temperature is not satisfactory). The static electricity of some formulations, such as ethylene vinyl acetate (EVA) or plasticized polyvinyl chloride (PVC), induces the accumulation of dust, which decreases transparency to light. The same effect is caused by scratches on some materials if they are not properly protected from the impact of winds carrying sand. Other aspects, such as the ageing of plastics (which affects how long they can be used) and their behaviour against fire, must be considered for optimum use.

In protected cultivation the solar radiation transmission properties of the plastic materials are of key importance for plant growing (see Chapter 4).

1.7 Summary

- The name 'protected cultivation' involves a series of techniques for the modification of the natural environment of plants, which totally or partially alter the microclimate conditions, with the aim of improving their productive performance.
- Among the protected cultivation techniques, it is worth noting windbreaks, mulches, low tunnels, direct covers, high tunnels and greenhouses.
- The main objectives of protected cultivation are, among others, to: protect the crops from harmful temperatures, wind, rain, hail and snow, as well as from pests, diseases and predators, creating a microclimate that allows for the improvement of their productivity and quality, contributing to a better use of resources.
- Protected cultivation has been used for many centuries; references to the use of protection date back to Roman times.
- The development of plastic materials has contributed to the widespread use of greenhouses and other protection techniques, from the last third of the 20th century all over the world.
- The estimated protected cultivation area worldwide in 2010 was 1,905,000 ha of greenhouses, 1,672,000 ha of low tunnels and floating covers and over ten million ha of mulches. The huge increase in area under protected cultivation in recent decades was due to the enormous spread in Asia, mainly in China.

2

The External Climate

2.1 Introduction

The local climatic conditions, to a large extent, determine the microclimate inside a greenhouse and its future management; therefore knowledge of the prevailing climatic conditions is necessary before designing and building a greenhouse.

The climate of a certain locality is the result of the radiative exchanges between the Sun and the Earth. In relation to greenhouses, the most important elements of the climate are: (i) solar radiation; (ii) atmospheric temperature and humidity; (iii) wind; and (iv) rainfall.

2.2 The Earth and the Sun

2.2.1 Introduction

The Earth's axis, around which our planet turns on itself, maintains a fixed inclination with respect to the plane of the Earth's orbit around the Sun, called the ecliptic (Fig. 2.1). The axis also maintains a fixed direction, that is, the Earth's axis points continuously to a fixed point in the sky. The 23° 27′ angle formed by the Earth's axis and the perpendicular to the plane of the ecliptic is called the obliquity (angle) of the ecliptic (Fig. 2.1), which determines the Earth's decline (see Appendix 1 section A.1.1).

2.2.2 The seasons

The movement of the Earth around the Sun determines the year's seasons. When the plane perpendicular to the ecliptic which contains the Earth's axis passes through the centre of the Sun, which happens twice a year, the summer and winter solstices occur (Figs 2.2, 2.3 and 2.4).

Between the two solstices we have the spring and autumn equinoxes (Fig. 2.5), moments in which an imaginary line linking the centre of the Earth to the centre of the Sun is perpendicular to the Earth's axis. During the equinoxes neither of the two poles is inclined towards the Sun (Fig. 2.5).

The winter solstice in the northern hemisphere (Fig. 2.3) corresponds to the summer solstice in the southern hemisphere. At the winter solstice at noon the Sun reaches an apparent elevation (maximum of the day) of around 30° in the south of Spain (latitude 37°N) and 90° in the Tropic of Capricorn. On the summer solstice (Fig. 2.4) the apparent maximum Sun elevation is around 76° in the latitude 37°N (south of Spain) and 90° in the Tropic of Cancer.

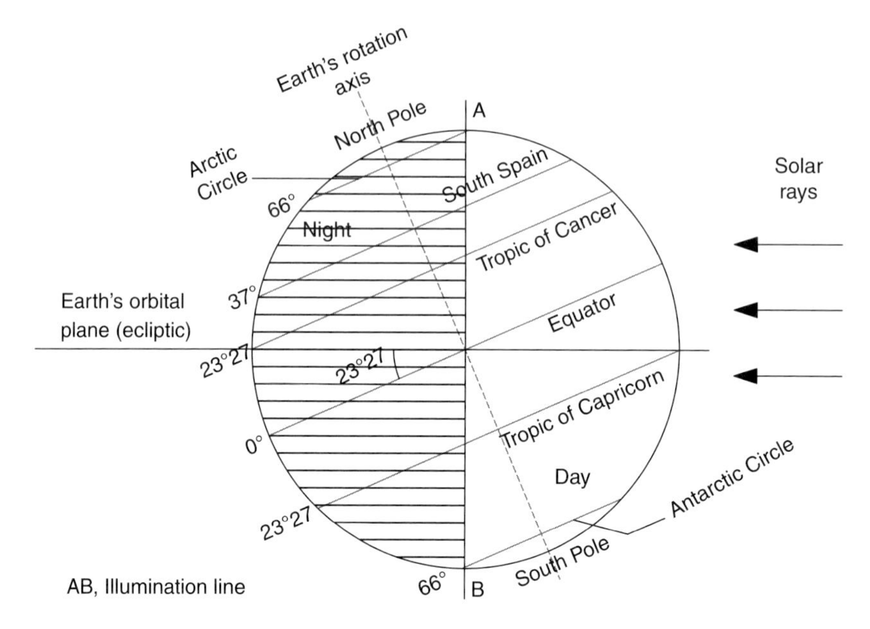

Fig. 2.1. The inclination of the Earth's axis of rotation with respect to the Earth's orbital plane.

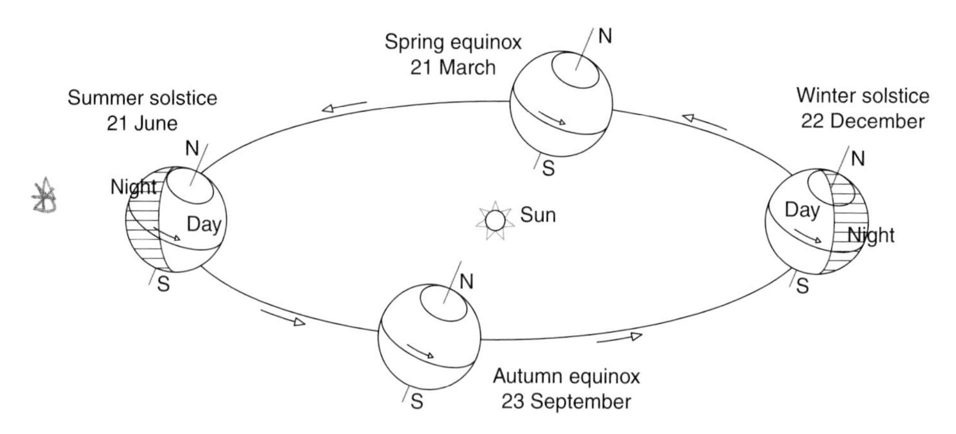

Fig. 2.2. The year's seasons in the path of the Earth's orbit around the Sun.

2.3 Day Length

The astronomic duration of the day throughout the year is variable, and the degree of variability is proportional to the latitude of the locality. The astronomic day is the period between sunrise and sunset, sunrise being the moment at which the solar disc appears on the horizon and sunset the moment at which it disappears. The differences in the duration of the day between different latitudes are greater in winter than in summer (Table 2.1).

In the equinoxes the duration of the day equals that of the night for each latitude, whereas in the solstices the duration of the day is maximum in the summer and minimum in the winter.

For medium latitudes, the natural light threshold which influences the photoperiodic phenomena is overcome during the 'civil twilight', the time after sunset and before sunrise when the Sun is below the horizon but not more than 6° below it (Berninger, 1989). The duration

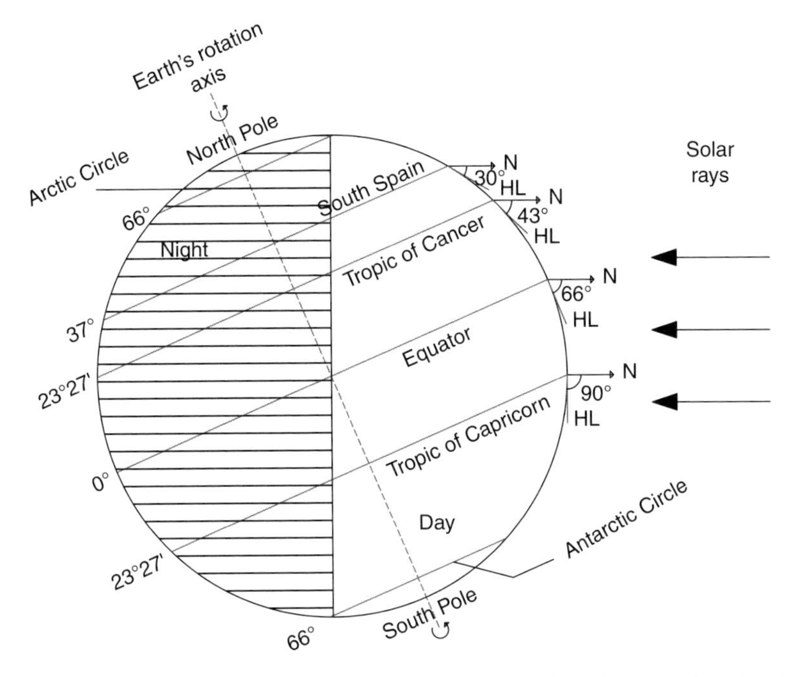

Fig. 2.3. The incidence of the solar rays on the Earth in the winter solstice in the northern hemisphere. N, Apparent position of the Sun at noon at different latitudes; HL, horizontal line on each latitude.

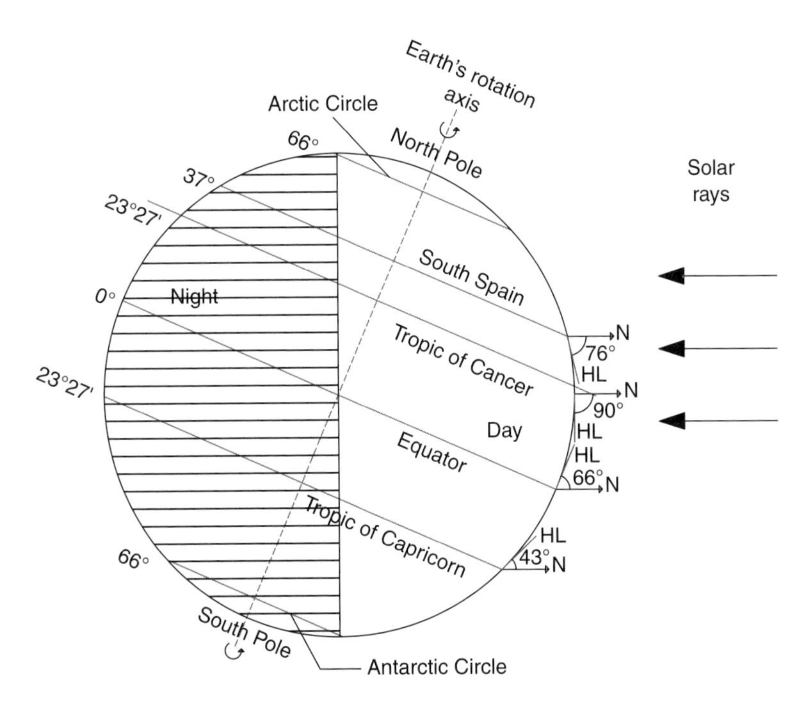

Fig. 2.4. The incidence of the solar rays on the Earth in the summer solstice in the northern hemisphere. N, Apparent position of the Sun at noon at different latitudes; HL, horizontal line each latitude.

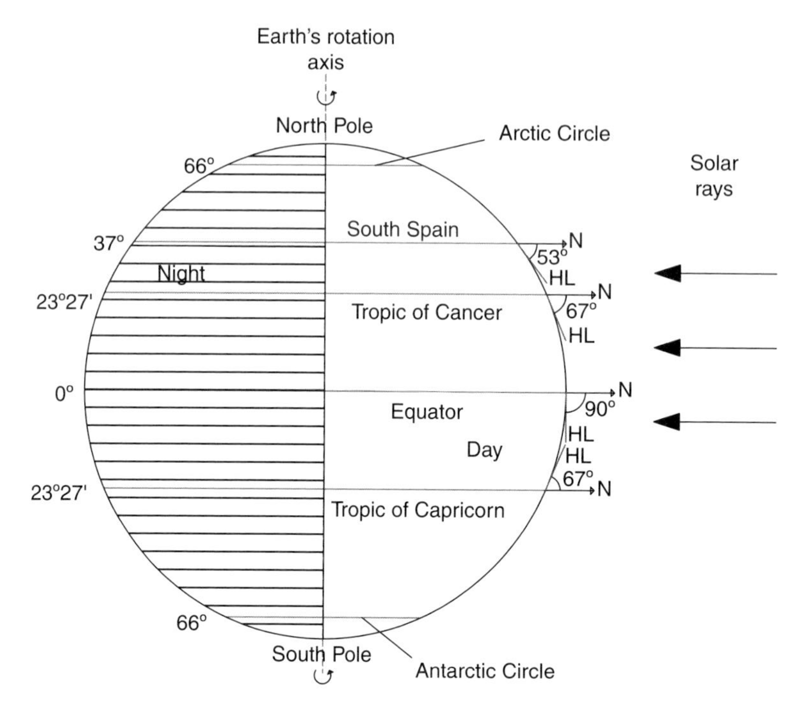

Fig. 2.5. The incidence of the solar rays on the Earth in the equinoxes. N, Apparent position of the Sun at noon at different latitudes; HL, horizontal line on each latitude.

Table 2.1. Duration of the astronomic day (expressed in hours and tenths of an hour) depending on the latitude.

	Month			
Latitude[a]	March	June	September	December
52°N	11.6	16.4	12.4	7.6
44°N	11.7	15.2	12.3	8.7
36°N	11.9	14.6	12.3	9.7

[a]Latitudes correspond to The Netherlands (52°N), the south of France (44°N) and the south of Spain (36°N).

of the 'photoperiodic day' for medium latitudes corresponds to 1 astronomic day increased by 40 min to 1 h (Berninger, 1989).

2.4 Solar Radiation

2.4.1 Introduction

All bodies emit radiation at wavelengths which depend on their temperature. The warmer the body, the greater the amount of energy emitted and the shorter the wavelength of emission (Wien's law, see Appendix 1 section A.1.4). The Sun, whose surface temperature is between 5500 and 6000°C, emits short-wavelength radiation, among which is the light (visible part of solar radiation). The Sun is very similar to a 'black body', a perfect emitter and receptor of radiation.

Radiation propagates through space as a wave, but it also exists as discrete energy packages called photons. Each type of radiation propagates in intervals of different wavelength (which is the shortest distance between consecutive waves) and with a determined frequency (or number of vibrations per second). The complete set formed by all the wavelengths constitutes the electromagnetic spectrum (Plate 1). The wavelength of a certain radiation and its frequency are related, their product being a constant (see Appendix 1 section A.1.5), thus the higher the frequency, the lower the wavelength and vice versa.

All the energy contained in the electromagnetic spectrum travels at the speed of light and it is called radiation. It includes the cosmic rays, gamma rays, X-rays, ultraviolet (UV), visible light (blue, green, yellow and red), infrared, radar and radio and television waves (Plate 1).

The solar energy, outside the Earth's atmosphere, changes very little and it is called the solar constant, but when crossing the atmosphere the radiation is partially reflected, absorbed or dispersed and suffers quantitative and qualitative modifications. The energy losses depend on the thickness of the atmosphere crossed and its characteristics (moisture content and gases, turbidity, cloudiness) (Fig. 2.6). The higher the Sun is over the horizon, the more energy reaches the Earth's surface. The maximum is received at noon when the Sun is in the point of maximum elevation and the sky is clear (Fig. 2.7).

With clear sky, the amount of solar energy which reaches a point of the Earth's surface depends on the Sun position, which varies with latitude, season and time of the day, besides the cloudiness and turbidity. The Sun's position at any moment is given by its coordinates: Sun elevation (h) and geodesic azimuth (γ). The geodesic azimuth is the angle (from 0° to 360°) of a certain direction in a horizontal plane, measured from the south direction (0°), following the movement of the clock hands (south-west–north-east) (Fig. 2.7). The topographical azimuth, used in topographical operations, is the angle measured from the north direction, following the movement of the clock hands. The zenith is the point of the sky located in the vertical that passes through the observer's head located at a certain point. The zenith angle (θ) is formed by the vertical at the zenith and the line formed by the solar rays (Fig. 2.8), so that: ($h + \theta = 90°$) (Dufie and Beckman, 1980).

In mid-latitudes the solar rays impinge more vertically in summer than in winter, implying that the intensity of radiation is higher in summer (Figs 2.8 and 2.9). Besides, the days are longer in summer, thus the total amount of solar radiation received each day is higher than in winter (Fig. 2.8) as the Sun covers a larger apparent trajectory (Fig. 2.10).

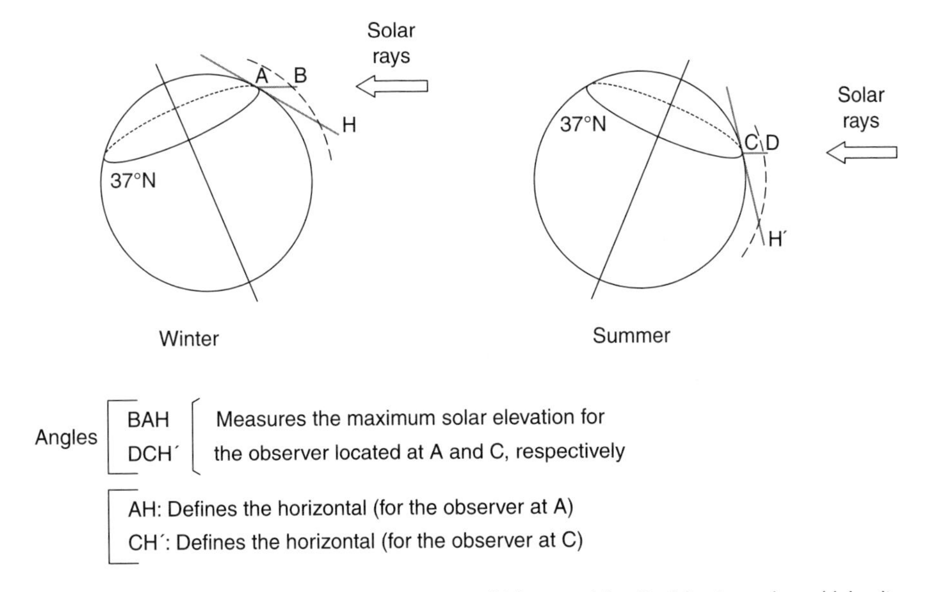

Fig. 2.6. In winter, the solar rays must cross a greater thickness of the Earth's atmosphere (doing it more obliquely, stretch AB) than in summer (stretch CD). The figure represents conditions at noon (the moment at which the thickness of the atmosphere to be crossed is lower).

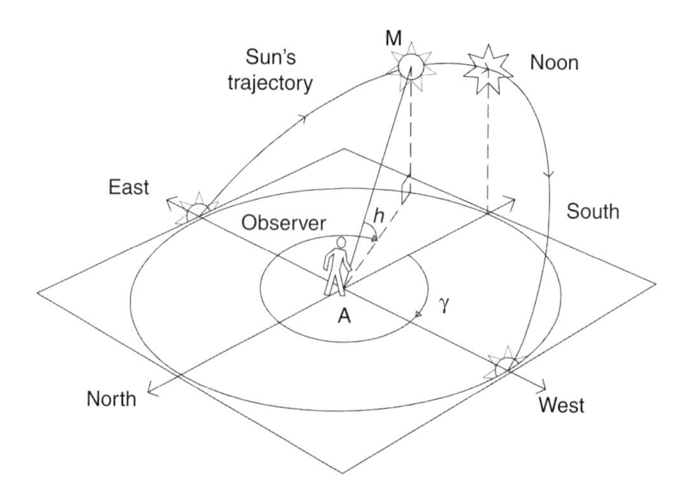

Fig. 2.7. The Sun's position at any moment (M) of the day is given by its coordinates: geodesic azimuth (γ) and solar elevation (*h*), for an observer located at A (adapted from Wacquant, 2000).

The relative differences of solar radiation received between regions are high if the latitudes are distant, especially in winter (Plate 2).

For greenhouse crop production, the most important factors are: (i) the total available solar energy; (ii) the duration of the day and of the night; and (iii) the quality of the radiation, because of its influence on the photosynthesis and in photomorphogenesis (Hanan, 1998).

Solar radiation acts on: (i) the plant, affecting photosynthesis (intensity, quality and amount of light), phototropism (relevant role of the red and the blue light), photomorphogenesis (photoperiod, i.e. the period of time per day that an organism is exposed to daylight), transpiration (the opening of the stomata), etc.; and (ii) the energy balance of the greenhouse, affecting the soil, water, air and plant temperatures, and the temperature of other objects absorbing radiation.

2.4.2 Quality of solar radiation

The incident solar radiation is composed of direct solar radiation, circumsolar radiation and diffuse radiation. Direct radiation comes directly from the solar disc, travels straight and its direction is determined by the latitude, day of the year and time of the day. The circumsolar radiation represents the radiation coming from the region close to the Sun. The diffuse radiation comes from all directions of the whole of the sky, due to reflections, deviations and scattering caused by the clouds, gases and aerosols present in the atmosphere (Hanan, 1998). The sum of direct solar radiation, circumsolar radiation and diffuse radiation is called global solar radiation.

In practice, the circumsolar radiation is considered together with the diffuse solar radiation, because the measuring method used integrates both of them (Day and Bailey, 1999).

The cloudiness, turbidity and transparency of the atmosphere have a great influence on the proportions of direct and diffuse radiation, as well as the solar elevation. On sunny days, with a clear atmosphere the percentage of direct solar radiation may reach a maximum of around 90% of the global daily radiation.

In urban or industrial areas, when the Sun is clear the direct solar radiation predominates on many occasions only if the Sun elevation is higher than 50°, due to the influence of the air's turbidity, whereas in coastal areas this predominance occurs at 30°, and at altitudes of 3000 m it occurs at only 6°, due to the great transparency of the air (Seeman, 1974).

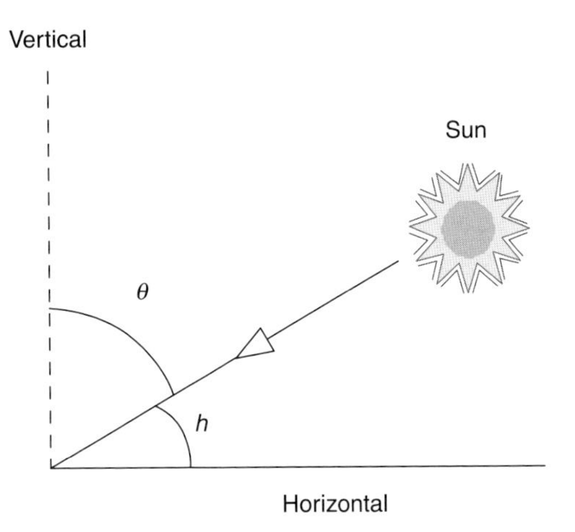

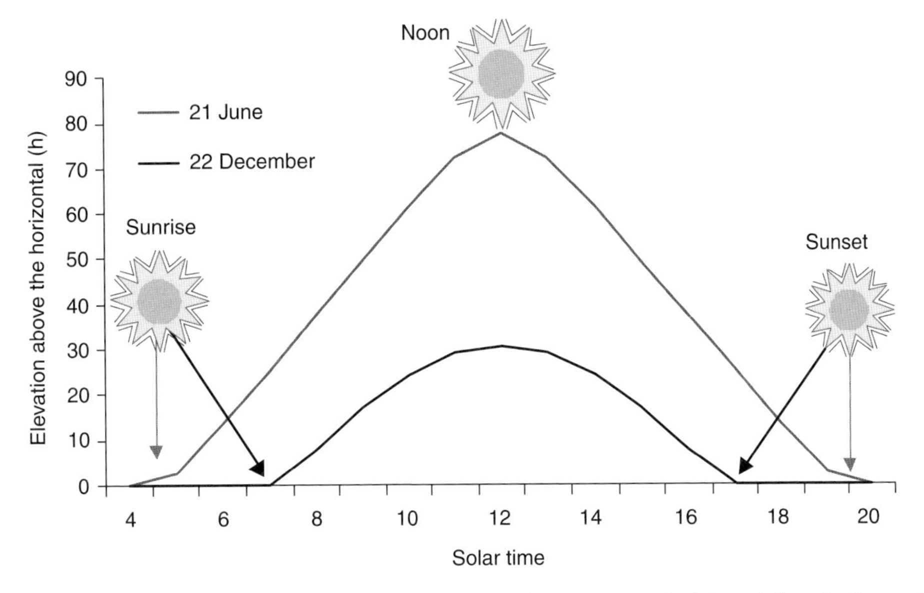

Fig. 2.8. Evolution of solar elevation (h) through the day in the summer and winter solstices for the south of Spain (latitude 37°N). θ is the zenith angle.

On completely cloudy (covered) days, the percentage of total direct solar radiation in relation to the global radiation is almost negligible.

Like all electromagnetic radiations, solar radiation propagates in the form of waves. The quality of solar radiation is characterized by its wavelength, measured in nanometres (nm). It can also be defined by its frequency (related to the wavelength). Plate 1 shows a sketch of the different radiations, by wavelength (or spectral composition).

The light is the radiation that stimulates the vision sensation in the normal human eye (photo-optical response). This response covers the wavelengths ranging from 380 to 720 nm, with a peak response around 550 nm. The colour, as a chromatic response of the human eye, ranges from 400 to 500 nm for the blue, 500–600 nm for the green, 600–700 for the red and 700–800 for the far red (Langhams and Tibbitts, 1997).

In the limit of the Earth's atmosphere, solar radiation ranges from 200 to near

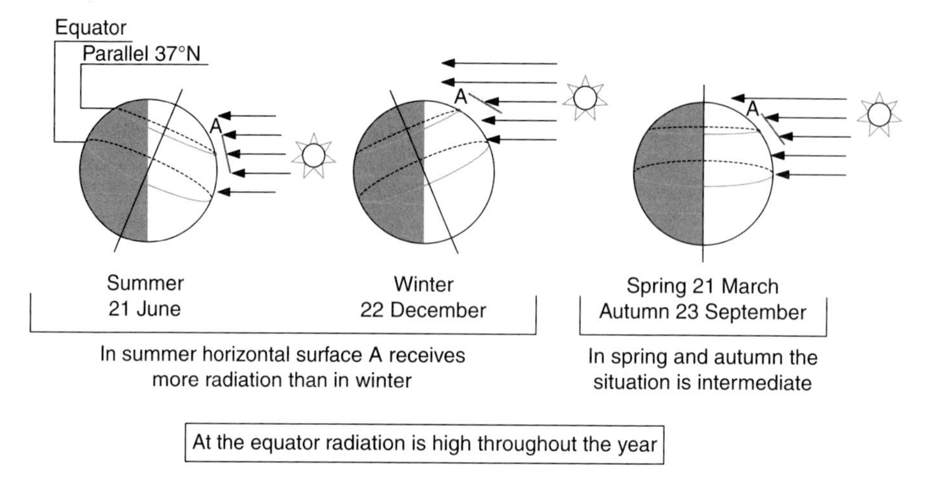

Fig. 2.9. The incidence of direct solar radiation at noon in the south of Spain (37°N) in the winter and summer solstices and in the spring and autumn equinoxes.

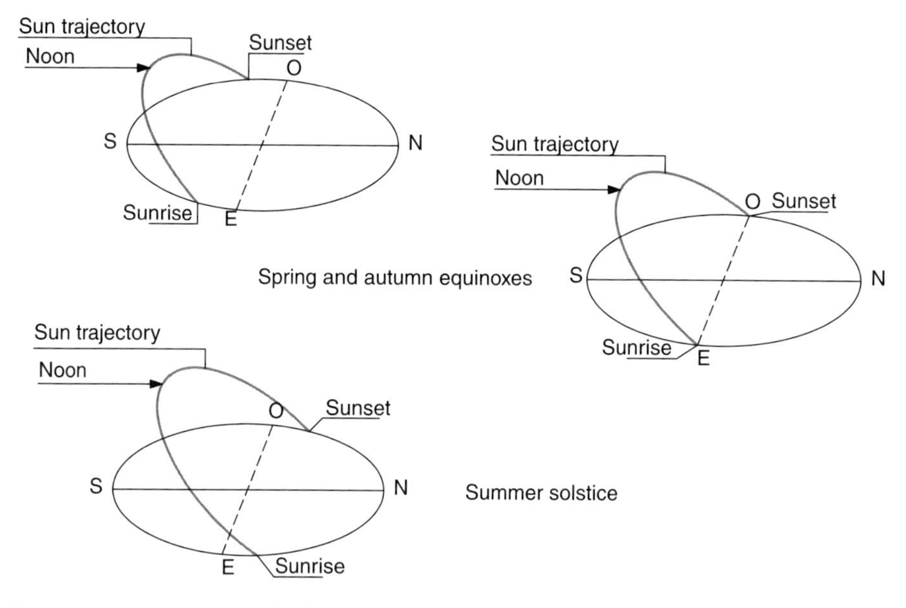

Fig. 2.10. Apparent trajectory of the Sun with respect to the horizontal plane in mid-latitudes of the northern hemisphere, in the winter solstice (top left), in the spring and autumn equinoxes (centre right) and the summer solstice (bottom left) (adapted from Fuentes, 1999).

5000 nm, with a maximum of emission at 470 nm (Fig. 2.11). During the crossing of our atmosphere, even with good weather conditions (clear sky), this radiation is mitigated and modified, due to the presence of water vapour, nitrous oxides, ozone, oxygen and other gases, which means that a large part of this radiation is scattered.

The majority of the global solar energy flux at the Earth's surface level (99%) is found between 300 and 2500 nm, made up of three categories of radiation as a function of the wavelength intervals they represent (spectral composition):

- *Ultraviolet radiation* (UV), below 380 nm. This radiation is scarce when the Sun's

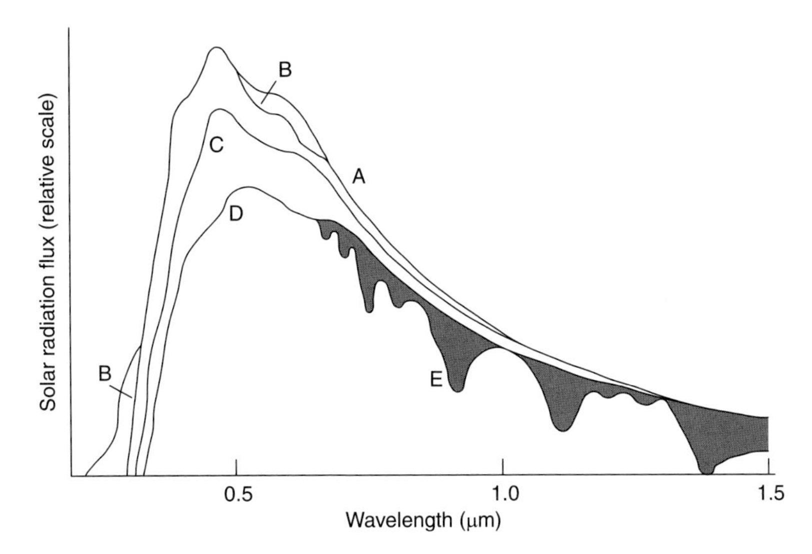

Fig. 2.11. Successive processes of reduction of the solar radiation flux when penetrating the Earth's atmosphere. A, Extraterrestrial radiation; B, after absorption by the ozone layer; C, after molecular diffusion; D, after the aerosol's diffusion; E, after water vapour and oxygen absorption (adapted from Monteith and Unsworth, 1990).

elevation over the horizon is low and at low altitudes. On the Mediterranean coast, its role is important in the ageing of plastic materials and for plant morphogenesis (Raviv, 1988). It amounts to 2–4% of the energy of the global radiation. The UV radiation may be subdivided into UV-A (higher than 320 nm) which is the one that tans the skin, UV-B (from 290 to 320 nm) responsible of the skin cancer and UV-C (from 200 to 290 nm) potentially dangerous but absorbed (Fig. 2.11) almost completely by the ozone layer (Monteith and Unsworth, 1990).

- *Visible radiation* to the human eye, from 380 (violet-blue) to 780 nm (red). This interval includes the PAR radiation (photosynthetically active or photoactive radiation; it amounts to 45–50% of the global radiation; Berninger, 1989).
- *Infrared solar radiation* (IR), from 780 to 5000 nm. It amounts almost to 50% of the energy of the global radiation (Berninger, 1989). The fraction of energy in the range from 2500 to 5000 nm is very low. Within the IR the NIR (near IR) is the band between 760 and 2500 nm.

The name PAR is used to designate the radiation with wavelengths useful for plant photosynthesis. It is accepted that the PAR radiation ranges from 400 to 700 nm (McCree, 1972), although some authors consider the PAR from 350 to 850 nm.

The composition of the radiation changes with time, as a function of the Sun's elevation and the cloudiness. When the Sun is low over the horizon, the short wavelengths are reduced (less UV and more red). The clouds reduce the amount of energy, greatly decreasing the NIR. The PAR proportion in relation to the global radiation increases with scattering (diffusion). It is lower with clear sky and in the summer (45–48%).

2.4.3 Quantity of solar radiation

The solar constant (solar radiation intensity at the outer regions of the Earth's atmosphere) is estimated to range between 1360 and 1395 W m^{-2}, as an average, measured in a perpendicular plane of the direction of the solar radiation flux (Takakura, 1989), although some authors estimate it as slightly lower.

Some authors consider more appropriate the term 'irradiance' than that of radiation intensity. Both designate it as the incident energy flux per unit surface, that is, the radiant flux density over a surface. The irradiance or radiation intensity is usually measured in watts per square metre (W m^{-2}).

The radiation intensity on a surface will depend on the inclination with which the radiation impacts on such a surface. If it impacts perpendicularly, the surface will receive the maximum radiation per unit area (Fig. 2.12). Lambert's cosine law dictates (Jones, 1983):

$$I = I_0 \cos i \qquad (2.1)$$

where:

I = Radiation flux density (irradiance or radiation intensity) impacting on the surface

I_0 = Radiation flux density (irradiance or radiation intensity) impacting on a surface perpendicular to the direction of the radiation

i = Angle of incidence between the radiation direction and the perpendicular to the surface. If the surface is horizontal, i is the zenith angle (θ) (Fig. 2.8).

At the Earth's surface, due to the absorption and dispersion of the radiation through the atmosphere, the radiation intensity is lower than the solar constant, being estimated as no more than 75% of it, under normal conditions (Monteith and Unsworth, 1990). The global radiation values are measured, by convention, over a horizontal surface, being admitted as a general rule that 48% of the global radiation is PAR radiation (Hanan, 1998).

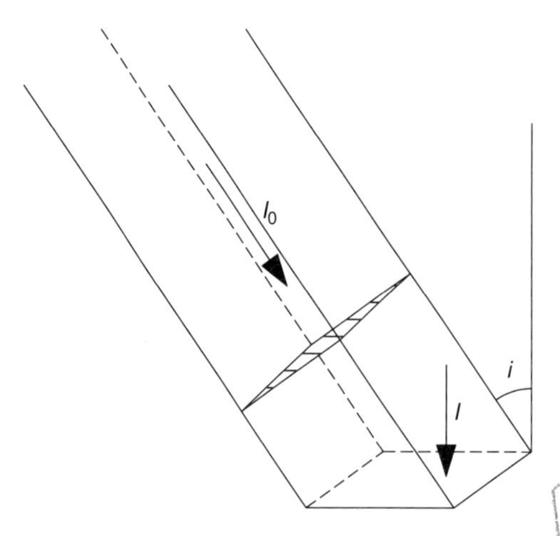

Fig. 2.12. The radiation impacting on a surface is calculated by the Lambert's cosine law (see text). The solar radiation values are measured, by convention, over a horizontal surface.

The proportions of direct and diffuse solar energy vary with the turbidity and transparency of the atmosphere, also influenced by the Sun's elevation. Through the day, if the sky is clear, the irradiance (radiation intensity) evolves in a very regular way, but if the sky is cloudy it varies a great deal (Fig. 2.13).

The global daily radiation varies greatly with the latitude, with higher differences in winter than in summer. It also varies with time throughout the year, depending on the irradiance (radiation intensity) and the length of the day, being higher during the summer months than during the winter months (Plate 2).

The global solar radiation is quantified by its irradiance or solar radiation intensity (instantaneous energy flux) in watts per square metre (W m^{-2}). The quantity of global solar radiation received or accumulated over a period is usually expressed in megajoules per square metre (MJ m^{-2}).

The PAR radiation (which ranges from 400 to 700 nm) can be quantified by its intensity in energy units or photosynthetic irradiance (W m^{-2}) or in photonic units (moles of photons). Within this range 1 W m^{-2} of PAR equals approximately 4.57 µmol m^{-2} s^{-1} under clear day conditions (Table 2.2).

The photometric units, adapted to human vision, do not have a constant equivalence with the energy and photonic units, so they are now disused.

The unit for the instantaneous light flux is the lux, which equals a lumen per square metre; 1 W m^{-2} of PAR radiation equals 247 lux, if the source of light is solar (Table 2.2), whereas it equals 520 lux, if the source of light is a low-pressure sodium vapour lamp (McCree, 1972).

The proportion of solar energy used in the synthesis of organic matter or in morphogenesis is minimal, and negligible in energy balance studies (Hanan, 1998). In greenhouse management it is worth mentioning that energy conversion phenomena

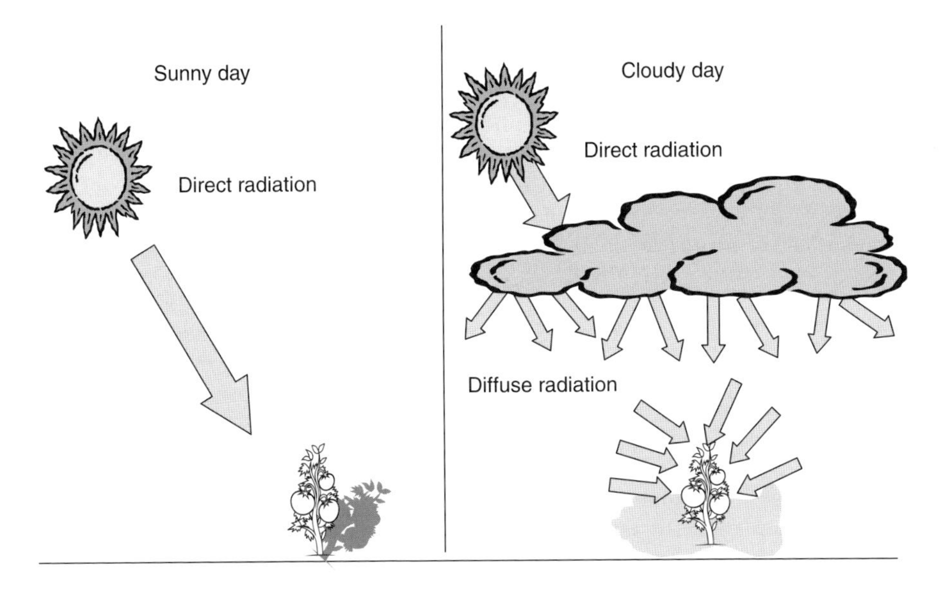

Fig. 2.13. On a sunny day direct radiation, coming from the Sun, predominates over the diffuse radiation which comes from the whole of the sky (dispersed by the clouds and the atmospheric gases). When the Sun's elevation over the horizon is low, on a sunny day, the proportion of diffuse radiation increases in relation to the direct radiation. The shadows on a sunny day are clear and well defined, whereas on a cloudy day they are not well defined.

Table 2.2. Equivalence of radiation and illumination units within the PAR range (400–700 nm) (adapted from McCree, 1972).

Light source	1 W m^{-2} (PAR) approximately equals:	1 W m^{-2} (PAR) approximately equals:	1 μmol m^{-2} s^{-1} approximately equals:
Sun and sky	247 lux	4.57 μmol $m^{-2}s^{-1}$	54 lux
Only clear sky	220 lux	4.24 μmol $m^{-2}s^{-1}$	52 lux
Incandescent lamp	250 lux	5.00 μmol $m^{-2}s^{-1}$	50 lux
Sodium vapour lamp (low pressure)	520 lux	4.92 μmol $m^{-2}s^{-1}$	106 lux
Sodium vapour lamp (high pressure)	408 lux	4.98 μmol $m^{-2}s^{-1}$	82 lux

are much related to the change from liquid water to vapour and vice-versa.

Once the solar radiation reaches the Earth's surface, part of the radiation is reflected. The so called 'albedo' is the proportion of the incident solar radiation which is reflected over a certain surface over the whole range of the spectrum (Monteith and Unsworth, 1990), although some authors relate it only to visible light. The albedo, expressed from zero (0) to one (1), ranges from 0.15 to 0.25 for herbal crops (Villalobos *et al.*, 2002).

2.4.4 Measurement of solar radiation

The effects of radiation on plants are determined by the quantity and quality (wavelength) of the radiation impacting on them. Another relevant factor is the direction of the radiation. It is assumed that the leaves act as a plane receptor, located horizontally (Langhams and Tibbitts, 1997). The radiation measurement sensors are constructed with filters which correct the radiation reflection when its incidence is not perpendicular to the sensor's surface.

The two most usual types of sensors are photoelectric and thermoelectric. The most-used photoelectric sensor is a photodiode sensor (photovoltaic). When exposed to radiation within its range of sensitivity, it generates a potential difference proportional to the radiation flux received.

The thermoelectric sensor uses a black body that absorbs the radiation impacting on the sensor and measures the resulting heat. To avoid temperature alterations due to air convective flows, they are covered with a plastic or glass dome, as a filter, whose transmission to different wavelengths limits the range of radiation to be measured.

There are spherical sensors (net radiometers) which measure the incident radiation (on their upper part) and the reflected radiation (albedo) on their lower part. The difference between them is the net radiation.

The difficulty in accurately measuring solar radiation, especially in the past, due to the high cost of the sensors, necessitated the estimation of global radiation by means of measurement of the number of sun hours, which varies throughout the year for each latitude (Table 2.1) and also depends on local cloudiness. The insolation or duration of sun hours during the months of autumn and winter is, in the absence of reliable data on solar radiation, used as an estimation of the aptitude index of a given area for protected cultivation (Nisen *et al.*, 1988).

Procedures to estimate radiation, when local measurements are not available, are described in detail in Appendix 1 section A.1.2.

2.5 The Earth's Radiation

Anybody whose temperature is above −273°C (i.e. absolute zero, on the Kelvin scale) emits radiation. This radiation will have a wavelength dependent on its temperature (see Chapter 5). The Earth's surface, therefore, emits long-wave radiation (far IR), which varies according to its temperature which in turn is influenced by the type of soil or vegetation (Fig. 2.14).

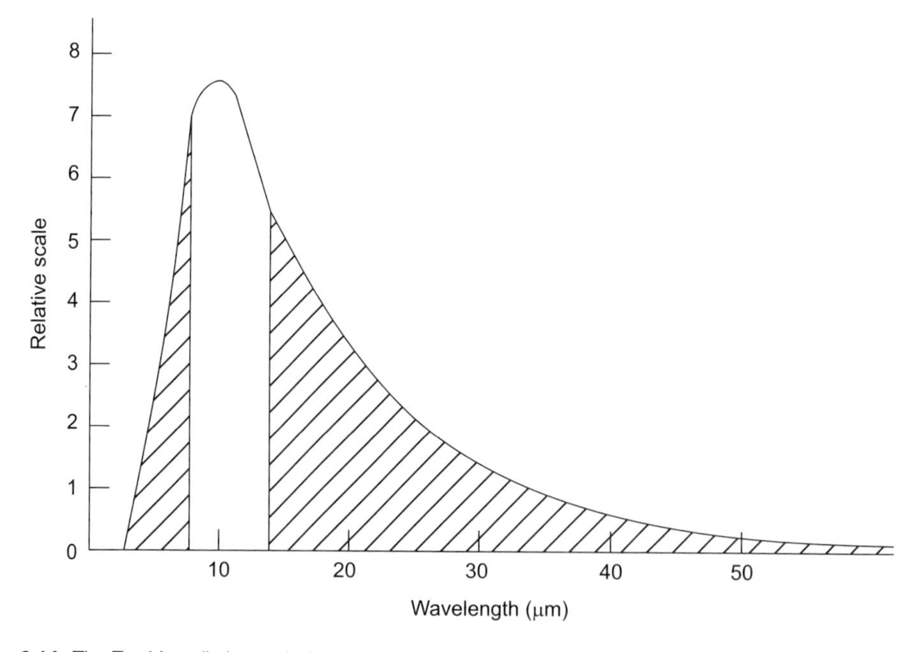

Fig. 2.14. The Earth's radiation emission according to the wavelength at the approximate Earth temperature (300 K, equivalent to 27°C). The hatched area below the curved line represents the radiation absorbed by the Earth's atmosphere and the clear area represents the remainder of the radiation that is lost into outer space under clear sky conditions. This latter part is known as the 'atmospheric window' (adapted from Rose, 1979).

In a similar way, the components of the atmosphere emit far IR radiation (in all directions), depending on the state of the atmosphere (i.e. humidity, turbidity, etc.).

The wavelength of these types of radiation ranges between 5000 and 100,000 nm, although the highest emission takes place between 8000 and 13,000 nm, as the temperature of the emitting bodies normally range between 0 and 30°C (equal to 273 and 303 K), an interval called the 'atmospheric window' (Rose, 1979) (Fig. 2.14), because within this interval, and if the sky is clear, the atmosphere is very permeable to radiation. The IR band between 2500 and 25,000 nm is usually designated as medium IR (MIR).

The balance of the Earth's radiation, which some authors call also atmospheric radiation, is always negative at night (i.e. the ascendant radiation surpasses the descendant radiation) and it is generally negative also during the daytime, if only the far IR radiation is considered (Berninger, 1989).

2.6 Net Radiation

The net radiation is the difference between the radiation flux received over a surface and the radiation flux emitted from the Earth's surface.

It involves the energy available to all physical and biological processes. The global solar radiation flux is always positive during the daytime and zero during the night, whereas the net radiation flux is negative at night (Rosenberg *et al.*, 1983).

2.7 Temperature

2.7.1 Air temperature

For each location, the air temperature varies with the evolution of solar radiation, in 24 h cycles and with changes caused by the seasons of the year. The average temperature follows, with a certain delay, the evolution of solar radiation (Plate 3). For instance, during the equinoxes, with the same solar radiation, it is colder by the end of March than by the end of September.

The measurement of the average temperature of the outdoor air is performed inside a meteorological box under normalized conditions, protected from solar radiation and with circulating air, located between 1.5 and 2 m above the ground. It ranges between a minimum (normally at dawn) and a maximum (usually 1 or 2 h after noon).

The temperature is normally measured in degrees centigrade or Celsius (°C). In the Kelvin scale, the temperature is expressed in Kelvin (K). In the Kelvin scale $T(°C) = T(K) - 273°C$. The Fahrenheit scale, which is much less used, expresses the temperature in Fahrenheit degrees (°F).

The conversion is:

$$T(°F) = \frac{9}{5}\,T(°C) + 32 \qquad (2.2)$$

$$T(°C) = \frac{5}{9}\,[T(°F) - 32] \qquad (2.3)$$

The measurement of the outside air temperature in greenhouses is sometimes performed over the cover (at several metres height) which alters how representative it is compared with normalized measurements. The altitude implies a decrease of the average temperatures of around 0.6°C per 100 m of elevation.

The 'actinothermal index' indicates the night temperature by means of a thermometer exposed horizontally in open air (Berninger, 1989). This situation permits radiative exchange without any limitations. With dry weather conditions and clear sky, without wind, the actinothermal index during the night reaches a lower value than the ambient air temperature. This index is more representative of the plant temperature than the air temperature during the night.

2.7.2 Soil temperature

The soil temperature determines the temperature of the subsoil organs. The surface layers of the soil act as thermal and seasonal heat sink, heating and cooling itself much more slowly than the surrounding air.

Oscillations in soil temperature throughout the year decrease with depth. At several metres deep, the temperature of the soil remains almost constant throughout the year.

Low soil temperature (below 14°C) may limit the growth of some crops (Berninger, 1989). In greenhouses the soil temperature is, obviously, higher than in open air.

2.7.3 The relationship between solar radiation and air temperature

The average values of global radiation and air temperature vary from month to month. The values follow sinusoidal curves, with a delay in the temperature curve with respect to the radiation curve (Plate 3).

If radiation and temperature are represented in a *xy* graph, the sequence of monthly values generates a climate diagram with an elliptic shape (Fig. 2.15). The daily differences in radiation and temperature are large, especially between cloudy and clear days.

2.8 Wind

Wind is the, mainly horizontal, displacement of air. It is characterized by its direction and velocity. If measured at different (standardized) heights, its velocity can be calculated at intermediate heights (Hellman's equation, see Appendix 1 section A.1.6). The direction is measured with a vane, and is expressed as the angle between north and the direction of the origin of the wind (in a clockwise direction).

The wind force is a basic consideration in the design of a greenhouse structure. Its direction and frequency are important when considering static (passive) ventilation. Its velocity is related to air renewal, even when the greenhouse is closed, and to the energy losses.

The wind velocity and direction patterns are more or less predictable for a certain location. For instance, in the coast of Almeria, the highest wind velocity values range between 10:00 and 17:00 (solar time) for the months from March to July (Plate 4), it being usual to have winds above 2 m s^{-1} during the daytime (Pérez-Parra, 2002). The dominant directions (Fig. 2.16) are from the west (west–south-west) and the east.

2.9 Composition of the Atmosphere

2.9.1 Water vapour content

From a crop production point of view, water vapour is one of the most important parameters of the atmosphere. The water phase changes (solid, liquid or vapour) involve the transport of large amounts of energy, which affects the crop and its surrounding air temperatures.

The atmospheric humidity and the availability of water, together with other factors, determine the rate at which plants transpire. The atmospheric air is never completely dry, containing some amount of water in the form of vapour. The water vapour content of air can be expressed in several ways:

1. By its concentration. This is the air absolute humidity (AH) which is expressed in kilograms of water vapour per kilogram of dry air.
2. By the water vapour pressure (e_a) expressed in units of pressure. From a free water surface in contact with the atmosphere part of the water molecules evaporate and others return to liquid, so that when those evaporating equal those returning to liquid we say that the atmosphere is saturated, a moment in which the partial water vapour pressure has reached its saturation value (e_s). The e_s value can be calculated (see Appendix 1 section A.1.7). The e_s value increases with temperature (Fig. 2.17).
3. By the relative humidity (RH) of the air, expressed as a percentage of the partial

water vapour pressure under certain given conditions, with respect to the saturation water vapour pressure (e_s). RH = $100(e_a/e_s)$. It changes with temperature.

4. By the vapour pressure deficit (VPD) or saturation deficit. This expresses the amount of water that the air, at a certain temperature, can still absorb before reaching saturation.

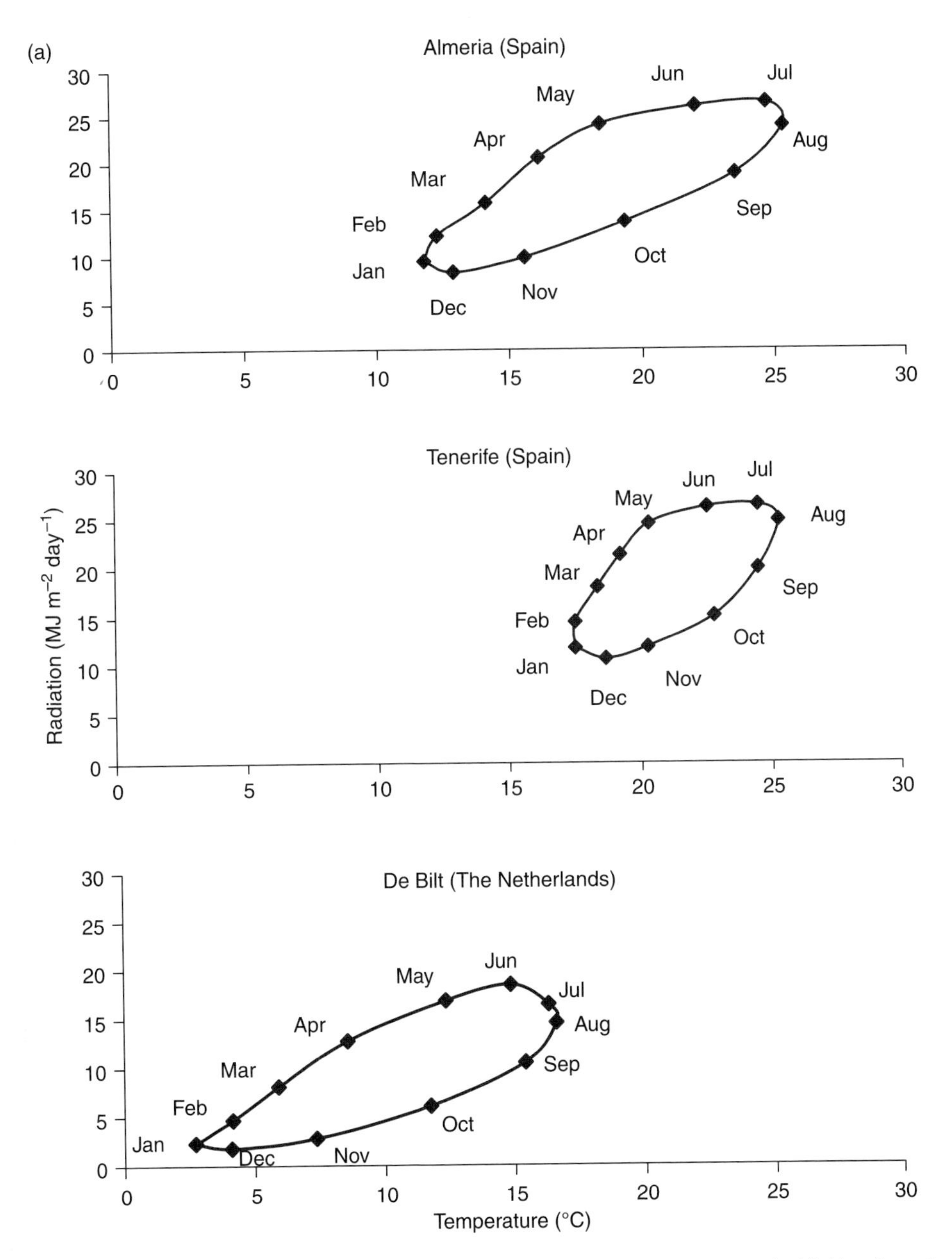

Fig. 2.15. Global radiation and average outside air temperature for an average year in (a) Almeria and Tenerife (Spain) and De Bilt (The Netherlands) (data from FAO; Kamp and Timmerman, 1996; Experimental Station 'Las Palmerillas', Almeria, Spain); (b) at Mexican locations of different altitudes: Culiacán (84 m), Ensenada (13 m) and Guanajuato (2050 m) (data from FAO).

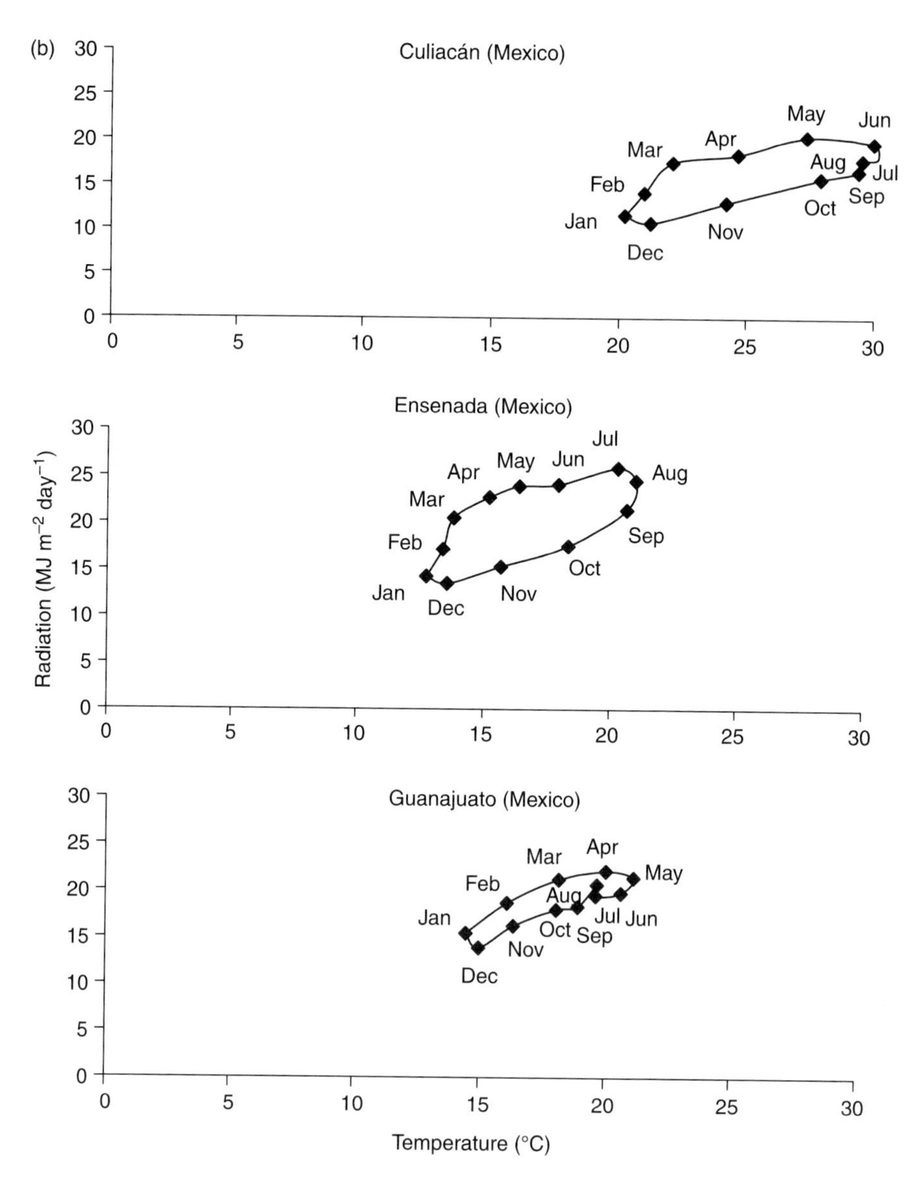

Fig. 2.15. Continued.

The expression VPD is the one that is used most often from a biological point of view. It is expressed in vapour pressure units and quantifies the 'drying power of the air'. The VPD notably influences transpiration and evaporation.

$$VPD = e_s - e_a \tag{2.4}$$

The atmosphere is normally sub-saturated, i.e. the e_a is usually lower at ambient temperature than its corresponding saturation value (e_s).

Dew point is the temperature at which the air water vapour starts to condense, due to reaching saturation. In greenhouses, this condensation starts on the coldest objects, initially on the greenhouse roof cover and walls.

The evapotranspiration (ET) integrates the water evaporated from the soil and that transpired by the vegetation. It will vary with the climate conditions

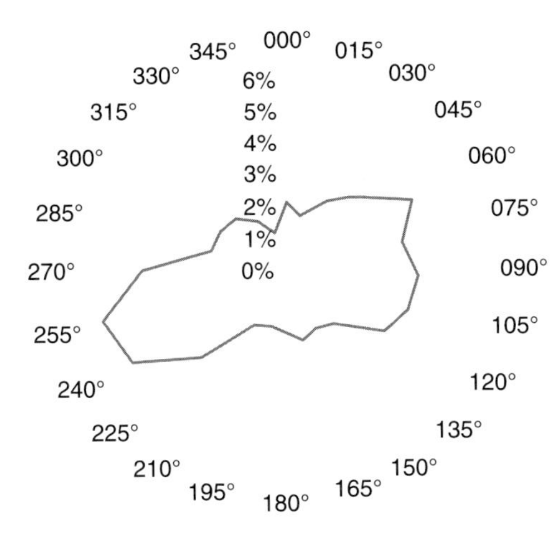

Fig. 2.16. Compass in Almeria (Spain) showing the predominant direction of the wind for the period 1996–2001 (Experimental Station of Cajamar Foundation-Cajamar; from Pérez-Parra, 2002).

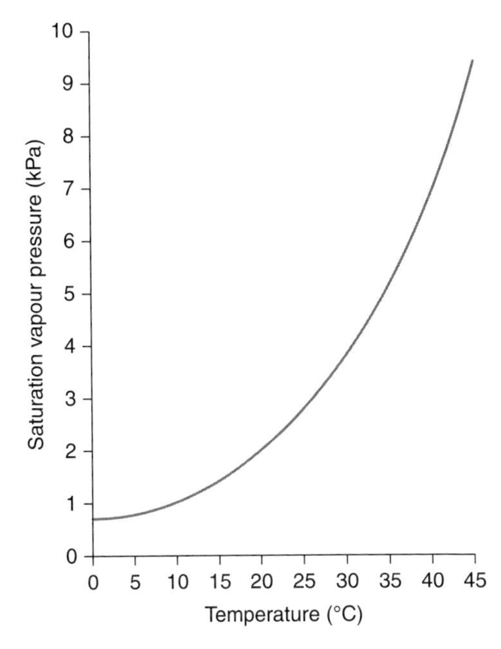

Fig. 2.17. The saturation water vapour pressure is dependent on temperature.

(radiation, air humidity, wind, CO_2) and with the water available in the soil or substrate. The ET contributes primarily to the water content of the greenhouse atmosphere.

2.9.2 CO_2 content

The CO_2 concentration in the air changed considerably during the last half of the 20th century, from 300 to 350 ppm (Allen, 1990); today, it increases about 1 ppm year^{-1} (Berninger, 1989). At the beginning of the 21st century the standard CO_2 level in the air was already 360 ppm. Its increase influences the atmospheric greenhouse effect, at a planetary level.

The most usual units to measure CO_2 are vpm (volumes per million) or ppm (parts per million), although it can also be defined as the partial pressure of CO_2 in the air, or the percentage of CO_2 in the air (e.g. 360 ppm equals 36 kPa, or 0.036% of CO_2; see Appendix 1 section A.7.1).

2.9.3 Atmospheric pollution

Atmospheric pollution caused by the presence of harmful gases (sulfur dioxide (SO_2), nitrogen oxides (NO_x) and ozone) affects the industrial parts and periphery of cities. It results in the corrosion of structures as well as having harmful effects on crops. Pollution increases the proportion of diffuse solar radiation.

2.10 Rainfall

The rain does not affect protected cultivation, except in special cases (e.g. greenhouses with perforated covers or screenhouses). It influences the design of the slopes, gutters, evacuation ducts and rainfall water storage systems.

Snow and hail are other phenomena to consider, due to the load they may exert and the resultant possible damage caused to the greenhouse structure and covering materials.

2.11 Altitude and Topography

The altitude influences the barometric pressure, which decreases as altitude increases.

Temperature also decreases with altitude, around 0.6°C for each 100 m of elevation (Jones, 1983). Solar radiation increases with altitude, decreasing the proportion of diffuse radiation on clear days (Seeman, 1974). In general, rainfall is higher at higher altitudes.

The topography plays a relevant role in the local microclimate mainly due to its influence on the wind, the rainfall, radiation and temperature (Photo 2.1). Of special importance are also the effects of the local topography on radiation, because of the shadows that may be produced, on wind velocity and direction, and on the thermal regime (Fig. 2.18).

2.12 Summary

- The local climate is a determinant for the greenhouse microclimate and its future management; therefore it must be evaluated, with the aim of choosing the right location for siting the greenhouse.
- In the context of a greenhouse, the available solar radiation and its qualitative

Photo 2.1. On the coast of Granada (Spain), greenhouses are located on the slopes and oriented to the south to benefit from better conditions of radiation in autumn and winter.

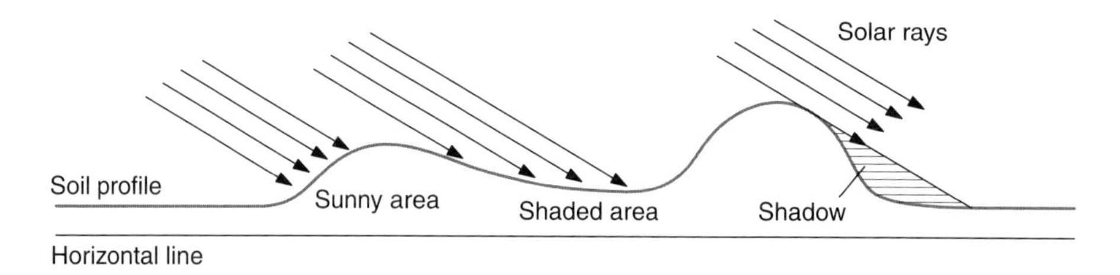

Fig. 2.18. In the northern hemisphere, the slopes oriented to the south receive much more radiation than those oriented to the north, especially in winter when the solar elevation is lower. If the slopes are very inclined, the shadows can also be important.

characteristics, as well as the duration of the day and night are primary elements of the climate.

- The qualitative characteristics of solar radiation, especially the proportions of diffuse and direct radiation, are important for greenhouses.
- Solar radiation or global solar radiation, is basically composed of: (i) UV radiation, which accounts for 2–4% of the total of the solar energy; (ii) PAR radiation (photoactive radiation) which accounts for 45–50% of the total solar energy and is indispensable for photosynthesis; and (iii) IR radiation which accounts for the remaining 50% of solar energy.
- The solar constant is the global radiation intensity outside the Earth's atmosphere and has a value between 1360 and 1395 W m^{-2}. At the Earth's surface, the intensity of radiation is only in the order of 75% of the solar constant, due to absorption, dispersion and reflection of the solar radiation in the Earth's atmospheric layer.
- Solar radiation varies greatly with latitude; throughout the year, it varies depending on the season, being minimum in winter and maximum in summer.
- The Earth emits long-wavelength radiation (far IR), known as the 'Earth's radiation'.
- Air temperature varies in 24 h cycles. The average temperature follows, with a certain delay, the evolution of solar radiation.
- Soil temperature follows, with smaller differences, the evolution of the air temperature. The daily oscillations of soil temperature decrease as the depth increases.
- Wind conditions are very important for greenhouses. The wind force is a basic consideration in the design of greenhouse structure while the wind direction and velocity are important for ventilation and the energy balance.
- The composition of the atmosphere is very important particularly regarding water vapour content and CO_2 content.
- The water vapour content of the atmosphere is usually represented by the RH of the air, but it is more precise to quantify it by the water vapour pressure of the air or its VPD.
- The CO_2 concentration in the air has risen in the last half century from 300 to 360 ppm.
- Rainfall influences the design of greenhouses as it affects the collection, evacuation and storage devices of rainfall water.
- Altitude affects the quality of the solar radiation (affecting the proportions of direct and diffuse radiation).
- Topography may affect solar radiation because of the shadows that may be produced, but it also affects the wind and thermal regimes.

3

The Greenhouse Climate

3.1 Introduction

Inside the greenhouse, the radiation, temperature and composition of the atmosphere are modified, and this results in a different microclimate from the one outside. The modifications depend essentially on the nature and properties of the cladding material, the air renewal conditions, the shape, dimensions and orientation of the greenhouse, but also on the plant canopy and the possibilities for evapotranspiration (Berninger, 1989).

This microclimate is not uniform and varies from the centre to the borders of the greenhouse, from the ground to the roof and from the limits of the canopy to its interior.

The so called 'spontaneous climate' is the one generated without important human or energetic intervention, especially without heating, forced ventilation or water spraying.

A greenhouse, normally, has a crop which is irrigated and its soil is wet. An empty and dry greenhouse is only of theoretical interest, as it is not representative of real conditions.

3.2 The Greenhouse Effect

The 'greenhouse effect' is the result of two different effects: (i) a 'shelter or confinement effect' (convective effect), derived from the decrease in the air exchanges with the outside environment, and which is perceptible even in greenhouses that are very permeable to the air; and (ii) an effect caused by the existence of a cover with low transparency to far IR radiation emitted by the soil, the plants and all the inner elements of the greenhouse exposed to sunlight (visible and short IR radiation, to which this cover is very transparent; Fig. 3.1). This second effect is sometimes referred to as the 'radiative greenhouse effect' or 'heat trap'.

According to Wien's law, the product of the temperature of a radiant surface (in K) by the dominant wavelength of the emitted rays (in microns) is constant and equal to 2897 (K μm).

The average temperature of the Sun's surface, assimilated to a black body, is 5800 K. The Sun emits radiation (ranging from 0.3 to 2.5 μm), with a dominant wavelength of $2897/5800 = 0.5$ μm (i.e. in the margin between green-blue and yellow of the visible range).

The materials that cover greenhouses are transparent to solar radiation, transmitting most of it but the plants and the soil absorb a large amount of it, of all wavelengths.

During the daytime, the majority of the solar radiation passes through the cover of a

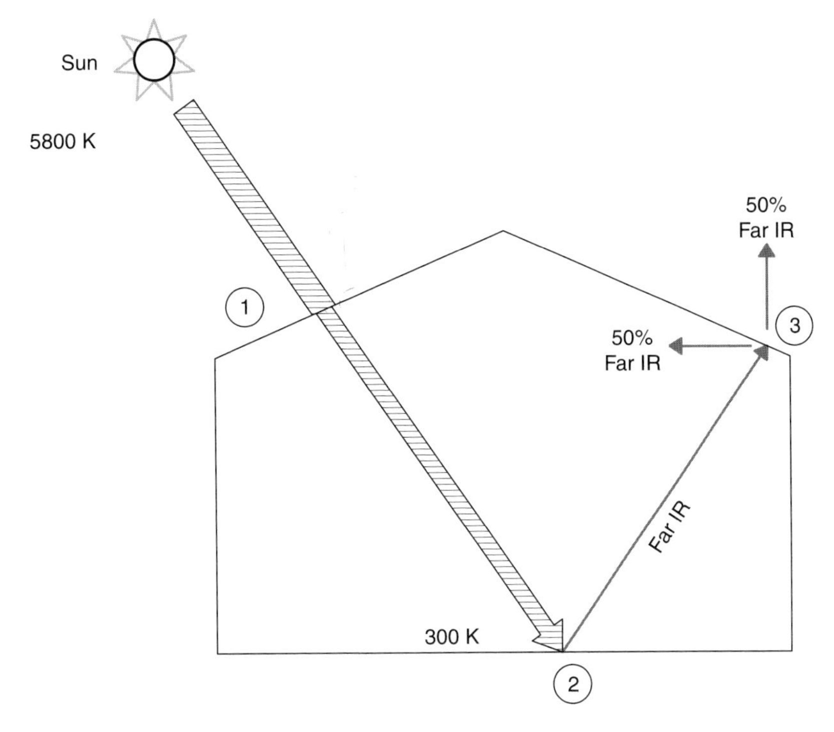

Fig. 3.1. The greenhouse effect. It is complemented by the shelter or confinement effect of the air inside the enclosure (see text). 1. The plastic lets a large part of the solar radiation pass through it (it transmits). 2. The greenhouse surfaces and the plants absorb the solar energy and re-emit (far IR) energy. 3. The greenhouse cover absorbs the energy (far IR) and re-emits it from its two sides, inwards and outwards.

greenhouse and is absorbed by the plants and soil. The plants and the soil are heated and re-emit energy, mostly with wavelengths of 10 μm but ranging from 2.5 to 25 μm (far IR range), according to Wien's law (because the temperature is about 300 K). This energy re-emitted by the plants and the soil is intercepted by the covering material (as the materials used are usually opaque to IR radiation), which is reheated and re-emits energy in turn outwards and inwards in similar proportions (Fig. 3.1). The greenhouse air is then heated, as it is confined and it is not renewed with outside fresh air.

These phenomena generate a temperature increase that is very evident during the daytime, in relation to the outside. This effect will vary depending on the specific conditions of transmission and absorption of the cover to radiation and depending on the ventilation and airtightness of the greenhouse.

At night, the temperature gradient with the outside is the result of a complex balance influenced, mostly, by the sky temperature and the temperature under the cover and the air exchanges between the greenhouse and the outside.

In an unheated greenhouse 'thermal inversion' may occur under certain conditions, depending on the cover type. For example, on nights when the sky is clear the energy losses in IR radiation to the atmosphere are very high (see Fig. 2.14). If the covering material is permeable to such radiation (as is the case with standard polyethylene (PE) films), on nights with no wind it may happen that the immobility of the air inside the greenhouse can cause a higher decrease in temperature than outside, resulting in thermal inversion. In other words, although a similar cooling process may occur outside, the free air movement and mixing with

Table 3.1. Brief description of the behaviour of the main microclimate parameters of the greenhouse in winter, depending on the weather conditions (adapted from Berninger, 1989).

		Greenhouse	
Type of sky	Outside	Day	Night
Clear sky	Large difference between day and night temperatures. High solar radiation (especially direct radiation). Low RH, especially if it is windy. At night, cold air, 'cold' sky	High solar radiation (direct and diffuse). High ventilation to limit temperature rise and avoid CO_2 depletion. High thermal storage. High evaporation	Possible heating to maintain temperature. High RH (without heating)
Cloudy sky	Stable temperatures. Weak solar radiation, diffuse. High RH. 'Warm' sky	Weak solar radiation, diffuse. Ventilation to limit the confinement (high RH, lack of CO_2). Scarce thermal storage. Low evaporation	Limited heating or may be unnecessary, except where there is a high plant disease risk associated with high RH

descending warmer air masses partially compensates for this cooling, resulting in a higher outside temperature at ground level.

In the past, the 'radiative greenhouse effect' was considered responsible for the greenhouse microclimate, but nowadays the importance of the convective effect has gained prominence, due to the air confinement, so the use of the expression 'greenhouse effect' must refer to both processes, radiative and convective (Papadakis *et al.*, 2000).

Table 3.1 summarizes the behaviour of the main microclimate parameters of the greenhouse in winter.

3.3 Solar Radiation in Greenhouses

3.3.1 Introduction

The solar radiation conditions in the greenhouse are very important from the point of view of production, not just quantitatively but also qualitatively. The first alteration which the greenhouse causes on the microclimate parameters is a decrease in available solar radiation (Fig. 3.2). The radiometric characteristics of the greenhouse cover may also significantly modify the quality of the radiation (distribution spectrum or proportion of diffuse radiation) affecting the crops, mainly in the efficiency of use of radiation and in its photomorphogenic effects (Baille, 1999), but also the insects and microorganisms in the greenhouse.

On single-span greenhouses and on the greenhouse sidewalls of multi-span greenhouses, an important part of the penetrating light is lost through the sidewalls. Therefore, the use of reflecting surfaces on the north sides of greenhouses (in the northern hemisphere) contributes to an increase in the available light (Day and Bailey, 1999).

Equally, the use of reflecting surfaces over the soil, to reflect the light not intercepted by the crop, allows for an increase in the light available for the crop.

3.3.2 Transmissivity to radiation

The fraction of global solar radiation transmitted inside the greenhouse is designated as 'greenhouse global transmissivity' (Zabeltitz, 1999). The limitations to productivity caused by low levels of radiation inside the greenhouse in autumn and winter in Mediterranean coastal areas on vegetable crops, which are highly light demanding, have been well documented (Castilla *et al.*, 1999; Gonzalez-Real *et al.*, 2003). Maximizing the radiation inside the greenhouse is in fact a desirable objective in all latitudes, especially during the autumn and winter seasons.

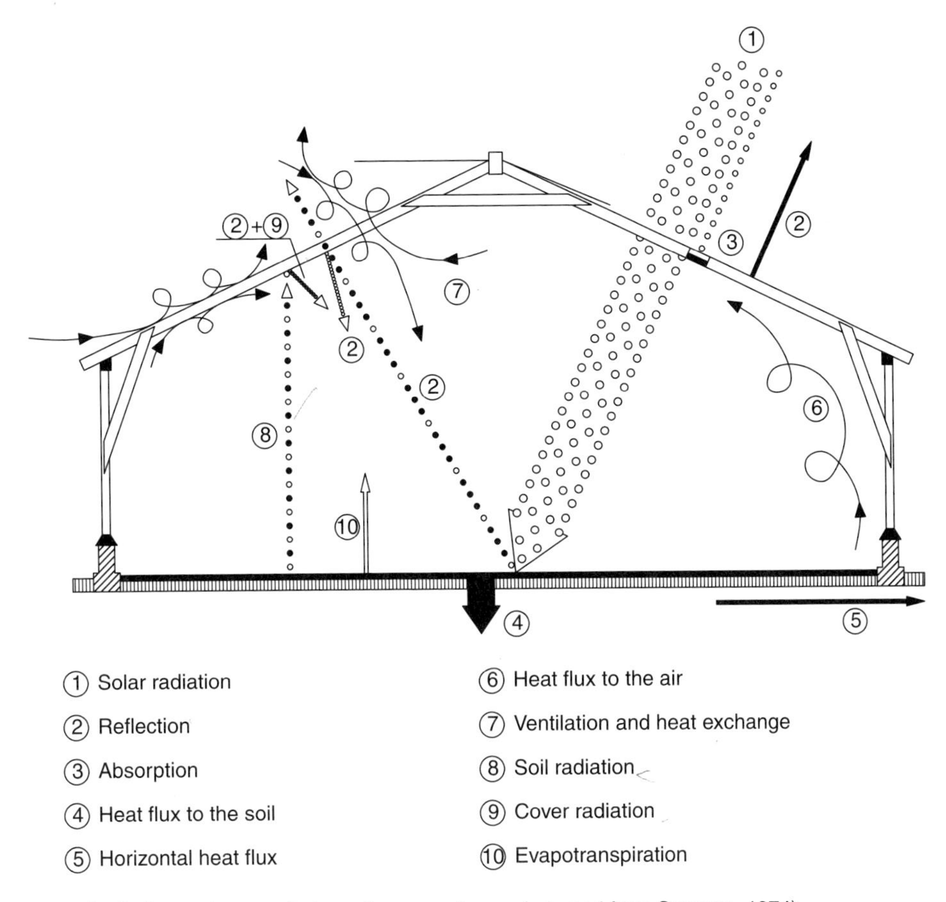

Fig. 3.2. Radiation and energy balance in a greenhouse (adapted from Seeman, 1974).

At latitudes higher than 30°, from the equator, the natural decrease of solar radiation is the most important uncontrolled limiting factor for crop growth inside greenhouses, and thus it becomes imperative under such conditions to strive for the maximum possible intensity, duration and uniformity of radiation (Giacomelli and Ting, 1999).

The above-mentioned transmissivity is a function, among other factors, of: (i) the climate conditions (cloudiness, mainly, which determines the proportions of direct and diffuse radiation); (ii) the position of the Sun in the sky (which will depend on the date and time of day and the latitude); (iii) the geometry of the greenhouse cover; (iv) its orientation (east–west, north–south); (v) the covering material (radiometric characteristics, cleanliness, water condensation on its inner surface); and (vi) the structural elements and equipment inside the greenhouse which limit, due to shadowing, the available radiation inside (Bot, 1983; Zabeltitz, 1999; Soriano *et al.*, 2004b). The transmissivity to direct solar radiation will vary depending on the angle of incidence (formed by the solar ray and the perpendicular to the greenhouse cover; Fig. 3.3), such transmissivity being higher, the smaller the angle is (i.e. when the radiation impinges on the greenhouse cover with high perpendicularity; Plate 5). The transmissivity to diffuse radiation, coming from every possible direction of the sky is scarcely influenced by the geometry of the greenhouse cover.

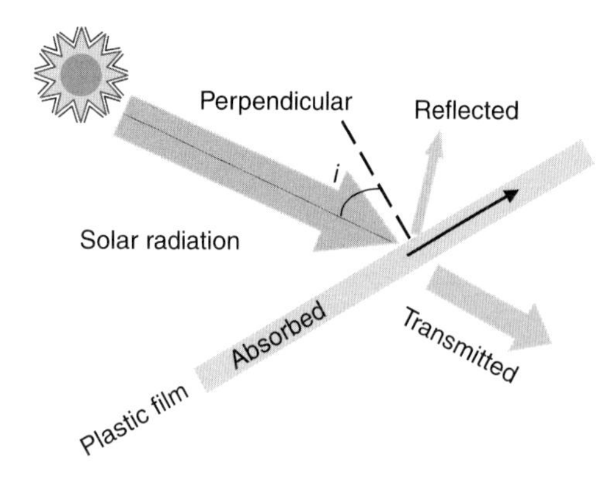

Fig. 3.3. The solar radiation which impinges on the plastic film is partly transmitted (crossing the film), reflected or absorbed by the film. The proportion of radiation transmitted (in relation to the incident) is known as transmissivity (or transmission) and varies depending on the physical and chemical characteristics of the film and on the wavelength of the radiation. When the radiation is direct it also depends on the angle of incidence (*i*).

On clear days, when direct radiation predominates, the average global transmissivity (fraction of global exterior radiation that penetrates inside the greenhouse) must be integrated as an average value for the whole greenhouse. This is because of the variability of radiation at different points throughout the greenhouse caused by differential shadowing of the structural elements of the greenhouse and of various pieces of installed equipment (Bot, 1983).

On completely cloudy days, when all the solar radiation is diffuse (i.e. when there are no defined shadows) the distribution of radiation is more homogeneous inside a greenhouse (Baille, 1999). The average instantaneous transmissivity of a certain greenhouse varies throughout the day, according to the position of the Sun in the sky and the characteristics of the radiation; normally, on a sunny day, it slightly increases from dawn until noon, and decreases later until dusk (Plates 6 and 7). When talking about global greenhouse transmissivity, it is normally understood as the daily average transmissivity (proportion of daily accumulated radiation which penetrates inside the greenhouse with respect to the outside), to distinguish it from the instantaneous values.

It is important to highlight the notorious differences that exist, from the point of view of radiation transmissivity, between single-span and multi-span greenhouses (even when spans have the same roof geometry) because of the shadows between spans (Fig. 3.4); consequently, transmissivity estimates obtained in single-span greenhouses cannot be extrapolated to multi-span types.

3.3.3 Orientation and transmissivity

The greenhouse orientation, which is designated by the direction of the ridge line (longitudinal axis of the span), at medium latitudes, clearly influences transmissivity, in autumn and in winter, under clear sky conditions (when direct radiation predominates). At latitudes higher than 30°, the north–south orientation results in less radiation being transmitted in winter than the east–west orientation, but in higher uniformity (Giacomelli and Ting, 1999); as the elevation of the Sun increases in spring, these differences notably decrease (Fig. 3.5). In greenhouses with roofs with a very low pitch (Fig. 3.5), the differences in transmissivity between the east–west and the north–south orientations are much smaller.

The uniformity of radiation in east–west oriented greenhouses (symmetrical with a roof pitch of around 30°) is less (on clear days) than in north–south oriented greenhouses, but their transmissivity in autumn–winter is higher, with differences of more than 10% of the outdoors daily global radiation around the winter solstice. However, these differences in uniformity between multi-span greenhouses oriented east–west and north–south are attenuated by: (i) the greater the height of the greenhouse (3.5–4.0 m at the gutters); (ii) the lower the span width; and (iii) the radiation diffusion characteristics of plastic films used nowadays.

Summarizing, at medium latitudes with a predominance of clear days in autumn and winter, such as in Mediterranean coastal areas, the east–west orientation is preferable to the north–south orientation, in greenhouses with roofs with a pitch of greater than 30°, whereas if the angle is low (e.g. low-tech greenhouses, with around 10° roof angle) the north–south

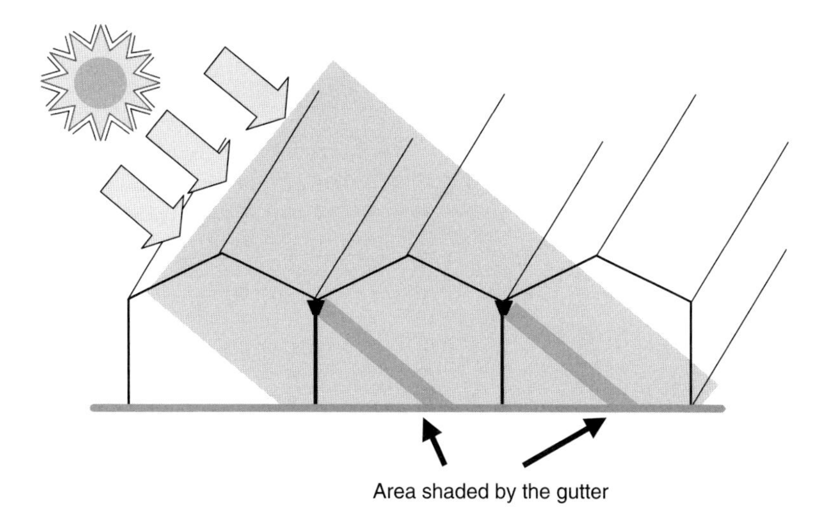

Fig. 3.4. The shadows of one span on the adjacent span have relevance in multi-span greenhouses (especially those that are east–west oriented). Shadows are caused by structural elements of the greenhouse, for instance gutters as well as roofs with a steep pitch.

orientation may be preferable (Castilla, 2001). In any case, if the priority in a greenhouse is to achieve the maximum uniformity of radiation (for instance, in a nursery) the north–south orientation would be preferable.

Another important aspect to consider when orienting the greenhouse is the direction of the predominant winds, which may become a primary consideration in choosing one or other orientation. The wind has a strong influence on the structure as a result of its mechanical effects, and because it has an indirect influence on the greenhouse indoor microclimate and energy balance (see Chapters 7 and 8). The wind increases the heat losses and the air infiltration leakage. Therefore, orientating the ridge parallel to the direction of the prevailing winds can, in certain cases, be advisable, but a reduction of ventilation must be expected.

The characteristics of the building plot (shape, slope, obstacles that generate shadows) may also limit the greenhouse orientation options.

3.3.4 Optimization of the transmissivity

Daily transmissivity values above 70% in simple cover greenhouses are very infrequent, because normally they range between 55% (winter) and 70% (summer), whereas in double-cover greenhouses they range between 50 and 60% (Baille, 1999). The average reflectivity (see Chapter 5) of a greenhouse ranges between 20 and 25%, and the absorptivity (see Chapter 5) for both the cover and the structure ranges from 15% with a simple cover to 25% with a double cover (Baille, 1999).

At canopy level, the 'radiation saturation level' has been defined as the value above which the radiation increments do not involve parallel increases of photosynthesis (see Chapter 6). This situation (widely studied in laboratory growth chambers) may occur in greenhouses during the high radiation months at midday, but only on the leaves located on the upper strata of the crop, exposed to higher radiation, whereas the leaves of the lower strata (shadowed by the upper leaves) receive much less radiation, and are far from the saturation level. Therefore, considering the plant as a whole, it is not usual to achieve radiation saturation in species of edible vegetables, even under Mediterranean conditions (see Chapter 6), and normally it does not seem justified to decrease the radiation in the greenhouse for this reason. It might be necessary, however, to limit radiation for other

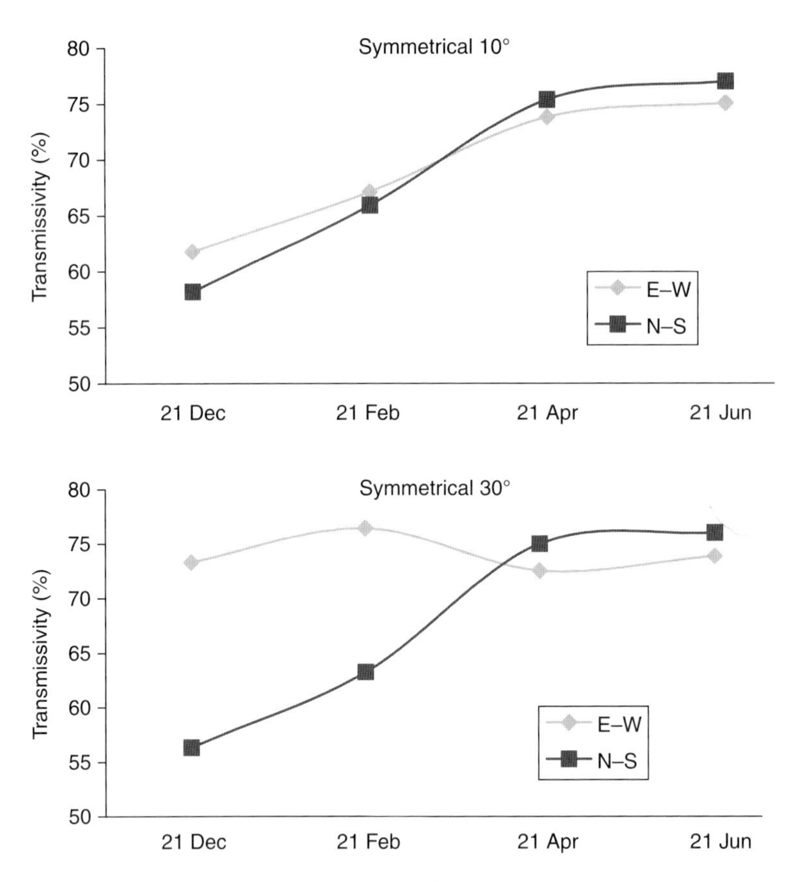

Fig. 3.5. Daily average global radiation transmissivity differences in a greenhouse with a symmetrical roof with a pitch of 10° (up), or 30° (down), depending on their east–west (E–W) or north–south (N–S) orientation on 21 December, 21 February, 21 April and 21 June. Data obtained by simulation, for a latitude of 37°N, for a low-tech parral-type greenhouse, with a new three-layer PE film cover, assuming only direct radiation (a hypothetical situation).

reasons (e.g. to limit temperature in insufficiently ventilated greenhouses, for fruit quality considerations, to improve the colour of the product, or to reduce water stress).

The anti-dripping effect of the inner side of a multilayer plastic film (once located over the greenhouse) prevents the formation of thick drops (when water vapour condenses on the film), reducing transmissivity (Jaffrin and Urban, 1990; Zabeltitz, 1999) and later contributing to water dripping on the crop, with negative effects on plant health (see Chapter 4).

Washing the plastic film covers and restricting greenhouse whitewashing as much as possible, together with a good selection of the plastic film, allow for a better availability of radiation inside the greenhouse (Montero *et al.*, 1985; Morales *et al.*, 2000). Other measures, such as limiting the shadows of the super-structure and of the installed equipment (thermal screens, ventilator's screens) and the outside windbreaks, are quite advisable.

The quality of radiation is affected by the soil particles deposited on the greenhouse cover, limiting the PAR even more than the IR radiation (Takakura, 1989).

We must also consider those crop management techniques which optimize the use of radiation (intercepting it) inside the greenhouse: (i) north–south orientation of the crop rows; (ii) plant density; (iii) plant training; (iv) pruning; and (v) use of mulching (Castilla, 1994). It is interesting to experiment with novel growing techniques, prior to their general adoption. In this respect, it is important to highlight the potentially negative influence in productivity of the use of white mulching in autumn–winter to increase the radiation intercepted by the crop, in unheated greenhouses under certain conditions, because of concomitant significant reductions in root temperature, both in crops grown in the soil or in artificial substrates (Lorenzo *et al.*, 1999, 2005; Hernández *et al.*, 2001).

3.4 Temperature

In an unheated greenhouse the main source of heat during the daytime is solar radiation, part of which is stored in the soil. During the night, the energy comes mainly from the soil, in the form of far IR radiation.

3.4.1 Air temperature

The air temperature inside the greenhouse is the result of the energy balance of the protection (Fig. 3.2). The greenhouse effect generally has two consequences:

1. At night, due to the limitation of IR radiation losses, the minimum temperatures are similar or slightly higher (1–3°C higher, depending on the covering material) than the outside (Plate 8). Nevertheless, on clear nights without wind, 'thermal inversions' may occur.
2. During the daytime, due to the 'heat trap' effect and the reduction in the convective exchanges (as the air is confined), the air temperature is higher indoors than outdoors, being possibly excessive when the radiation is high and the greenhouse is not efficiently ventilated (Plate 8).

The measurement of the air temperature must be performed in a representative location of the greenhouse, protected from direct sunlight and below a flow of air.

There are temperature differences inside a greenhouse, the east, west and north borders usually being colder (in the northern hemisphere), which can be minimized with double cladding (Berninger, 1989).

The thermal inversions in unheated greenhouses may occur on calm nights, with a clear sky, when the radiation losses towards the atmosphere are larger in the interval known as 'atmospheric window' (Rose, 1979; Fig. 2.14), if the greenhouse covering material is permeable to radiation in that interval (Day and Bailey, 1999).

3.4.2 Plant temperature

A thermometer located (without protection) inside a greenhouse during the night may provide a reading different to the actual air temperature. By approximation, we call it the 'radiative temperature' or 'actinothermal index' (see Chapter 2). These differences are larger with a normal PE cover than with a glass cover. This 'radiative temperature' better represents the plant temperature than the air temperature, during the night.

During the daytime, there are large differences in plant temperature, with respect to the air and also between parts of the plant, depending on the radiation intercepted, the water evaporation and the air movement, among other factors. The temperature of the flowers and the fruits depends greatly on their colour, which influences the absorption of radiation. For instance, green fruit is usually colder than red fruit.

Plant temperature has traditionally been assessed on the basis of the air temperature, corrected with the temperature of the greenhouse walls and the ground surface, and on the rate of evapotranspiration (Berninger, 1989). However, now technology is available for the direct measurement of plant temperature and the theory/philosophy of the 'speaking plant' in greenhouse crop management is widely accepted (Takakura, 1989; Challa and Bakker, 1995).

The recommended values for the greenhouse air temperature, for the majority of the species, range between 10 and 30°C, with daytime values higher than the night values. It is preferable (for economic reasons) to maintain lower temperatures at night, not only because the largest greenhouse energy losses happen at night (around 75%), but also because the lower temperature will also reduce respiration losses (Hanan, 1998).

In crops such as tomato, a night temperature around 15°C limits the losses by respiration and may be considered optimal. However, when the night temperature is much lower than 15°C, as the case may be when the greenhouse is located in low altitude subtropical latitudes, it becomes a limiting factor (Jensen and Malter, 1995).

The optimal temperatures usually decrease with the age of the plant, being higher during germination and the first stages of development. Different parts of the plant might have different temperature optima: for example the tomato plant growing point (at the top) would benefit from a higher temperature whereas the fruit (below) would rather be at a lower temperature. When the available radiation or the air CO_2 content increase, the optimal temperature, from a photosynthetic point of view, also increases. Other physiological processes may have different optimal temperature values, such as, for instance, the distribution of assimilates.

When temperatures are lower than optimal, normally, the quality of the product decreases, which may occur in winter in unheated greenhouses.

The capacity of most horticultural species to integrate the temperature on 24 h periods, or longer, within a range of 10–25°C, means that if the average temperature of the period is maintained (24 h) the growth won't change (Hanan, 1998), which allows for flexible heating management to reduce costs.

The concept of the thermal integral, applied to longer periods, allows for the prediction of the crop's phenology with the aim of scheduling the harvest (Mauromicale *et al.*, 1988). The thermal integral is based on the hypothesis that the lower the temperature the slower the growth rate and development of the plants will be (see Appendix 1 section A.2.1).

Within the range of horticultural species, we may distinguish three types of thermal requirements: (i) low demand, such as for lettuce, strawberry, endive, carnation, whose day/night thermal levels are around 10–25°C during the day/7–10°C during the night; (ii) medium demand, such as for tomato, beans, pepper, aubergine, courgette, with day/night thermal levels around 16–30°C during the day/13–18°C during the night; and (iii) high demand, which require values of 20–35°C during the day/18–24°C during the night, such as cucurbitaceous crops (melon, watermelon, cucumber) and some ornamentals.

3.4.3 Soil temperature

Close to the surface, the soil temperature follows a very similar pattern of development to the air temperature (i.e. sinusoidal shape and slightly delayed in relation to the air; Plate 8). The extreme values are buffered with the depth of the soil.

The type of irrigation system used influences the soil temperature; on the one hand by the water temperature itself and on the other hand by its effect on water evaporation from the soil and plants, and, therefore, the energy balance (Berninger, 1989).

3.4.4 Thermal inertia in the greenhouse

The soil, as well as the substrate in soilless crops, or the pots in ornamentals, are the heat sink of the greenhouse, that is they are centres of thermal inertia. The crop has little importance with regard to thermal inertia compared with the soil. A 10 cm soil layer has five to eight times more thermal capacity than the mass of a normal crop (Berninger, 1989).

During the night the soil returns part of the energy it has stored during the day back to the greenhouse. Use of white mulch, used to reflect radiation, limits the daytime heating of the soil and, thus, reduces the thermal inertia of the soil.

Mulching with organic matter, which limits the surface water evaporation and contributes CO_2 when decomposing, modifies the thermal inertia function of the soil in a similar way.

3.5 The Wind Inside the Greenhouse

Cladding the roof and sidewalls of the greenhouse with a film cover results in an enormous decrease of wind velocity with respect to the outside (Day and Bailey, 1999). This wind reduction has a great effect on crop physiology (see Chapters 6 and 9) and on the greenhouse microclimate, due to air being confined inside the structure (part of the greenhouse effect).

When greenhouse ventilation is limited and/or when stratification of the indoor air layers occurs, it becomes imperative to move the air with destratification fans (see Chapter 9).

3.6 The Greenhouse Atmosphere

A closed greenhouse is not completely airtight, because complete tightness is never technically achieved. The air exchanges with the outside depend greatly on the outside wind. The hourly exchange rate (R) is defined as the quotient between the outside incoming volume of air in 1 h and the total volume of the greenhouse (see Chapter 8), or, the average air volume exchanged per ground unit surface in 1 h (cubic metres of air per square metre of ground surface per hour; $m^3\ m^{-2}\ h^{-1}$).

3.6.1 Greenhouse ventilation

Greenhouse ventilation is justified for three main reasons: (i) to avoid excessive heating during the daytime: (ii) to ensure adequate levels of CO_2; and (iii) to control the humidity.

In large greenhouses, an hourly renewal rate (R), during the summer, of 30–60 volumes h^{-1} (i.e. volumes of greenhouse per hour) must be achieved (see Chapter 8). Normally, the static ventilation is not sufficient, except with large vents and constant wind.

3.6.2 Air humidity

The most important contribution to the water vapour exchanges in a greenhouse comes from crop transpiration, although water evaporation from the humid soil also contributes, as well as condensation of water vapour on the different greenhouse surfaces when they get cold (Day and Bailey, 1999).

During the daytime, the RH decreases in the greenhouse when temperature increases (Plate 8), although the absolute humidity has increased due to transpiration. When ventilating, the outside air (colder and dryer) which enters the greenhouse decreases the RH, because this outside air gains heat faster than moisture. As a general rule, VPD values larger than 1.1 kPa in the winter, or 2.7 kPa in the spring, should be avoided (Berninger, 1989).

At night, as the greenhouse gets colder, the RH increases and may reach saturation (Plate 8), and then condensation occurs over the greenhouse surfaces, starting at the colder ones, such as the cover. If the greenhouse has enough slope in the roof, the condensate will slide down, being collected in proper gutters, if available. With a low roof slope the water will drip over the crop.

If the greenhouse has a double cover, the inner cover will be warmer (than with a simple cover) and it will take longer for the water vapour to condense.

The water transpiration by the crop has a great effect on air humidity. A well-developed crop evaporates water actively, shades the soil and limits the warming of the greenhouse during the daytime. A greenhouse without a crop and without irrigation will be much warmer on a sunny day and the day/night variations of temperature and RH will be larger.

3.6.3 CO_2 content

The ventilation, photosynthesis and respiration of the plants and the CO_2 generation in the soil (by root respiration and decomposition of organic matter), influence the CO_2 content of the greenhouse air.

At night, the accumulation of CO_2 due to plant respiration increases the greenhouse concentration above the values outside (Fig. 3.6). During the daytime, due to photosynthesis, the CO_2 content decreases with respect to the outside value (Fig. 3.6). In a closed greenhouse, on a sunny day, the CO_2 content may decrease below 200 ppm, it being a limiting factor for crop production (Lorenzo *et al.*, 1997a, b).

The goal of ventilation is to avoid CO_2 depletions of more than 30 ppm, with respect to the normal atmospheric content (360 ppm). The ventilation required to decrease the temperature, which ranges, at least, from 20–30 renewals h^{-1}, is usually sufficient to maintain appropriate CO_2 levels (Fig. 3.6). In winter, at noon, the normal consumption is 1.5–2 g CO_2 m^{-2} h^{-1} increasing in spring to average values of 3 g CO_2 m^{-2} h^{-1} in the Mediterranean area (Berninger, 1989).

3.6.4 Pollutant gases

In greenhouses, pollutant gases (besides those that originate from plant protection treatments) are the result of using inappropriate fuels (with excess impurities) or defective burning, when the combustion gases are injected inside the greenhouse. Combustion defects mainly generate carbon monoxide (CO), ethylene (C_2H_4) and nitrogen oxides (NO_x). The most usual impurities (presence of sulfur) generate sulfur dioxide (SO_2) (see Chapter 7).

The plastizicers present in some materials (e.g. silicone) may also be toxic to the plants, but their concentration in normal greenhouses does not generate problems, which may occur otherwise in small growth chambers (Langhams and Tibbits, 1997).

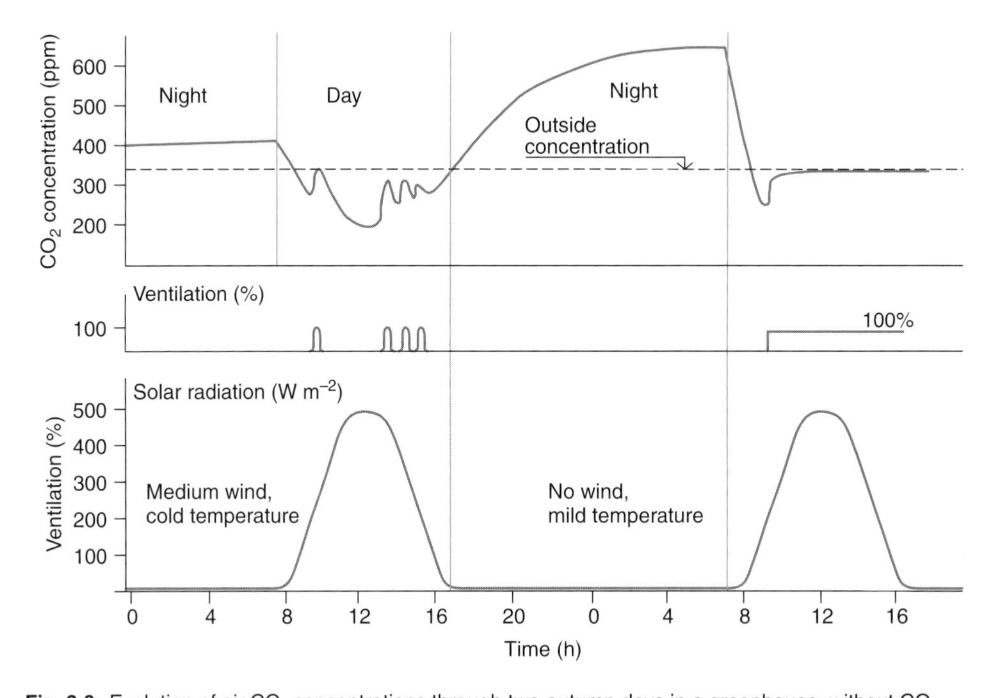

Fig. 3.6. Evolution of air CO_2 concentrations through two autumn days in a greenhouse, without CO_2 enrichment and a tomato crop, as influenced by ventilation and outside global radiation (adapted from Wacquant, 2000).

3.7 Summary

- Inside a greenhouse, radiation, temperature and atmospheric composition are modified generating a different microclimate.
- The 'greenhouse effect' is a consequence of two different phenomena: (i) a confinement effect (convective effect), due to the reduction in the air exchanges with the outside atmosphere; and (ii) a radiative greenhouse effect, caused by the existence of the greenhouse cover, which is a screen that is transparent to the Sun's rays but has low transparency to far IR radiation (emitted by the soil, the vegetation and the inner elements of the greenhouse).
- The first modification which a greenhouse generates in the climate parameters is a decrease of solar radiation, due to the presence of the greenhouse cover.
- The transmissivity of a greenhouse is the fraction of solar radiation which penetrates inside the greenhouse. It is usually expressed as a percentage.
- The transmissivity of the greenhouse depends, among other factors, on: (i) the cloudiness; (ii) the position of the Sun in the sky; (iii) the geometry of the greenhouse cover; (iv) its orientation; (v) the covering material; and (vi) the structural elements of the greenhouse.
- When direct radiation conditions prevail, on sunny days, the geometry of the cover (shape and pitch of the roof) and the orientation greatly influence the greenhouse transmissivity. When diffuse radiation prevails (cloudy days) they have little influence.
- In the climatic conditions of the Mediterranean coast, with an abundance of sunny days, the east–west greenhouse orientation transmits more radiation in autumn–winter, than the north–south orientation, on gabled greenhouses with a certain roof pitch (around 30° roof angle). If the roof pitch is lower (around 10° roof angle) the differences in transmissivity due to orientation (north–south or east–west) are small.

- There is less uniformity of radiation inside a greenhouse that is east–west oriented than inside a north–south oriented greenhouse. However, these differences in uniformity between multi-span greenhouses oriented east–west and north–south are attenuated by: (i) the greater the height of the greenhouse; (ii) the lower the span width; and (iii) the radiation diffusion characteristics of plastic films used nowadays.
- Transmissivity to solar radiation must be maximized in the greenhouse, because an increase in solar radiation involves a parallel increase in yield. This requires careful consideration when choosing the type of greenhouse, the covering material, and the crop management technique (crop rows orientation, plants density, pruning, training, etc.) that will be used.
- The air temperature in the greenhouse is, normally, higher during the day than the outside air temperature and similar or slightly higher during the night than the outside air temperature. However, in unheated greenhouses, during clear nights (without clouds) with no wind, 'thermal inversion' may occur (the temperature of the air in the greenhouse being lower than outside).
- The plant temperature varies greatly during the day (with respect to the air temperature and between different parts of the plant) depending on the intercepted radiation, transpiration and the air movements, among other factors.
- The soil has great thermal inertia and during the night it returns part of the energy it has stored during the day back to the greenhouse. The crop, due to its low mass, has little importance in greenhouse thermal inertia.
- The wind is very limited inside the greenhouse, in relation to the outside. Therefore, it may be convenient to move the inside air, because the lack of air movement negatively affects

photosynthesis and, as a consequence, the yield.

- Greenhouse ventilation is necessary to avoid excessive warming during the daytime, to maintain minimum acceptable CO_2 levels and to avoid excessive air humidity.
- At night, as the greenhouse gets colder, the RH increases and may reach saturation, and then condensation occurs over the greenhouse surfaces.
- Transpiration by the crop has a great effect on air humidity and the decrease of temperatures.
- The air CO_2 content during the daytime, if there is no renewal of the inside air, may decrease to values of the order of 200 ppm, negatively affecting the yield. During the night, CO_2 levels are higher than in the outside air, due to the CO_2 supplies from plant respiration and from the soil.

4

The Plastic Greenhouse

4.1 Introduction

The development of plastic materials has been one of the most decisive factors in the great expansion of greenhouse industries around the world. During the second half of the 20th century, the lower cost of plastic-covered greenhouses in relation to traditional glasshouses, promoted their use without heating in many mild climate areas, but advanced designs also became commonplace in northern colder countries, because of the significant savings in energy use.

The light weight of plastic materials compared with glass have allowed for a notable reduction in the supporting structures for the cladding materials and, as a consequence, of the building cost.

It is difficult to define the distinction between high tunnels and greenhouses, but in this book high tunnels (i.e. those structures that are high enough to allow workers to walk inside them and work on the crop) are included within the term 'greenhouse'. However, facilities which do not use sunlight as the main source of radiation (e.g. growth chambers, phytotrons, etc.) are not included in the term.

A greenhouse is a structure which allows for the delimitation of a crop compartment, in which the climate differs from that outdoors, due to modifications caused by the cladding material in exchanges between the soil, the substrate and the canopy with the surroundings (Villele, 1983).

The new regulation UNE-EN-13031-1 (*Greenhouses: design and construction*; the Spanish version of the European regulation EN-13031-1) defines the greenhouse as a structure used for growing and/or to provide plant and yield protection, optimizing the solar radiation transmission under controlled conditions, to improve the crop environment and whose dimensions allow people to work inside.

4.2 Evolution of the Greenhouse Concept

In greenhouses, the main function of increasing the temperatures in relation to the open field, as a consequence of the 'greenhouse effect', was previously the main consideration. However, this is not always the main consideration now, although it remains the most important factor during short periods of low outside temperatures, as is the case in many Mediterranean greenhouses. Indeed, in some regions what is more important is the 'shading effect' in the season of high radiation, or the 'windbreak effect', at least during some periods of the year.

The covering of greenhouses with a screen (net) instead of with a plastic film, is a recent introduction at low latitudes. It limits the radiation and the outside wind without increasing the temperature. In very arid or desert areas, the insulation that the greenhouse provides in relation to the outside environment (which is very dry and hot) represents a different concept to that of conventional protected cultivation, as it provides for an increase in the ambient humidity and a reduction in the temperature, provided the crop is well irrigated. This has prompted talk about the 'oasis effect' (Sirjacobs, 1988; Photo 4.1).

In a greenhouse, the reduction of the radiation with respect to the open air means there is a decrease in the irrigation requirements (as evapotranspiration decreases), which together with a significant increase in yield results in a much more efficient use of the irrigation water (Stanghellini, 1992). This is relevant in regions with scarce water resources.

Photo 4.1. A banana crop grown in a greenhouse. In very arid or desert areas, the greenhouse generates a certain 'oasis effect'.

In tropical and subtropical regions of high rainfall, the 'umbrella effect' predominates in greenhouses (Photo 4.2), because the primary purpose of their use is to avoid the harmful effects of frequent and heavy rains over the crops. Here the greenhouse effect is normally undesirable, as the natural thermal conditions are sufficient, or even excessive, for the development of the crops (Garnaud, 1987).

4.3 Geographical Production Areas

In the past, greenhouses were located near cities, were the produce was destined for sale. Until the middle of the 20th century, transport difficulties confined the use of greenhouses to the areas adjacent to the centres of consumption.

Nowadays, the location of the production areas is influenced, mainly, by climate conditions and by the type of species to be cultivated, which determine the costs and quality of the yield, as well as the transport costs to the consuming markets. Other technical factors (e.g. availability and quality of irrigation water, soil characteristics) and socio-economic factors (e.g. commercial channels, levels of technology in the area, financing possibilities, communication and electricity infrastructures, farm size, availability of inputs for horticultural production) also have an influence at a time marked by the internationalization of trade and, in general, the globalization of the economy.

The displacement of greenhouse production areas that occurred during the last decades of the 20th century are a good example of this. In Europe, greenhouse vegetable production has moved from the north to the Mediterranean area (Photo 4.3), where the better climate conditions in autumn and winter enable production at notably lower costs, without using heating (Tognoni and Serra, 1988, 1989). Similar movements occurred in the USA when carnation cultivation partly moved to California and Florida (with a better climate) from the traditional production areas (East Coast) and from Colorado, where an important sector had emerged due to its

Photo 4.2. In tropical and subtropical regions of high rainfall, the protection against the rain becomes, in many cases, the main purpose of using greenhouses.

Photo 4.3. The vicinity of the sea provides good temperature conditions for greenhouse cultivation in Mediterranean coastal areas.

good light conditions which induced a product of better quality (Nelson, 1985; Wittwer and Castilla, 1995). Afterwards, Colombia partially displaced some carnation growers, due to its lower growing costs and their good quality (due to the climate conditions), despite the high transport costs to the North American markets. The displacement of the production areas to regions with milder climate has enabled the use of less sophisticated greenhouses and therefore, those of cheaper construction which enabled production at more competitive costs (Castilla *et al.*, 2004).

However, it must be pointed out that the most technologically advanced and most productive greenhouse industries are centred in northern colder countries (e.g. The Netherlands, Belgium, Canada, Japan). The main reason for the survival of the greenhouse industries of the north in an increasingly competitive world market is that, despite the high heating costs, and contrary to public misconception, it is cheaper to heat a space than to cool it and this offers better opportunities for year-round production in the north than in the south. In reality, it is not possible to control the greenhouse air temperature within acceptable levels during the summer months in hot climates with present greenhouse cooling systems. In turn, the good prospects for year-round production (and much higher yields, of stable product quality) allow for investment in more expensive (and more sophisticated) greenhouses in the north.

Pot plants are more limited in where their production areas can be located due to their high transport costs. In a similar way, species such as roses, which need strict climate control to obtain high quality flowers, must be cultivated in greenhouses with a good level of technological equipment. Sometimes the commercial aspect, such as when produce is sold directly to the final consumer, may dictate the construction of the greenhouse in specific locations, such as peri-urban areas.

The trend to suppress political and geographical borders, and the consequent restriction on barriers for export (agreements of the World Trade Organization, WTO), will have implications in the long term with regard to the prevalence of horticultural areas with higher production efficiency, which are able to adapt themselves to the requirements of the market. The phytosanitary barriers (limits placed on the export of produce due to pesticide residues, or the possibility of introducing pests or pathogens from the exporting country) are expected to become even more relevant than they are today in this sector (Castilla *et al.*, 2004).

Given the need for export, and the great importance that the external market has in countries with a well-developed horticultural sector, it is expected that in the future international competition will increase and will become an even greater challenge.

4.4 Climatic Suitability for Greenhouse Vegetable Production

4.4.1 Introduction

Today, technology allows greenhouse cultivation of any horticultural species in any region of the world, provided that the greenhouse is properly climatized, but a profitable cultivation of the target crop requires a much more strict selection of the region, depending on its climatic conditions and the requirements of the selected horticultural crop.

Solar radiation is the first climate parameter to be considered when evaluating the climate suitability of a region for protected cultivation. The length of the day and the global solar radiation intercepted by a horizontal surface through the daytime hours (see Chapters 2 and 3) determine the total daily radiation (global solar radiation integral in that period). The ambient temperature is another basic climate parameter to consider.

The stability of both values (radiation and temperature) through the different months of the year enables the representation of their average monthly values (obtained by averaging data sets for several years) for a given location, in a climate diagram with an elliptic shape, which represents the location's climate (Fig. 2.15).

Other climate parameters, such as soil temperature (which is linked to the air temperature), wind, rainfall and air composition (humidity and CO_2), should also be considered, although their influence is not as important, when evaluating climate suitability.

Depending on the climate characteristics of a region and the requirements of the crops to be grown it will be necessary to choose a certain type of greenhouse. For instance, in a region with a tropical humid climate, where the defence from the rain is the main objective of the greenhouse (prevalence of the umbrella effect), it will be necessary to choose a different type of construction from that used in a region of semi-desert or with a Mediterranean climate.

4.4.2 Climate requirements of vegetables

The most usual cultivated species in greenhouses are vegetables with medium to slightly high thermal requirements (tomato, pepper, cucumber, melon, watermelon, marrow, green bean, aubergine) with the aim of extending the production season beyond the conventional growing season. Nowadays, in cases of very high product prices, greenhouse production in geographical areas that do not have perfectly suitable climate conditions force a notable and expensive artificial intervention over the climate parameters. In any case, the economic results will determine where greenhouses are located.

The cited horticultural crops are, essentially, warm season species, adapted to average ambient temperatures ranging from 17 to 28°C, and whose limits we can establish as between average temperatures of 12°C (minimum) and 32°C (maximum) (Nisen *et al.*, 1988). They are sensitive to the cold, and suffer irreversible damage with frosts. The persistence of temperatures below 10–12°C for several days, as well as temperatures above 30°C, in the case of dry air, or higher than 30–35°C in cases of high air humidity, affect their productivity (Nisen *et al.*, 1988). In the past, it was accepted that a daily variation between day and night average temperatures (thermal periodicity, between 5 and 7°C) was necessary for proper physiological functioning (Nisen *et al.*, 1988). Currently, it has been stated (Challa and Bakker, 1995; Erwin and Heins, 1995) that this day/night temperature difference, known as DIF, is not a requirement but a tool for the management of greenhouse crops (see Chapter 7, section 7.4.6).

The minimum daily requirements of radiation of these species are estimated at around 8.5 MJ m^{-2} day^{-1} (equivalent to 2.34 kWh m^{-2} day^{-1}) in the three shortest months of the year (November, December and January in the northern hemisphere; May, June and July in the southern hemisphere). This means around 6 h of light day^{-1}, which translates to a minimum of 500–550 h of light in these three months (Nisen *et al.*, 1988). The duration of the day and the night is influenced by the geographical latitude together with the time of the year. The photoperiodic requirements of some crops are linked to the duration of the night.

Other desirable climate parameters for these species would be a soil temperature above 14°C and an ambient relative humidity of 70–90% (Nisen *et al.*, 1988).

4.4.3 Obtaining the required climate conditions

The lack of possibilities to practically increase, at a reasonable cost, the natural radiation conditions (except in very sophisticated greenhouses and with high value crops), has made it necessary to design and locate greenhouses with the aim of maximizing the interception of solar radiation during the months of autumn and winter. Therefore, the natural radiation availability is a critical limiting factor when considering the establishment of greenhouses.

Due to the interdependence between air and soil temperatures (even with less oscillation inside a greenhouse than outside), achieving a suitable air temperature involves proper soil temperature values as well.

According to the methodology proposed by the FAO (Nisen *et al.*, 1988), protected cultivation in greenhouses or high tunnels enables daytime thermal increases in relation to the outside air, mainly depending on: (i) the characteristics of the cladding material; (ii) the outside wind velocity; (iii) the incident solar radiation;

and (iv) the transpiration of the crop that is grown inside the greenhouse, possibly reaching very high values (Fig. 4.1). By contrast, night temperatures are only slightly increased in relation to the outside (2–4°C, at most) and, in some cases, are lower (thermal inversion).

The maximum thermal increases vary with the latitude and, for each specific location, with the time of the year, as the solar radiation changes (Fig. 4.2).

To increase the low temperatures the most usual solution is to heat the greenhouse, which is not always profitable. In some cases, a highly isolating system can avoid the temperature decrease during the night, as is done in the 'lean-to greenhouse' type in China, where a special curtain of canes and wood materials is manually placed over the greenhouse cover at sunset, and removed at sunrise, to avoid relevant temperature decreases at night, a highly labour-intensive activity.

To limit thermal excesses, the renewal of the interior air by means of ventilation is the classic and most economic tool (see Chapter 8).

The hourly air renewal rate needed to limit the temperature gradient to an acceptable value, depending on the maximum predictable solar radiation (Table 4.1), can be very high (Fig. 4.3), and this may not be attainable in practice without mechanical ventilation or evaporative cooling.

4.4.4 Climate suitability

The fundamental requirements of the thermophilic horticultural species previously cited as candidates for out-of-season cultivation (e.g. tomato, pepper, melon, watermelon) would be:

1. A minimum global radiation of 8.5 MJ m^{-2} day^{-1} (equivalent to 2.34 kWh m^{-2} day^{-1}).
2. Average ambient temperatures between 17 and 27°C in coastal areas, and 17–22°C in inland areas far from the sea. This distinction is based on the higher daily thermal oscillations observed in continental climates (around 20°C) than in marine-type climatic regions (10°C), and by setting the upper threshold of air temperature at 32°C (Nisen *et al.*, 1988).

Considering the usual unacceptably high cost of active intervention (e.g. heating systems) on the microclimate in the case of unsophisticated greenhouses, the minimum greenhouse temperatures are usually similar to those of the open air. The maximum temperatures, with passive normal ventilation would be in some cases around 10°C higher than outside. This would involve an

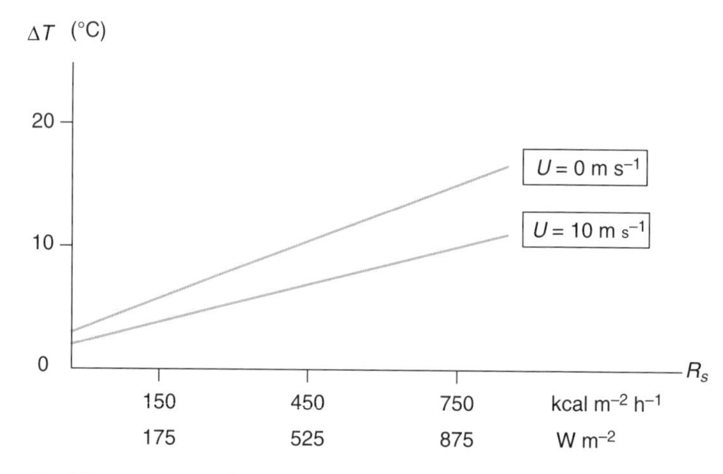

Fig. 4.1. Example of the temperature increase (ΔT) in a closed greenhouse, well irrigated, as a function of the solar radiation intensity (R_s) and wind velocity (U) (adapted from Nisen *et al.*, 1988).

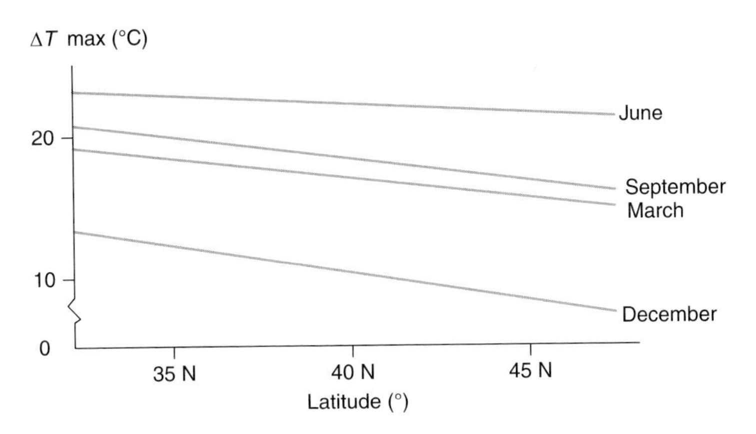

Fig. 4.2. Example of the maximum amplitude of the temperature increase (ΔT max) in a closed greenhouse, well irrigated, at different times of the year and at different latitudes (adapted from Nisen *et al.*, 1988).

Table 4.1. Values of maximum global solar radiation intensity (W m^{-2}) predictable as a function of latitude, at noon (northern hemisphere). (Source: Nisen *et al.*, 1988.)

Latitude	December	March	June	September
32°N	550	915	1050	855
38°N	455	845	1025	780
44°N	355	770	995	685

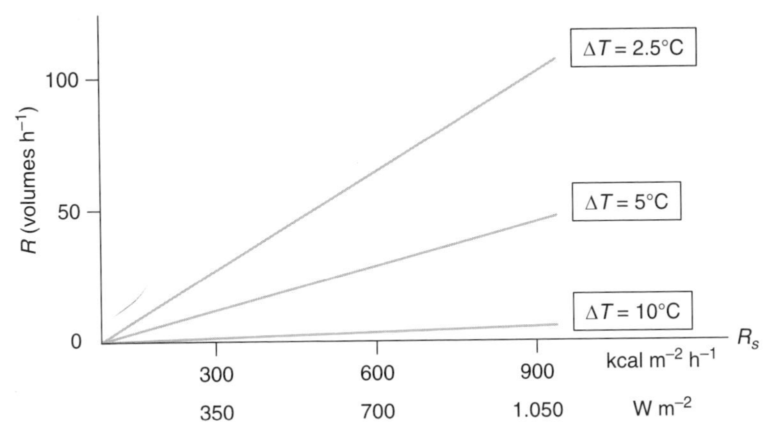

Fig. 4.3. Estimation of the hourly air exchange rate R in a well-irrigated greenhouse, to maintain ambient warming to a given air temperature (ΔT) value with respect to the outside air as a function of solar radiation (R_s) (adapted from Nisen *et al.*, 1988).

increase of the average temperatures of around 5°C.

In view of these predictions, the thermal climate limits for protected cultivation, without active climate control equipment, would be between 12 and 22°C of average temperatures in coastal areas and 12–17°C in inland areas. Outside these limits, protected cultivation would require active climatization systems: heating and mechanical ventilation or misting for cooling.

Figure 4.4 represents the climate diagram for Almeria. The solar radiation in December is at the minimum limit. The temperatures are slightly lower than 12°C (minimum threshold) in the month of January, which denotes that it would be necessary to heat the greenhouse during this month. Except for the summer months (June, July, August and September), the remaining months have thermal conditions which are suitable for protected cultivation (between 12 and 22°C), with efficient ventilation. In summer, it is necessary to limit the thermal excesses to cultivate inside greenhouses.

Obviously, this method constitutes only a primary approach to evaluate the climate suitability of a region for the cultivation of thermophilic vegetable species. In a similar way, we can evaluate the climate suitability of a certain location for greenhouse cultivation of other less thermally demanding species (e.g. lettuce, Chinese cabbage).

The use of screens instead of plastic films as a covering material, which do not create a greenhouse effect and cause a shading effect and a windbreak effect, is an option for protected cultivation which is being used in low rainfall areas with mild winter temperatures (e.g. the Canary Islands or tropical American areas) or during the summer season in the uplands of medium latitudes (i.e. the province of Granada, Spain).

4.5 The Plastics

4.5.1 Introduction

The name 'plastics' comes from the main characteristic of these types of material which is their plasticity, or capacity to be moulded or shaped. The plastics are composed of macromolecules (polymers), made by the union of other smaller molecules (monomers), obtained in the industrial process of polymerization, and various additives. If the polymer is made of just one

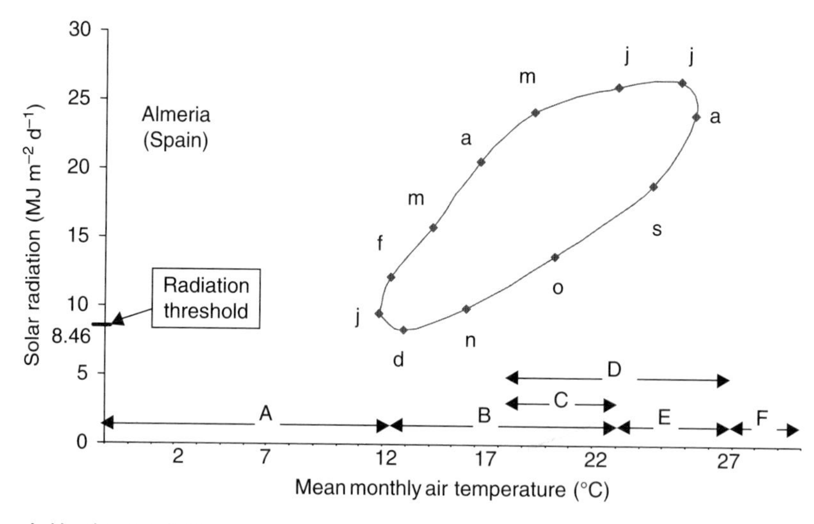

A. Heating required
B. Protected cultivation possible without heating but with natural ventilation (passive)
C. Open air cultivation possible (inland areas)
D. Open air cultivation possible (coastal areas)
E. Need to use techniques to decrease temperatures
F. Excessive temperatures

Fig. 4.4. Estimation of the climate suitability of Almeria (Spain, 37°N) for the cultivation of thermophilic vegetable species in greenhouses, from the monthly average values of air temperature and solar radiation (adapted from Nisen *et al.*, 1988).

monomer it is called a homopolymer, and when it is made from two or more polymers it is called a copolymer (Díaz *et al.*, 2001). Almost all the monomers used in the plastics industry come from petroleum.

4.5.2 Plastic materials commonly used in agriculture

The plastics have a great variety of uses in agriculture, not just as films but in many other forms (tubes, boxes, threads).

The plastic films have a low weight, as compared with glass. For instance, 1 m^2 of polyethylene (PE) film, 0.2 mm thick, weighs a bit less than 200 g, while 1 m^2 of glass, 2.7 mm thick, weighs 6.5 kg. This characteristic has favoured the use of PE as greenhouse cladding material.

The most common plastic materials used as agricultural films (Table 4.2) are the low density polyethylene (LDPE, with a density less than 0.93 kg m^{-3}), the copolymer of ethylene and vinyl-acetate (EVA) and the plasticized polyvinyl chloride (PVC). The more common materials, used as rigid panels, are polycarbonate (PC), polymethyl methacrylate (PMMA), rigid PVC and polyester reinforced with fibreglass.

4.5.3 Plastic additives

Additives are incorporated in the process of plastic making in order to provide certain properties or improve their characteristics, without altering the molecular structure of the polymer. They can be processing additives (that ease the fabrication process) or functional additives (that give certain qualities to the plastic), reaching up to 10% of the final weight of the product (Díaz *et al.*, 2001).

The most important functional additives are: (i) the photostabilizers; (ii) the anti-acids; (iii) the long and short-wave IR radiation blockers; (iv) additives that modify the surface tension; and (v) the luminescence additives (Díaz *et al.*, 2001).

Photostabilizers prolong the life of the material by delaying ageing of the material; ageing being precipitated by polymer degradation due to UV light of the Sun's rays. Common photostabilizers are the nickel quenchers and diverse organic compounds, such as the 'HALS' (hindered amine light stabilizers) (Díaz *et al.*, 2001). The latest generation of 'HALS' are known as the 'nor-HALS'. The latter are more stable than the conventional HALS, in the presence of pesticides and other aggressive acid conditions, and are being used for long-life films and in those greenhouses where sulfur is commonly applied (e.g. rose culture).

The nickel quenchers give a yellowish-green colour to the films and their use is diminishing, due to their residues. The HALS additives are non-coloured.

The anti-acid additives improve the resistance of the HALS, increasing their useful life. Zinc oxide, used as anti-acid, gives the films a light-diffusing effect.

Table 4.2. Materials commonly used for greenhouse cladding.

	Plastics		
Glass	Rigid panels	Flexible films[a]	Screens
	Fibreglass reinforced polyester	Low density polyethylene (LDPE)	
	Polycarbonate (PC)	Conventional	
	Polyvinyl chloride (PVC)	Long life or UV	
	Polymethyl methacrylate (PMMA)	Thermal or IR	
		Ethylene vinyl-acetate copolymer (EVA)	
		PVC	
		Others: Polyvinyl fluoride (PVF) (Tedlar), Mylar, Polyester	

[a]May be used as multilayer films.

The IR radiation blockers have allowed the fabrication of thermal PE films. These additives can be mineral products of very diverse type: natural or synthetic silicates, aluminium or magnesium hydroxides (Díaz *et al.*, 2001).

The additives that modify the surface tension can be of two types: (i) the anti-static ones, that have an anti-dust effect, avoiding the electrostatic attraction of dust on the outside of the film; and (ii) the anti-drip ones, that avoid the generation of water droplets in the inner side of the film (increasing the surface tension of the film) in such a way that water vapour is condensed as a water film (Díaz *et al.*, 2001).

The anti-static additives reduce the static electricity, avoiding the attraction of dust particles. However, they induce the film surface to become sticky, and dust that does settle by gravity or brought on the wind cannot easily be washed away by rainfall. Therefore, these additives are not practically useful.

The IR radiation blockers known as NIR (i.e. near IR, with wavelengths between 760 and 2500 μm) reduce the heat load of greenhouses covered with films with these additives. They are used in tropical areas to limit the greenhouse effect (Verlodt and Verchaeren, 2000).

The luminescence additives (fluorescent or phosphorescent) can alter the 'quality' of the light transmitted by films that contain them. These additives absorb wavelengths of less interest for plant growth and transform them in other wavelengths more effective for photosynthesis. Normally, this effect is reached converting UV radiation (that is absorbed) in visible radiation or converting green radiation into red radiation that is more efficient in the photosynthesis process (Yanagi *et al.*, 1996).

Other interesting additives for greenhouse plastic films are those that alter the red/far red (R/FR) ratio. These can have a great influence in plant photomorphogenesis, through the plant pigment 'phytocrome' (see Chapter 6). Some luminescence and IR radiation-blocking additives also have this R/FR-altering effect.

The additives that block UV solar radiation affect the behaviour of some insects, limiting their vision capacity, and the development of some fungal diseases.

4.5.4 Properties of plastic films

When considering greenhouses, the most important properties of plastic films are the radiometric and mechanical characteristics as well as their durability and their behaviour when water vapour condenses on the film.

Radiometric properties

For greenhouse use, the relevant radiometric properties of plastic films are their transmissivity to solar radiation, in the different wavelengths (UV, PAR, NIR and to long IR wavelengths). In addition, the reflection and adsorption characteristics are relevant (see Chapter 5) (Fig. 4.5).

A material which lets a great proportion of the radiation pass through it is called transparent, whereas a material which prevents radiation from passing through it is called opaque. When crossing a translucent material, radiation is dispersed (scattered) in all directions and this dispersed (diffuse) radiation generates shadows that are less defined and less sharp than those generated outside.

The perfect covering material, besides having good insulating properties should allow 100% transmission in the PAR range (Papadakis *et al.*, 2000), which is unattainable. Normally, a good covering material must be transparent to NIR solar radiation and be as opaque as possible to long-wave IR, to have a good greenhouse effect. In addition, a good covering material must be diffusive (Fig. 4.6), and have good insulation characteristics and behave well with regard to condensation.

The transmission characteristics to PAR radiation change depending on the type of radiation, direct or diffuse, and it is important to avoid the confusion between transmissivity to diffuse radiation of a film and the proportion of direct radiation

(a)

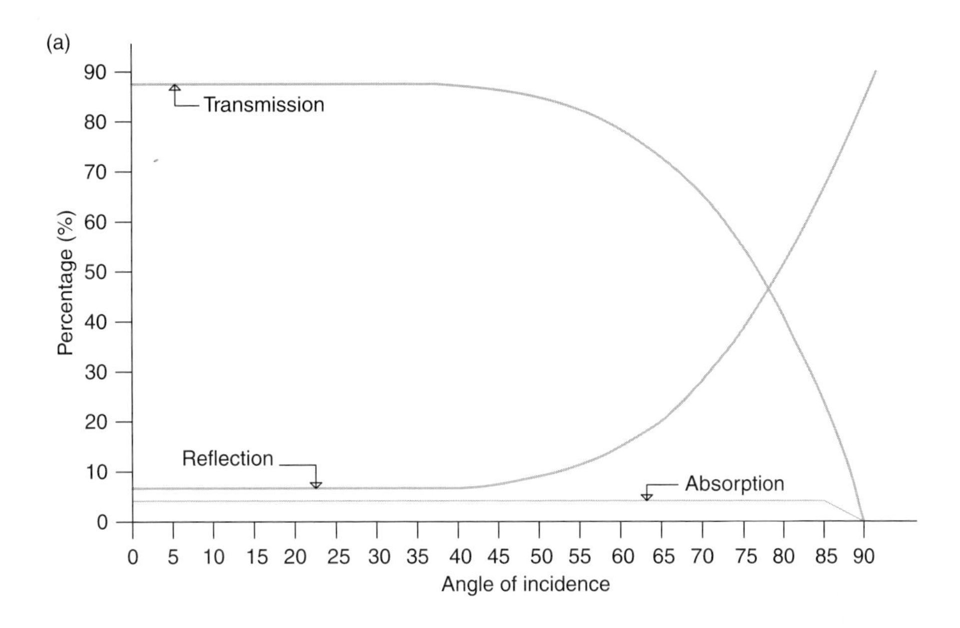

(b)

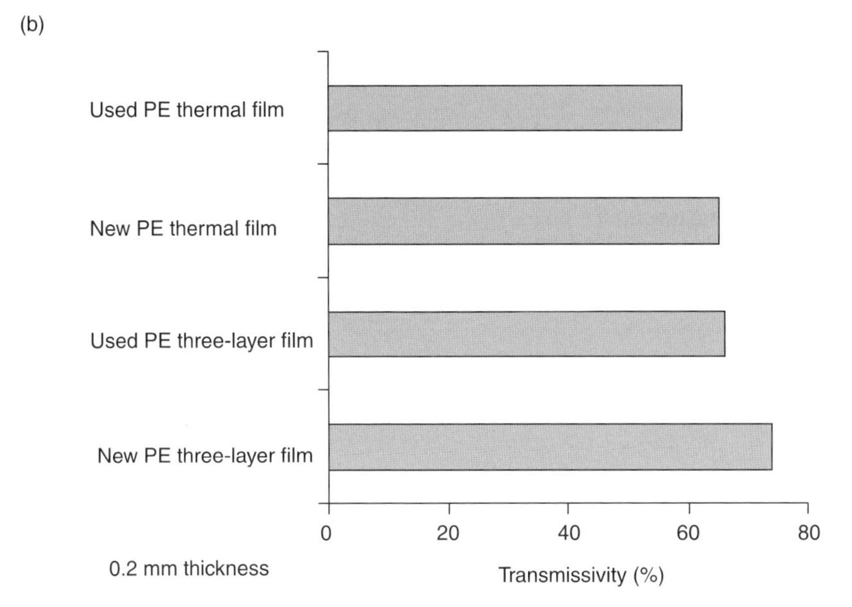

Fig. 4.5. (a) Influence of the angle of incidence of direct radiation on the transmission, reflection and absorption of a greenhouse covering material. (b) Transmissivity of diffuse solar radiation of different greenhouse covering materials (Montero *et al.*, 2001).

that is converted into diffuse radiation when passing through a film (Papadakis *et al.*, 2000). On sunny days, with a diffusive film (Fig. 4.6), the diffuse radiation inside a greenhouse can be three or four times higher than the outside diffuse radiation (Baille *et al.*, 2003; see Appendix 1 section A.3.1).

The transmission characteristics are altered by the presence of condensation and water vapour on the film, depending on the shape that the condensed vapour

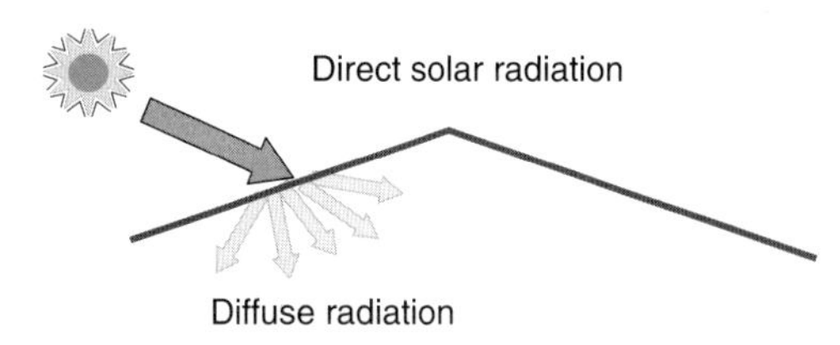

Fig. 4.6. Some greenhouse-covering plastic films have the power to diffuse solar radiation, increasing the proportion of diffuse radiation inside the greenhouse. Therefore, on a sunny day, inside a greenhouse covered with this type of film the shadows are less defined and sharp than outside.

acquires (droplets or water film). The condensation on the inner face of the plastic film can notably reduce the transmissivity to solar radiation, depending on the shape of the droplets (Fig. 4.7). Indeed, condensation reduces the transmission to long-wave IR. The dust and dirt accumulated over the film decrease its light transmissivity, in the same way as the ageing of the material which specifically limits the radiation (Matallana and Montero, 1989; Montero and Antón, 2000b; Papadakis *et al.*, 2000).

When evaluating the transmissivity of new materials it is important to relate the results to the prevailing light during the test (light quality) (Kittas and Baille, 1998; Kittas *et al.*, 1999).

Mechanical properties

The mechanical characteristics of the film depend both on the intrinsic factors of the material (i.e. type of raw material) as well as on the conditions during their transformation into a film (homogeneity in the distribution of additives, proper and uniform thickness). The degrading action of solar radiation also affects these properties, mainly depending on the exposure time.

From the point of view of the grower, the most relevant mechanical properties are resistance to traction, tearing and impact (Briassoulis *et al.*, 1997b; Marco, 2001). The resistance to traction, which evaluates the capacity of the film in the greenhouse to withstand tensile forces, is important during the assembly of the film in the greenhouse and for withstanding strong winds. The resistance to tearing of the plastic is important to avoid tears due to accidental cuts of the film, which are not unusual in the low-cost 'parral'-type greenhouses. The resistance to impact is needed so that the film can withstand hail and strong winds.

Durability

The durability of an agricultural film is defined as the shelf life during which the film retains, at least, 50% of its initial mechanical properties (Díaz *et al.*, 2001).

The degradation of the film in the greenhouse occurs mainly by the action of UV radiation from the sun, which degrades the polymer (photodegradation). In addition, durability is also influenced by other factors, for example: (i) climate conditions (temperature and radiation mainly); (ii) the additives used; (iii) the thickness of the film; and (iv) the management of the greenhouse (pesticides used, assembly of the film) among others (Briassoulis *et al.*, 1997a, b). Both halogenated and sulfurated pesticides attack the photostabilizers (HALS and nickel quenchers), shortening the shelf life of the films (Barahona and Gómez-Vázquez, 1985; Gugumus, 2000).

The artificial ageing of films in the laboratory using special lamps, which reproduce similar conditions (but more intense) to the natural solar radiation, allow for a quick estimate of the longevity characteristics of a film.

Behaviour with regards to condensation

Anti-dripping additives increase the surface tension in the film, so that the water vapour condenses in the shape of a film without producing droplets, so compared with the same film without anti-dripping additives there is an increase of light transmission due to the reduced reflection of light (Pearson *et al.*, 1995; Von Elsner *et al.*, 2000a). However, the durability of the effect is normally shorter than the shelf life of the film (Papadakis *et al.*, 2000).

When anti-dripping additives are used in monolayer films, the external surface of the film (once it is fixed on the greenhouse)

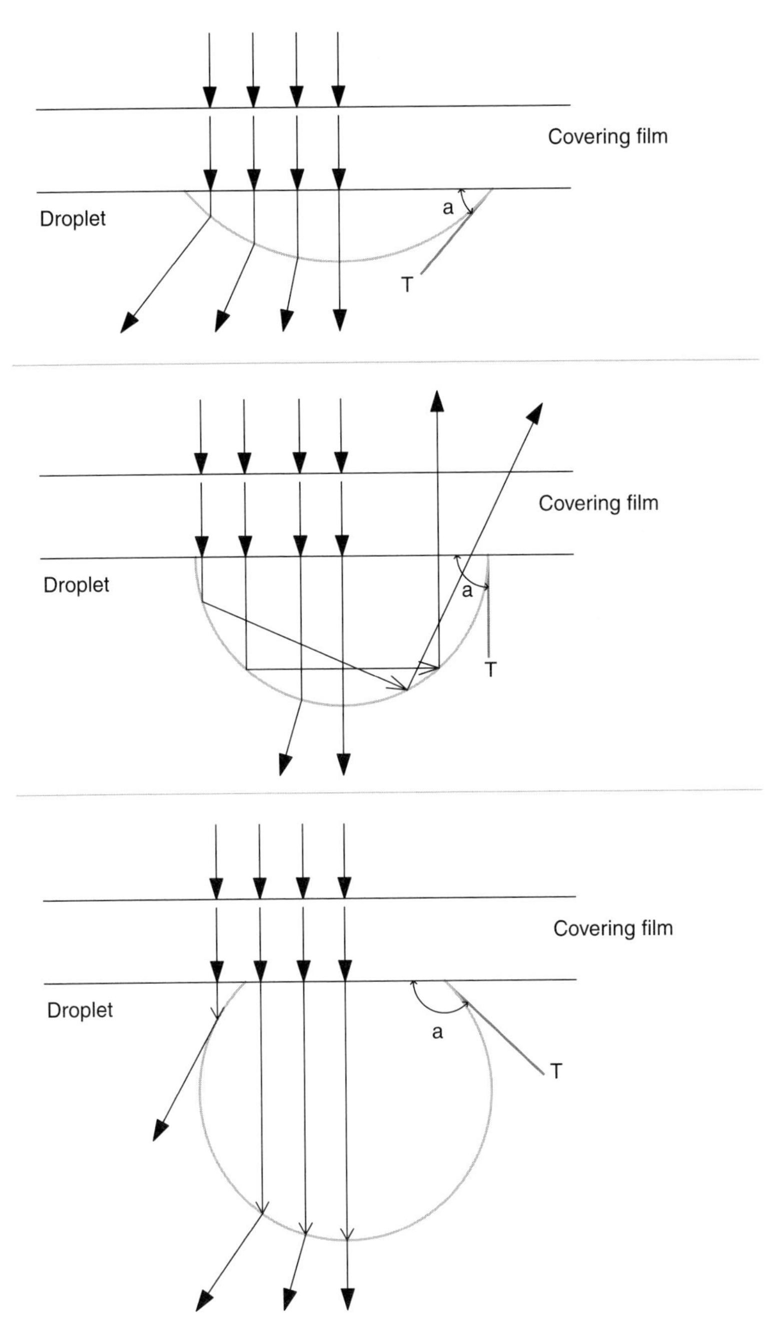

Fig. 4.7. The optical behaviour of the light rays when crossing a water droplet condensed on the inner surface of a greenhouse covering film depends on the angle (a) of contact of the droplet with the covering material; T is the tangent to the droplet at the contact point (adapted from Díaz *et al.*, 2001).

has a great adhesion for dust, which is avoided by using these additives only on the internal surface of multilayer films (Salmerón *et al.*, 2001). The effect of these additives usually lasts less than the lifespan of the film.

The condensation of water on the plastic film improves the insulation conditions, decreasing the global heat transfer coefficient (Table 5.2). However, condensation in the form of droplets is harmful, resulting in burn on the leaves as each droplet acts as a lens (magnifier) to solar radiation and as the droplets fall over the crop this favours the development of cryptogamic diseases (Papadakis *et al.*, 2000). The size of the droplet depends on the contact angle (Fig. 4.7), decreasing the transmissivity if the angle is less than 40° and reaching a higher reduction with contact angles of 90°, which is the case for PE-composed films (Papadakis *et al.*, 2000; Díaz *et al.*, 2001).

The condensation of water vapour on the inner surface of the plastic film needs a minimum angle of the roof of 20–25° for the droplets to slide and not fall over the crop (Von Elsner *et al.*, 2000a). The snow slides over the roof if the angle is equal to or greater than 26.5°; equivalent to a 1:2 slope (Von Elsner *et al.*, 2000b).

4.5.5 Plastic films most commonly used in greenhouses

Normal films

The plastic films most commonly used in greenhouses are LDPE, EVA and their derivatives, frequently incorporating several layers of different materials in multilayer films (Robledo and Martin, 1981). The plasticized PVC films are seldom used, except in eastern Asian. Equally, polyester and polyvinyl fluoride (PVF) films are seldom used.

All these films can be manufactured in small thicknesses and this allows for low weights (less than 200 g m^{-2}) minimizing the loads on the greenhouse structure. Their mechanical properties are good, withstanding hail better than glass, although they soften at high temperatures and are fragile at very low temperatures. Under normal climate conditions they are stable. Due to static electricity they accumulate dust on their surface, especially EVA and PVC, which is usually countered by adding anti-dust additives. Their chemical resistance is generally good, but the use of some pesticides (containing sulfur or halogens) may decrease their lifespan (affecting the HALS additives).

The thermal dilatation coefficient of plastic materials is higher than in metals, which must be considered during the assembly of the greenhouse.

The ageing of plastic films involves a decrease of their mechanical and radiometric properties, as they degrade with time.

The *polyethylene* used in greenhouses is of the low density type (LDPE), obtained by radical polymerization in high pressure processes. Its transmission of solar radiation is good, although decreases with time as the film gets dirty and old, but its thermal behaviour is mediocre, due to its transparency to long-wave IR when there is no condensation on the film (Table 4.3). The incorporation of thermal additives solves this problem.

To improve their shelf life, additives are incorporated which protect the degrading action of the UV rays, calling them long duration or long-life polyethylene (LD-PE) or ultraviolet polyethylene (UV-PE).

Table 4.3. Characteristics of several flexible materials for greenhouse covers (adapted from CPA, 1992 and Tesi, 2001).

	PE	UV-PE	IR-PE	EVA	PVC
Thickness (mm)	0.10	0.18	0.18	0.18	0.18
Weight (g m^{-2})	92	165	173	179	230
Direct PAR transmissivity (%)	91	88–90	85–86	90	90
Diffuse PAR transmissivity (%)	90	86	86	76	89
Long-wave IR transmissivity (%)	68	63–65	≤25	18–27	10–15
Durability under non-aggressive climate (cropping seasons) (years)	1	3 or more	3	3	2

EVA copolymer is a copolymer obtained by the same polymerization system as the LDPE. Its optical properties are slightly different from those of PE having in general a higher PAR transmission and a lower turbidity but thermal properties are better (Table 4.3). These thermal performances (transmissivity to long-wave IR) depend on the content of vinyl acetate, being 12–14% in most cases. Its main disadvantage is its higher tendency to fluency when cold ('creep'), which is quite inadequate for windy areas (the film stretches but does not recover well to the initial shape). As it gets dirty easily, it is usually incorporated in multilayer films. The EVA films can be very transparent to light (crystal variant) or if it is of the opaline type (translucent) it has great diffusing power.

PVC film has similar optical properties to EVA, but better thermal properties (Table 4.3). It attracts dust, as does EVA. Its widths are limited (6.5 m with extrusion and 2 m with calendering) and it has very little resistance to tears which is solved by incorporating a fabric, preventing it from tearing and breaking. After its use it is difficult to dispose of and for this reason, and its cost (which is higher than PE), its use is limited in the Mediterranean area.

PVF film has excellent optical and mechanical properties and is very durable, lasting for a long time, but its price is really high, so it is seldom used. *Polyethylene terephthalate (PET)* also has excellent optical properties but is very expensive. Both are difficult to fix to the structure.

Special films

LONG DURATION FILMS. The incorporation of photostabilizing additives (see section 4.5.3) is a common practice in the manufacture of films to extend their lifespan. LD-PE (long duration PE), also called UV-PE, may last three growing seasons, in high annual radiation areas. In areas of lower radiation it will last longer. The guarantee usually given for the durability of these films is on the condition that there will be no abusive use of pesticides, quantified as a maximum permissible content of chloride and sulfur on the film.

THERMAL FILMS. To improve the transmissivity to long-wave IR the standard PE films are enriched with additives that block the long-wave IR, resulting in thermal PE or infrared polyethylene (IR-PE) that prevents thermal inversion in the greenhouse, in relation to standard PE (Fig. 4.8). These additives slightly decrease the PAR transmissivity (Table 4.3).

Nowadays, additives are being used (new thermal loads) which can vary the diffusing power of light (Salmerón *et al.*, 2001).

Another way of increasing the opacity to long-wave IR in the EVA films is to increase the proportion of vinyl acetate.

A film can be considered as thermal when its transmissivity to long-wave IR (7–14 μm) is lower than 25% for 200 μm thickness films (European regulation EN-13206).

ANTI-DRIPPING FILMS. The anti-dripping films are hardly used as monolayer films, but as a part of multilayer films, being usually placed on the inner surface. There are also products which, when applied to the plastic film as a spray, provide an anti-dripping effect.

MULTILAYER FILMS. With the aim of incorporating in a single film the best characteristics of several types, the multilayer films were created, consisting of two or more plastic films (layers) welded by coextrusion (a manufacturing process).

Nowadays the use of the three-layer type is increasing. One of the most commonly used has an EVA film in the middle of two PE films, the external film containing UV additives (for longer durability) and the internal film containing anti-dripping additives. In addition, it may contain other specific additives.

The multilayer films have displaced the use of anti-dust monolayer films.

PHOTOSELECTIVE FILMS. Initially, photoselective films were based on the 'waterfall' effect (fluorescence) by means of luminescence additives which converted UV radiation in blue light and green into red (see section 4.5.3),

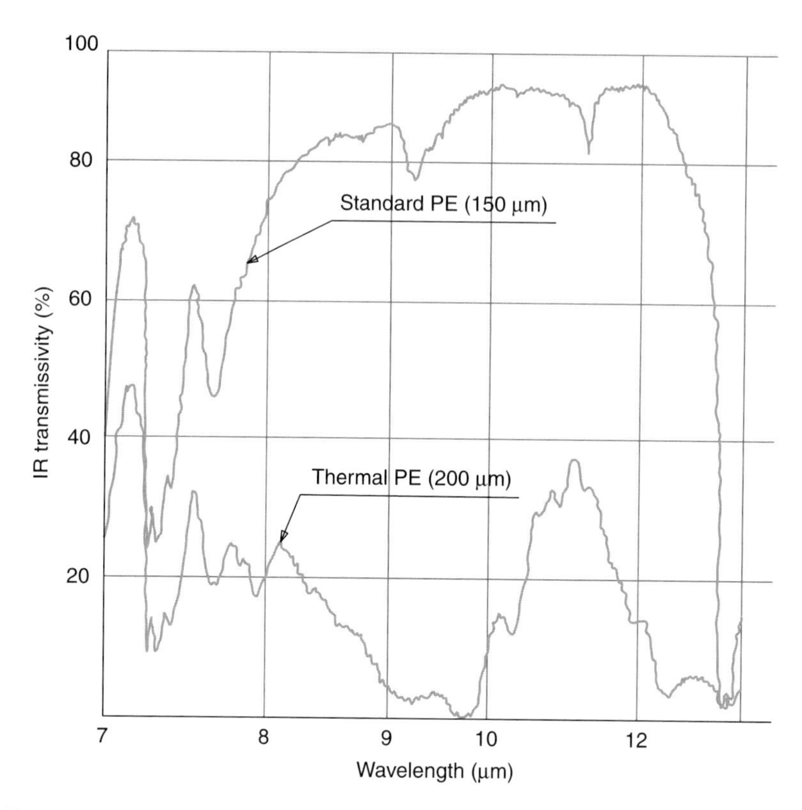

Fig. 4.8. Transmissivity of standard and thermal PE films in the range of long-wave IR (adapted from Castilla *et al.*, 1977).

but nowadays there are photoselective films that filter the radiation selectively (Papadakis *et al.*, 2000).

There are very different types of photoselective films. The *antithermal films* or *static thermal shields* filter the NIR radiation (near infrared) of the solar spectrum, therefore limiting the input of energy inside the greenhouse, and decreasing the temperature (Verlodt and Verchaeren, 2000). They have application in warm climate areas or during the summer, but their use has been limited because of the effect they have on decreasing daytime temperatures (2°C, on the average) and because of their high cost.

A variant of these thermal shields are the *thermocromic films* or *dynamic thermal shields* that are now under development, which filter the NIR radiation depending on the temperature (Díaz *et al.*, 2001). Therefore, during the summer, when the temperatures are high, they would filter the NIR radiation limiting the warming of the greenhouse, whereas during the winter they would let it enter the greenhouse contributing to its warming.

Fluorescent films alter the proportion of R/FR light (red/far red) affecting morphogenesis but they have a transmissivity to PAR lower than the standard PE as the process of fluorescence is not very efficient (Pearson *et al.*, 1995). Their application to agriculture is not yet very clear (Kittas and Baille, 1998).

Anti-pest films are photoselective materials that block part of the UV solar radiation reducing the development of pests or diseases caused by fungi or viruses that are transmitted by insects, which are sensitive to the decrease or absence of this type of radiation (Salmerón *et al.*, 2001).

The best known example of this is the reduction of *Botrytis*, when blocking the radiation in the range from 300–400 μm (Jarvis, 1997). In a similar way, the suppression of UV radiation around 340 μm limits the sporulation of *Sclerotinia* and *Alternaria* (Jarvis, 1997).

White fly (*Bemisia tabaci*), vector of the tomato yellow leaf curl virus (TYLCV), and thrips (*Frankliniella occidentalis*), vector of tomato spotted wilt virus (TSWV), are examples of insects which need UV light for their visual organs. The lack of UV light limits their mobility and decreases their presence, reducing the incidence of disease caused by these viruses (González *et al.*, 2003; Monci *et al.*, 2004).

The normal behaviour of pollinating insects, bumblebees and bees, used in greenhouses is slightly affected by the use of these anti-pest films, partially affecting their vision. However, the use of these films is compatible with pollination by these bees (Salmerón *et al.*, 2001). Further studies are being conducted in this area.

All these variants of photoselective films are usually used as part of a multilayer, with the aim of incorporating other interesting properties into the film.

OTHER FILMS. The diffusing effect of a film is induced by specific additives or with additives used for other aims. These diffusive films increase the proportion of diffuse light inside the greenhouse. They have a whitish appearance, are opaline and their diffusive power is quantified by their turbidity (European regulation EN-13206).

Other types of films incorporate fabrics inside them or are interlaced with fabric to improve their mechanical properties, such as in those used in retractable roof greenhouses.

Normalization of plastic films used as greenhouse covers

The Spanish regulation UNE-EN-13206 (*Thermoplastic films for covers used in agriculture and horticulture*; the Spanish version of European regulation EN-13206) has replaced Spanish regulation UNE-53328 (*Polyethylene films used as greenhouse covers*).

Regulation UNE-EN-13206 covers three types of films: (i) standard; (ii) clear thermal; and (iii) diffuse thermal.

Previously, regulation UNE-53328 (now superseded) defined the durability, in agricultural years, taking solar radiation in Almeria as a point of reference, whose annual value was estimated at 148 kLy (kilo Langley) year^{-1}; 1 kLy = 41.84 kJ m^{-2}).

At the time of writing, the existing regulation (UNE-EN-13206) defines three climate areas as a function of the solar radiation expressed in kilo Langleys per year: (i) between 70 and 100; (ii) between 100 and 130; and (iii) between 130 and 160. The anticipated durability of the film under real conditions is estimated by means of accelerated artificial ageing, establishing a correlation between artificial and natural ageing, depending on the annual solar radiation of each site (Tables 4.4 and 4.5). Therefore, a film that lasts 2 years in a climate area with 130–160 kLy year^{-1} of solar radiation could last 4 years in a less sunny climate area of between 70 and 100 kLy year^{-1}.

Regulation UNE-EN-13206 defines, among other properties, the characteristics and tolerances with regard to: (i) resistance to traction, tearing and impact; (ii) the thermicity (transmissivity to long-wave IR radiation, which is emitted by the Earth's surface); and (iii) the transmissivity to visible light (regulated by European regulation EN-2155-5).

Regulation UNE-EN-13206 is applicable to LDPE, to linear LDPE and the mix of both, to EVA and to the ethylene and butyl acrylate copolymers, as well as to their mixtures with LDPE.

The designation of a film or covering film, according to this regulation, must include: (i) the type (standard, clear thermal and diffuse thermal); (ii) the polymer used in its manufacture; (iii) a reference to the regulation; (iv) the width and thickness; and (v) the accelerated artificial ageing data. For instance, the designation of a LDPE film, of 5000 mm width, 105 μm thickness with an accelerated artificial durability of 1700 h would be: COVER WITH STANDARD FILM EN-13206...5000 105 A.

Table 4.4. Greenhouse covering films: classification based on their longevity estimated by means of accelerated artificial ageing (regulation UNE-EN-13206).

Film type[a]	Durability (h)
N	≥400
A	≥1700
B	≥3200
C	≥4600
D	≥6000
E	≥7300

[a]Film classifications. For example a film classified as C means that it will last 4600 h or more (in the artificial ageing test) and this number can be correlated with the expected commercial life use, as detailed in Table 4.5.

Table 4.5. Correlation between artificial and natural ageing (regulation UNE-EN-13206).

Annual radiation (kLy)	Artificial ageing (h) for an anticipated shelf life of:			
	1 year	2 years	3 years	4 years
70–100	1700	3500	5300	7100
100–130	2800	5600	8400	–
130–160	3900	7800	–	–

In addition, films must be marked at the ends of the films (and on rolls of the film) with the manufacturing date, together with other data about the brand and commercial name.

4.5.6 Rigid plastic materials

Rigid plastic materials, also known as organic glass, can be presented in the form of rigid or semi-rigid simple panels (PVC, polyester, PC) or double- or triple-wall alveolar or cellular sheets (PC and PMMA).

Physical and mechanical properties

The dimensions of the panels are larger than those of glass, being limited by their resistance to flexion (which is equal or greater than that of glass). They are lighter than glass and less rigid (Table 4.6), which allows them to adapt to curved shapes. Their rigidity is improved giving them fluted or corrugated shapes, by incorporating a fibreglass armour (fibreglass reinforced polyester).

Their resistance to impact is much greater than that of glass, being able to withstand hail (especially PC).

On the other hand, their resistance to abrasion is worse than that of glass, so it is important to avoid scratching. As their thermal dilatation coefficient is high (higher than glass) dilatations must be prevented by using oval fixation holes and fixing only one side of the sheets to allow it to slide in tracks (Wacquant, 2000).

The rigid panels vary as to how flammable they are (Table 4.6); flammability must be considered during their assembly and use.

The global heat transfer coefficients (see Chapter 5) are similar to those of glass for simple walls and they are much more insulating in double or triple walls. PMMA and PC are permeable to water vapour, which must be taken into account when they are used in double or triple panels (cellular).

The optical properties are similar to those of glass but their durability is lower. As they age their transmissivity is altered, the amount of change depending on the assembly conditions and the local climate.

Frequently, plastic rigid panels are used for the sidewalls and front walls of greenhouses (covered with flexible film) because they last longer and are versatile to use as they can be curved, cut or perforated.

More commonly used rigid panels

Polyester reinforced with fibreglass that is used in greenhouses is known as 'horticultural quality' and has higher transmissivity to global solar radiation (80% instead of 60–65%) than the one used in the building industry. It retains its optical properties for 10 years, with losses below 20% (Wacquant, 2000). A surface coating on the external face protects it and prevents ageing from UV rays. In order to provide rigidity to the sheets, they are corrugated or reinforced.

Its light transmission, which is lower than that of glass, is good, having a great

Table 4.6. Characteristics of several rigid materials used as greenhouse covers (adapted from CPA, 1992).

	Horticultural glass	Reinforced polyester	Bioriented PVC	Double PMMA	Double PC
Thickness (mm)	4	1	0.9	16	6
Weight (kg m^{-2})	10	1.5	1.45	5.0	1.4
Inflammability	No	Easy	No	Medium	Medium
PAR transmissivity (%)	90	80–85	79–85	82	75–80
Long-wave IR transmissivity (%)	1	4	1–2	2	5
Durability (years)	≥ 20	10	10	20	10

light diffusive power. The transmissivity to short-wave IR is low (which is an advantage during the summer). Although it is a good insulator, the shape of the sheets increases the exchange surface. Its resistance to impact is slightly higher than that of PMMA.

Its main advantages are the light weight, the ease of handling, its high diffusion of light and its high resistance to impact. Its main disadvantage is its lower transmissivity to diffuse radiation and its combustibility.

Bioriented PVC is made as corrugated sheets and is protected against UV light. Its light transmission is similar to that of polyester, although it varies depending on the spectra. It is a better insulator than glass. It withstands hail well, but withstands very high temperatures poorly.

Polycarbonate (PC) is less transparent than glass, but has a great resistance to impact. It is used in alveolar sheets, mainly of double-wall types but also as simple sheets. Its light transmission decreases when ageing. A surface coating protects it against UV rays ensuring after 10 years, minimum transmission and resistance values (Wacquant, 2000). It is used when there is serious danger of hail, and in passages, doors or as separating walls.

Methyl polymethacrylate (PMMA) is presented as alveolar panels of double or triple walls in thicknesses of 8–16 mm (known as Plexiglass, Altuglass). It behaves in a similar way to glass and retains its properties well with time. Its permeability to water vapour requires some precautions in its use. Its high price and the thermal dilatations limit its use in commercial greenhouse; besides, it is easily combustable.

4.6 Greenhouse Construction

4.6.1 Introduction

The prime aim in constructing a greenhouse must be obtaining higher profitability. Therefore, the best design must be the result of a compromise between technical requirements, some of which conflict with one another, while obtaining the highest profitability.

Obviously, the type of greenhouse that is constructed will vary depending on the main purpose of the greenhouse. A greenhouse that must protect crops from the rain (rain-shelter), as is the case in some tropical areas, will be conceived in a different way from a conventional greenhouse in an area with a Mediterranean climate, or to one in the northern latitudes or in semi-desert climates.

Besides, the local climate of each region, the specific type of greenhouse selected will depend on the bioclimate requirements of the species to be grown (the type of greenhouse required for vegetable production will be different from that required for growing ornamental species as they have different light and heat requirements). The socio-economic conditions will also affect the type of greenhouse to be chosen. For instance, limits on the initial investment in many cases restricts options available for the type of greenhouse.

The availability of labour will determine the level of mechanization and automation. Whether the crop is sedentary (species whose cycle is set in the same place) or mobile (pot plants) is another aspect to consider, especially with regard to optimizing the use of the space.

4.6.2 Greenhouse types

There are different classifications of the greenhouse types. According to their architectural shape we can distinguish two basic types: (i) a single-module or monospan greenhouse, consisting of a greenhouse isolated from other structures; and (ii) a multi-module or multi-span greenhouse, formed by a series of gutter-connected greenhouses. The multi-span types allow for better use of the soil, are cheaper to construct and to heat, and having a lower ratio between the cladding surface and the soil surface.

With regards to roof geometry there are many variants (Fig. 4.9). Curved roofs may adopt a semi-circular, semi-elliptic or ogival (Gothic arch) shape, among others. The difficulty to mount glass on curved structures has limited its use until the appearance of semi-rigid plastic panels or flexible films. The use of wood in the structures has also restricted the use of curved geometries, due to the difficulty for its construction.

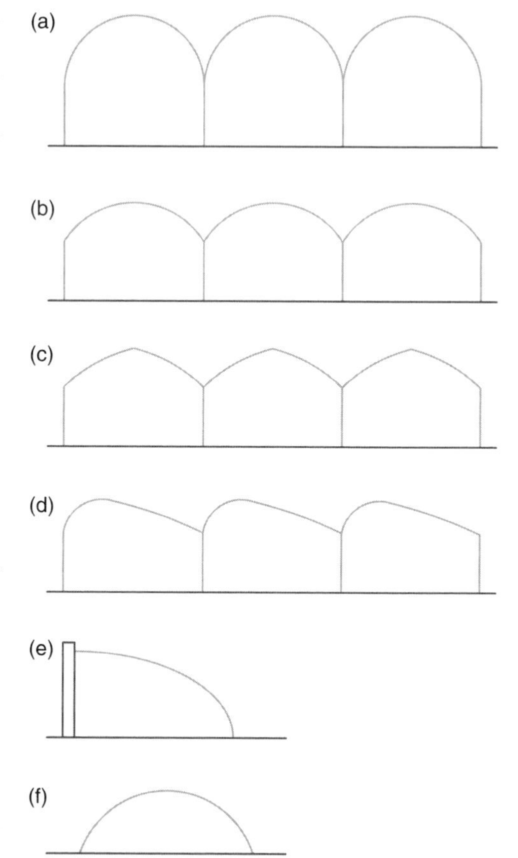

Fig. 4.9. Some common types of curved-roof greenhouses: semi-circular (a), semi-elliptic (b), Gothic arch or ogive (c), asymmetric (d), attached (lean-to) single-span (e), and single tunnel (f).

In the case of gable-roof greenhouses (Fig. 4.10), the roof may be symmetrical or asymmetrical, with great diversity of angules, depending on the latitude and local conditions. The flat-roof greenhouse, initially used in the parral-type greenhouse (typical of the south of Spain), is less and less used. The sidewalls, in single-span or multi-span greenhouses, can be vertical or slightly inclined, the latter being more advantageous for their higher light transmissivity; its importance decreases as the greenhouse width increases. However, the disadvantage is that they limit the cultivation of vertically trained species by the sidewall, as the useful height is reduced.

Other types of greenhouses that are seldom used are inflatable greenhouses, in which the cladding plastic film is kept in position by air pressure, notably reducing the structural elements. Fans must supply, continuously, the necessary pressure to keep the plastic in position. This idea induced the use on conventional greenhouse structures double clad with inflated plastic films, by means of air pressure. The double cover decreases the heating costs and better withstands the wind than the single layer, but limits the transmissivity to radiation in relation to the simple film.

Other variants of structures have very little commercial interest. Greenhouses made with hanging structures to decrease the weight of the roof have problems of resistance against strong winds, and as a consequence, their use has expanded very little.

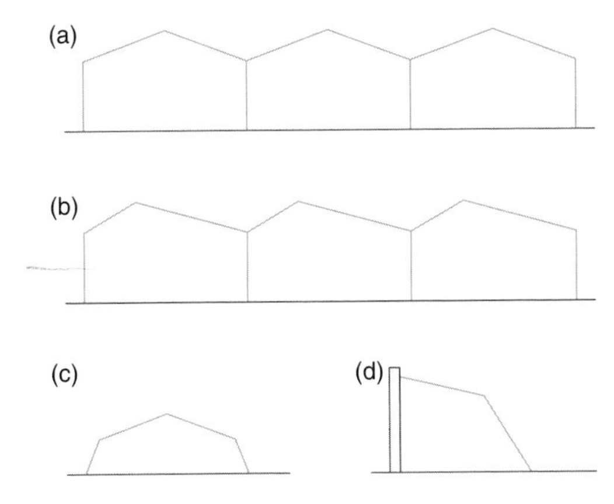

Fig. 4.10. Some common types of gable-roof greenhouses: symmetric multi-span (a), asymmetric multi-span (b), simple single span (c), and attached (lean-to) single span (d).

Although greenhouses are usually located on flat terrain, in some cases (south coast of Spain, for instance) they are built in south-oriented slopes (in the northern hemisphere), profiting from more favourable winter radiation conditions than if they were built on flat terrain. Although this location limits the possibilities of mechanization, the roof and sidewall surfaces exposed to the north are lower, with the subsequent advantage from the thermal point of view during the winter. In a similar way, the soil surface (or soil and terrace wall, if the greenhouse is terraced) is higher with a positive incidence on the greenhouse thermal inertia. In addition, the ratio of plastic cladding area:soil area is reduced, limiting the thermal losses. The disadvantages of this type of greenhouse, besides the already mentioned difficulty for mechanization, are the observed temperature stratification, both during the daytime (which requires the proper location of ventilators) and during the night, due to the higher weight of the cold air, which stratifies on the lower parts.

Another type of greenhouse, common in home gardens and as commercial greenhouses in China (solar greenhouse), is the 'lean-to' greenhouse where the greenhouse leans against a wall. In the northern hemisphere, the greenhouse is oriented towards the south and the wall is on the northern side (Fig. 7.5). The first greenhouse constructed years ago was of this type, the northern wall providing thermal inertia. This type (passive solar collection) has also been used in recent decades as useful complementary equipment for solar heating of buildings.

Greenhouses that are buried or semi-buried in the ground allow for natural thermal regulation with much lower oscillations than in conventional greenhouses. Their use has been limited to the cultivation of ornamental species that need high temperatures and humidity, or to greenhouses for plant propagation. The areas where they are built must not have a shallow water table or be prone to flooding.

Depending on the cladding material we may distinguish the following greenhouse types:

- Glass
- Plastic:
  - Flexible film
  - Rigid panel:
    - Simple
    - Alveolar
  - Screens

Although today it is not a commonly used form of classification, greenhouses can be classified according to the minimum temperature level that they can maintain, so we can distinguish: (i) '*cold greenhouses*' (5–8°C); (ii) '*temperate greenhouses*' (12–15°C); and (iii) '*warm greenhouses*' (20–25°C).

4.6.3 Structure materials

The rigidity characteristics of the cladding materials determine, to a great extent, the shape of the greenhouse and the materials from which it is made. The rigidity of glass limits its use on curved-roof structures, so it is normally used on straight-roof structures, such as gable roofs. Panels, depending on the type, only permit small curvatures, so their use on curved-roof greenhouses is rare.

In a conventional greenhouse, a functional structure that complies with the primary purpose of holding the cladding material and the predictable loads (snow, wind, trained crops and attached facilities) must, primarily, avoid shadows and be as cheap as possible, in terms of building and maintenance costs, and be consistent with the agronomic requirements.

In general, the cheaper cladding materials (flexible films) have a short lifespan, so they must be replaced more frequently than the long-life materials (semi-rigid panels, glass) which are more expensive and heavier, requiring, therefore, more expensive structures. As indicative figures, for the Spanish Mediterranean coast conditions, a glasshouse would need a metallic structure, which weighs 15 kg m^{-2}, whereas a structure for a semi-rigid panel would weigh around 12 kg m^{-2} and for a PE film around 7 kg m^{-2}. From the grower's point of view, the main aspect to consider when choosing between the different types of greenhouses is the annual cost per unit greenhouse surface (amortization plus the maintenance costs) depending on the performance of the greenhouse. The cost and availability of labour also notably influence the periodic replacement of the plastic film.

The greenhouse structure includes the foundations and the elements to hold the cladding material and support the structure itself.

To hold the cladding material, in glasshouses or in greenhouses with semi-rigid panels, aluminium has the advantage of being able to be used to create difficult profiles, but it is expensive. In plastic greenhouses, such as the low-cost parral-type greenhouse, wire (steel or galvanized) fulfils this role well.

The use of wood, nowadays, is limited to straight-roof structures (Photo 4.4), due to the difficulty and high cost associated with its use on curved sections. Metal, on the contrary, adapts well to curved structures, and due to its better resistance characteristics, bears strong forces. Today, wood is used almost only for plastic-film greenhouses, or artisan construction.

In plastic-film greenhouses, the choice of structural materials is linked to: (i) their availability and cost; (ii) their technical characteristics depending on the greenhouse to be built (use of wood, steel); (iii) the performance required by the greenhouse depending on the crops to be grown; (iv) the local climate; and (v) the local conditions in terms of experience and creativity (the parral greenhouse originated when a plastic film was adapted to cover a table-grape growing structure, 'pergolato').

Photo 4.4. Wood has frequently been used in greenhouse structures in many areas.

In general, cheap greenhouses originate from the use of affordable and available materials within the local context.

In the tunnel greenhouses of the Mediterranean area, the use of galvanized steel prevails, normally with tubes of circle or oval sections, for widths of up to 10 m and heights of up to 4 m.

For multi-tunnel greenhouses, metallic structures prevail (a predominance of galvanized steel, due to the high cost of aluminium; Photo 4.5) or a mixture of materials (wood–wire, steel–wire, steel–wood, steel–concrete) are used over wooden structures.

Steel structures, which are normally more expensive than wooden structures, allow for a reduction in the number of interior pillars (relative to wood), easing the interior manoeuvrability (passage of machinery, implementation of thermal screens) and creating fewer shadows than wood, increasing the available light. In addition, steel structures are easier to assemble than wood, have more accessible roof ventilation mechanisms and are more airtight, although the higher heat conduction of metal weakens these advantages. Reinforced concrete structures are not common.

4.6.4 Covering materials

Types of covering materials

Until the introduction of plastic materials, glass was the only greenhouse cladding used. The rigidity of glass limited the use of curved shapes in greenhouses so the predominant roof geometry was that of gable-roof greenhouses.

The availability of flexible films or semi-rigid panels broadened the range of possibilities in terms of greenhouse design choices and notably decreased the carrying weights (1 m^2 of 4 mm thickness glass weighs approximately 10 kg, whereas 1 m^2 of PE of 0.2 mm thickness weighs 0.2 kg). The lower construction costs of plastic greenhouses, especially those covered with flexible film, allowed for an extension of greenhouse cultivation to many regions of the world.

A usual classification of greenhouse cladding materials is that indicated on Table 4.2. Glass was, as already mentioned, the first material used to cover commercial greenhouses. Glass, as a greenhouse cover, is a material of excellent optical and thermal characteristics (Table 4.6). It bears ageing and pollution well and it is not

Photo 4.5. The steel structure is used in multi-span greenhouses.

flammable. Its main disadvantage is its low resistance to impacts (sensitive to hail). In addition, it is heavy and expensive. Among its variants we can find horticultural glass, cathedral glass (of greater light diffusive power) and glasses of low emissivity (which improves the insulation). All of them are very expensive and require sophisticated greenhouse structures.

Plastic materials in the form of rigid panels have a long durability (more than 10 years) with a low reduction of light transmission. They have good thermal characteristics and weigh less than glass, so they require lighter structures than glass (Table 4.6). See section 4.5.6.

The most commonly used rigid panel is polyester reinforced with fibreglass, whose durability improves with external coatings of PVF film. The simple PC and polyester sheets have been the most used, because the PVC sheets deform at high temperatures. In general, their use has been preferred for high value crops, and same as glass, when a great durability of the greenhouse cover is required.

The alveolar panels, composed of two layers of rigid material, cross-linked at regular intervals for higher strength, are an excellent solution for avoiding thermal losses, but reduce the light transmission and their cost is high. PMMA and PC have been used mostly for alveolar sheets. Simple semi-rigid panels PVC (bioriented), PMMA and PC have had limited spread.

In summary, the great expansion of greenhouses during the last quarter of the 20th century has been based on the use of flexible plastic films, the use of PE prevailing clearly in the Mediterranean area.

The use of standard PE films was progressively substituted by the use of long-life PE and thermal PE (see section 4.5.5) initially and more recently by multilayer plastic films.

In the use of cladding materials it is common to distinguish between roof and sidewalls, because in the roof the criteria of maximizing the transparency to solar radiation must prevail. In the sidewalls (especially in those oriented to the north, in the northern hemisphere) the insulation function must prevail.

Therefore, the choice of the covering material of a greenhouse must combine an economic price (which does not always mean the cheapest) with longevity according to its characteristics and price. Table 4.7 summarizes the costs of several materials. A good covering material must have maximum transparency to global solar radiation, especially within the PAR range, and be as opaque as possible to long-wave IR radiation. It must have a global heat transfer coefficient (K) as low as possible, to achieve a good greenhouse effect and avoid the 'night thermal inversion'.

To avoid limiting their transparency, the covering material must not retain dust (by electrostatic attraction), limiting solar light, and it must be easily cleaned by washing. In addition, the material must avoid condensation in the shape of large droplets (the formation of continuous condensed water films being preferable) and have a good resistance to abrasion, especially in arid and desert areas (where strong winds carrying sand are frequent), as the scratching of the film will increase reflection, limiting its transmissivity.

Table 4.7. Average cost of different cladding materials, according to Muñoz *et al.* (1998).

Material	Euros m^{-2}	Relative cost[a]
Rigid:		
Glass (4 mm)	18.0	60.0
Corrugated PC (0.8 mm)	15.0	50.0
Alveolar PC (6 mm)	19.2	64.0
Polyester (1.2 mm)	14.4	48.0
PVC (0.8 mm)	9.0	30.0
Flexible:		
Standard PE (0.1 mm)	0.30	1.0 (base)
Long-life (UV) PE (0. 18 mm)	0.36	1.2
Thermal (IR) PE (0.2 mm)	0.42	1.4
EVA (0.2 mm)	0.50	1.7
Three layers (0.2 mm)	0.54	1.8
PVC (0.2 mm)	0.48	1.6
Reinforced PVC (0.3–0.5 mm)	1.50	5.0

[a]The relative cost indicates the relation with the cheapest (standard PE).

The development of quality regulations adapted to the regional or national conditions of use is necessary for an appropriate transparency (in this case in the economic sense) of the market.

The use of double layers to reduce the thermal losses is not limited to alveolar panels, because the idea is also used in the case of glass (double glass) and flexible films, using a double layer (which plays the role of thermal screen) in some cases or, in others, as an inflated double layer (by means of air pressure, coming from a fan).

It is worth highlighting the recent expansion of screens (nets) as cladding material of greenhouse structures (Montero *et al*, 2009), when temperature increases are not the goal, but rather protection against wind or shading or improving the ambient humidity. The structural requirements of these greenhouses are lower than for those of film, as the screens offer less resistance to the wind and are permeable to water.

The so called 'screenhouses' are effective and economical structures for shading crops, protecting them from wind and hail, improving the temperature and humidity regimes, saving irrigation water and excluding insects and birds (Tanny *et al.*, 2006).

There is a relationship between the porosity of a screen and its transmission of solar radiation, but other parameters also influence diffusion effects on the incident radiation and, consequently, on shading and transmission levels (Sica and Picudo, 2008). Screens contribute to increasing the diffuse fraction of the transmitted solar radiation through them at different levels, depending on the structure, texture and colour of the screens (Abdel-Ghany and Al-Helal, 2010; Moller *et al.*, 2010).

Dust deposition on the screen can widely alter its light transmission (Santos *et al.*, 2006). Screenhouses can save around 30% of the annual irrigation water required for outdoor conditions, without any loss of yield and even improving quality (Tanny *et al*, 2006).

The use of coloured screens, instead of the conventional white or black screens, to manipulate the vegetative growth of the crop and to improve the yield and quality, has been recommended (Oren-Shamir *et al.*, 2001). In coloured screens, the spectral manipulation is aimed at specifically promoting desired photomorphogenic/physiological responses, while light scattering improves light penetration into the inner canopy (Shahak *et al.*, 2009). In order to limit the visual environmental impact of screenhouses, the colour of the material should be chosen carefully (Castellano *et al.*, 2008a).

Net or screens are characterized, among others, by their porosity (which influences their shading effect) and their permeability to the air (Castellano *et al.*, 2008b). Porosity is determined by the diameter and physical characteristics of the thread and the density of the screen (number of threads per centimetre), which also affects its durability, overall weight, strength and elasticity (Castellano *et al.*, 2006). The most common prime raw material used for making agricultural screening is high density polyethylene (HDPE). In Europe, screens are characterized by the number of threads per centimetre in each direction (e.g. a 10 × 20 screen has ten threads in one direction and 20 in the other) (Teitel, 2006). See Chapter 8 (sections 8.4.6 and 8.4.7).

Assembly of the cover

In the case of flexible films, the mechanical characteristics of the film and its assembly determine the life of the film and the degree to which the greenhouse performs properly (how airtight it is, thermal losses, etc.).

The form in which the film is assembled depends on the dimensions and the characteristics of the ventilation system (roof, side, mixed) and is limited in principle by the width of the film. It must be assembled at high temperatures (which will allow the plastic to expand) so the film remains as taut as possible (Fig. 4.11). The fixing must avoid, if possible, nailing or drilling the film (which is sometimes very difficult in artisan greenhouses such as the parral), because these drillings are the point of entrance for rainfall water which will directly drip over the crops, and also

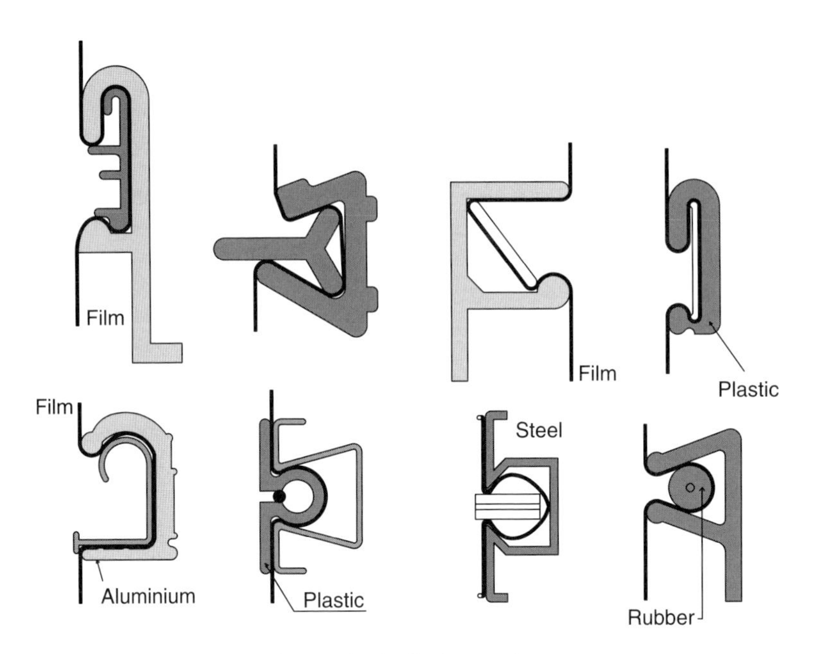

Fig. 4.11. Different systems of how to fix the plastic film in greenhouse structures.

the drillings may rip the film. The metallic surfaces of the structure exposed to the sun will degrade the film as they warm up; therefore, it is advisable to protect the film in those contact areas with painting that reflects the sun rays.

The ageing of films depends, besides their intrinsic characteristics, on the conditions of use, particularly the climatic conditions such as solar radiation (intensity, length and degree of exposure to the weather). With ageing, the mechanical characteristics deteriorate and the optical characteristics are altered (transmission reduction) loosing quality from the point of view of their agronomical use.

4.6.5 Greenhouse screens

Introduction

Screens have been traditionally used to shade or obscure greenhouses. Later, following the energy crisis of the last third of the 20th century, their use was expanded to reduce energy costs (thermal screens), especially in heated greenhouses. More recently, the use of screens has allowed for an improvement of the climate control in sophisticated greenhouses.

Nowadays, the different types of screens must be integrated in the greenhouses structures, minimizing the light losses.

Reasons for the use of screens

Basically, screens are used to obscure, to shade, to decrease energy losses and for climate control. A new application is the use of insect-proof screens which prevent insects entering the greenhouse (Chapter 8).

Darkening screens are mobile and are used to limit the length of the day. They must have a minimum light transmission (<0.1%) to achieve the short-day treatments in crops sensitive to photoperiod (see Chapter 9).

Shading screens are used to decrease the light intensity and to limit direct radiation, with the aim of restricting thermal excesses, improving the quality of the produce or avoiding water stress. They can be used inside or outside the greenhouse. Whitewash, usual in hot climates, is a type of low-cost fixed shading.

Energy-saving screens, if used only during the night, do not need to be transmissive to PAR radiation. But if they are also used during the day it is necessary to have a good PAR transmission, besides their insulating and anti-condensation properties.

Climate control screens are a combination of shading and energy-saving screens.

Screen assembly

Depending on their mobility, screens can be fixed, semi-fixed or mobile.

The fixed energy-saving screens decrease the light transmission permanently and increase the humidity of the air. The most common are perforated films of high transmissivity to light, to palliate these defects, and their use is limited to short periods in the winter.

Semi-fixed screens decrease the problems of fixed screens but do not avoid them, so the use of mobile screens has expanded, with different systems employed to move the screens (rolling, sliding, folded sheets) (Fig. 8.16).

Mobile screens are usually extended horizontally, although sometimes the shading ones are implemented parallel to the roof cover. When folded they must keep shading to the minimum. If the folded screen adopts a north–south orientation it will distribute the shadows more uniformly than in the other direction. The insect exclusion screens are usually extended vertically or inclined, covering the vents.

Nowadays, there are several automatic mechanisms to deploy and retract the screens (traction wires, displacement bars, rolling tubes) which allow for different implementation options, depending on the greenhouse structure. In windy areas, an anchorage system in the upper part of external screens must be provided to avoid the wind lifting them and possibly destroying them.

Materials for the screens

All types of screens must be hard wearing and resistant to scratching, as well as to ageing by UV radiation, temperature, humidity and chemical agents. Additives are usually incorporated into the plastic materials used as screens in order to prevent dust particles sticking to the screens and to increase their shelf life.

The screen materials must have dimensional stability, and not to be prone to stretching beyond 2% (Bakker and Van Holsteijn, 1995). They must allow for an easy deployment and retraction.

Nowadays, screens are made using PE, polyester or acrylic materials as the raw materials because other materials such as polypropylene, polyamide and cellulose have limited use mainly due to their lower durability (Bakker *et al.*, 1995).

Regarding texture, the screens adopt the shape of plastic films, of woven or braided fabric, and canvas or aluminized sheets, associating aluminium sheets with other materials in one or both surfaces (Urban, 1997a). There are non-woven screens too, such as agro-textiles, which are bonded plastic filaments spun together as a fabric.

The PE sheet screens usually only last for two cropping seasons, with the proper thicknesses and additives (UV). The screens built with polyester or acrylic materials may last for more than 5 years.

Darkening screens are usually built of black PE film and black woven tissues, which absorb condensation. These materials may be coated with aluminium, on their inner face, to fulfil a complementary function of energy saving (Bakker *et al.*, 1995).

Shading screens may incorporate white or aluminium sheets joined to the base material, in an open structure (with drillings), to allow the passage of air. If little shading is desired, aluminium sheets are not used. If the screens are of the closed type (no drillings), that do not allow the passage of the air, they may be used to save energy. To allow for ventilation with these screens, they are not completely deployed so that there are small open sections.

The energy-saving screens used at night must be made of PE film, coated with aluminium, a material that has no transmissivity and great reflectivity (see Chapter 5), or canvas, either woven with aluminium sheets or not.

To avoid high humidity levels and be able to remove water vapour, up to 0.25% of the surface of PE films can be perforated (Bakker *et al.*, 1995). The modern woven materials allow for the absorption of a certain amount of condensed water droplets, avoiding them dripping over the crop.

If energy-saving screens are to be used during daylight hours, it will be necessary that they have good PAR transmission.

The climate control screens are preferably of woven or joined materials, based on PE or polyester, depending on their intended use. They can be used with aluminium sheets to achieve different degrees of shading and energy saving, with white translucent sheets to decrease the direct radiation and increase the diffuse component of radiation, or coloured sheets to alter the light spectrum in the greenhouse (Bakker *et al.*, 1995).

4.7 The Selection of the Greenhouse: Options

When choosing a greenhouse the main aspects to consider must be: (i) the transmission to PAR radiation which determines the productivity potential; (ii) the solidity and durability; (iii) the functionality and ease of maintenance; (iv) the energy economy, if heating must be used; and (v) the price.

Limiting ourselves to the most common types, the choice is between the artisan and the industrial greenhouses. Artisan greenhouses use flexible plastic films and are of low cost. Of the industrial greenhouses, the most common are the Venlo greenhouses, of Dutch origin, which use glass as cladding, and the multi-span greenhouses, with a curved-roof shape which are very popular in regions with a mild climate and which normally use a flexible film as cladding, although they may use rigid panels in the cover (Figs 4.12–4.16).

The artisan greenhouse covered with flexible film and little equipment constitutes the low cost and lower agronomic performance option, whereas the sophisticated glasshouse is at the other extreme. Between these two extremes there are a number of intermediate options. Chapter 14 summarizes the information on costs of greenhouse construction.

The choice of a greenhouse of the appropriate technological level must be the result of a compromise between the technical and economic requirements, in order to achieve sustainable production.

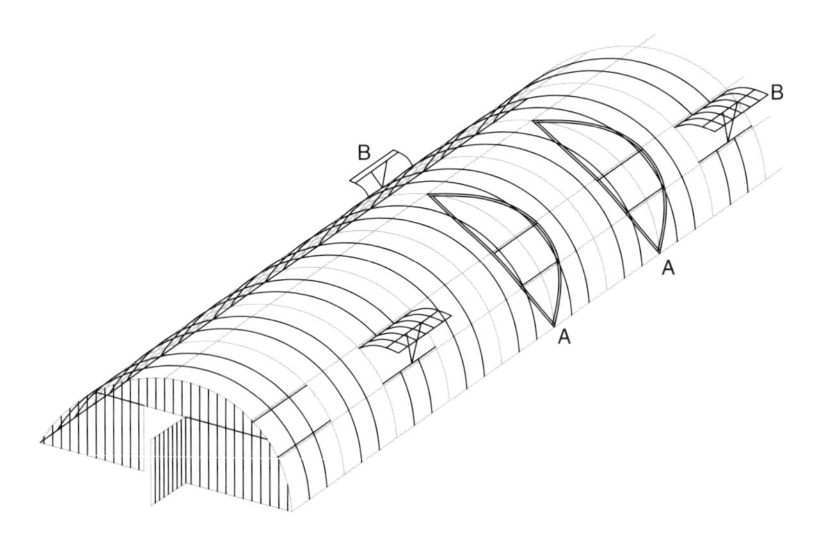

Fig. 4.12. Curved-roof tunnel greenhouse, common in the Mediterranean area. A, Opening separating the plastic film; B, vent.

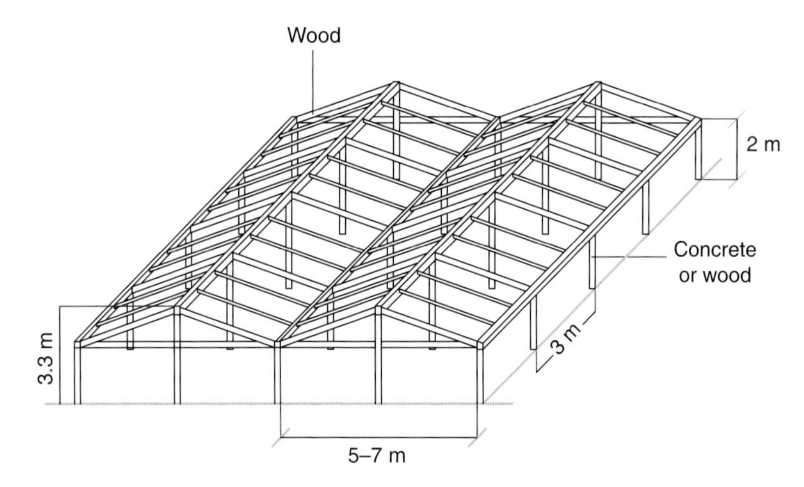

Fig. 4.13. Wooden-structure greenhouse, common in Italy.

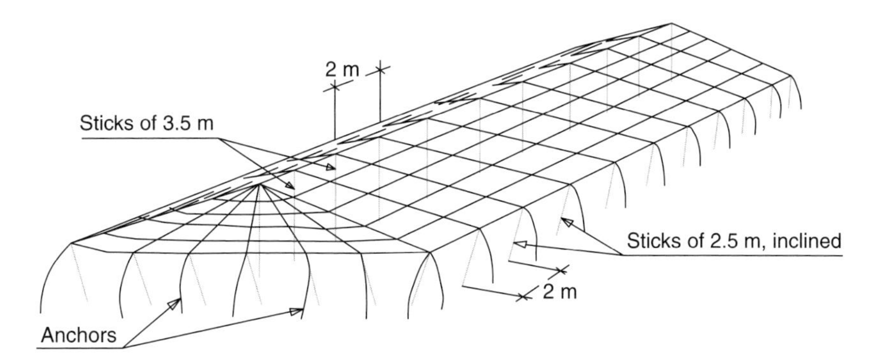

Fig. 4.14. Parral-type greenhouse structure, common in Spain.

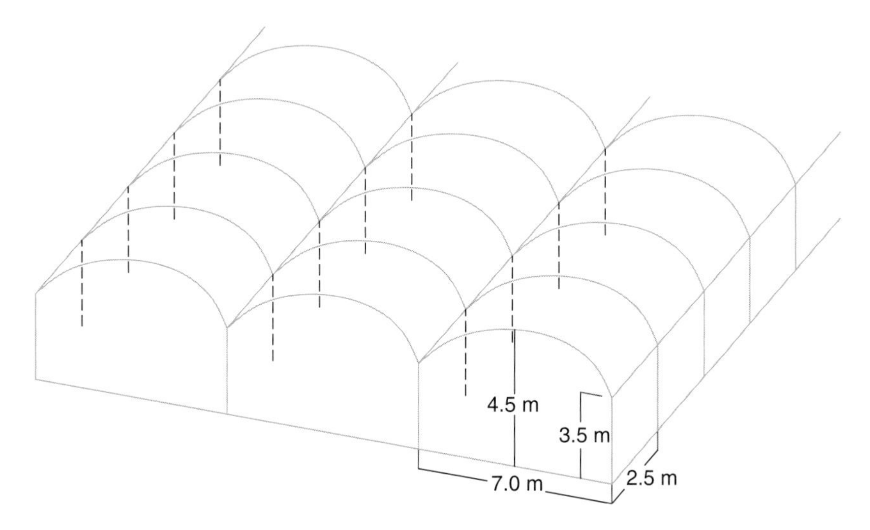

Fig. 4.15. Multi-span greenhouse structure with a curved roof.

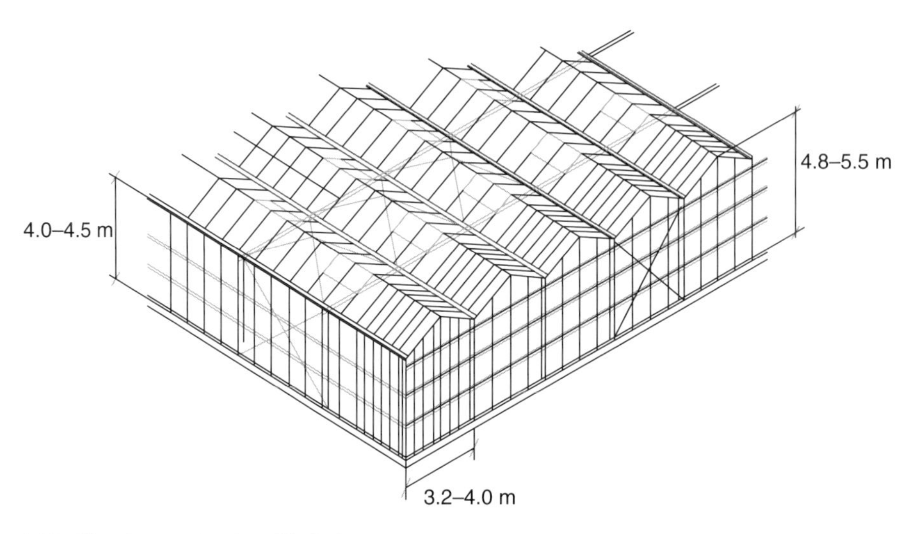

Fig. 4.16. Glasshouse structure, Venlo type.

4.8 Greenhouse Site Selection

The specific selection of the location of the greenhouse must take into account aspects such as:

- Topography. In principle, the place must be flat in the width direction, with a slope along the main axis between 0 and 0.5% (never more than 1–2%, which would involve it being terraced). In some cases, however, an inclined plot oriented towards the south (in the northern hemisphere) may be of interest, if the greenhouse type chosen adapts well to the plot, although it would be difficult to use mechanization in such a situation (such as with low-cost parral greenhouses on coastal slopes of the south coast of Spain). Normally, on steep terrains, it would be preferable to build several separate greenhouses with their axes parallel to the contour lines. The evacuation of rainfall water must be considered, avoiding it collecting in hollows.
- Microclimate of the selected site. There should be proper drainage of cold air for calm nights, and areas that frequently experience fog should be avoided. The site should be well illuminated and without shadows (from hills or buildings).
- Protection from cold winds (usually from the north in the northern hemisphere), using windbreaks or taking advantage of the topography. If snow is to be expected, the greenhouse must be far away from trees to avoid snow accumulation.
- Supply of irrigation water, in sufficient amounts and with the required quality, for the crop to be cultivated.
- Good drainage conditions of the selected plot. This aspect is especially important in regions of high rainfall. Places with a high water table must be avoided.
- Good soil characteristics for horticultural cultivation, either if plants are going to be cultivated directly in the soil or if the soil is going to be used to fill pots or containers.
- In the case of greenhouses located near to cities, it is important to evaluate the air pollution, not just by its incidence on the plants but also by residues that may be deposited over the greenhouse, limiting solar radiation (dust from factories) or that may be harmful for the greenhouse cladding material.
- Space for future expansion or auxiliary facilities (e.g. a water reservoir for collection of rainfall water or storage of irrigation water) and buildings (e.g. for handling or as stores or offices).

- Availability of labour.
- Closeness to transport networks (roads, railways), communication (telephone) and energy (gas, electricity).
- The orientation, besides avoiding shadows from hills or neighbouring buildings, must be considered in relation to the dominant winds, depending on the shape and slope of the greenhouse roof. Normally the orientation would be selected that captures the maximum amount of light in the greenhouse (see Chapter 3).

4.9 Criteria for the Design and Construction of Greenhouses

4.9.1 Introduction

Depending on the local climate and the bioclimatic requirements of the species to be cultivated, once the proper site has been selected, it will be necessary to choose the cladding material, the type of structure and the architectonic shape of the greenhouse. If the predictable climate generated by the greenhouse is not appropriate complementary facilities and equipment for climate control will have to be considered.

4.9.2 Criteria for the design of plastic-film greenhouses

Greenhouse design is very much influenced, in practice, by the local climate and the latitude of the site (Von Elsner *et al.*, 2000a), and in many cases is limited by the availability of materials for the construction.

No design is perfect, thus it is necessary to prioritize in each case, the criteria to follow, these being: (i) the maximization of the light (the main goal to be achieved; Bailey and Richardson, 1990; Giacomelli and Ting, 1999; Swinkels *et al.*, 2001); (ii) minimizing, if possible, the structural elements to avoid shadows (Briassoulis *et al.*, 1997a); (iii) ensuring good insulation which decrease the heat losses (Swinkels *et al.*, 2001); and (iv) affordable costs (Bailey and Richardson, 1990). Greenhouses with retractable roofs were conceived to maximize light (Photo 4.6).

The physical and mechanical properties of the covering materials and their

Photo 4.6. Retractable roof greenhouse (Cravo type) that permits the complete retraction of the plastic cover to maximize the solar radiation.

availability limit the options when building a greenhouse (Briassoulis *et al.*, 1997a), so there is a certain trend among growers to build traditional greenhouses (Von Elsner *et al.*, 2000b).

Relative to plastic-film greenhouses, the most important aspects to achieve are those detailed in the following paragraphs (Zabeltitz, 1990, 1999; Von Elsner *et al.*, 2000a, b).

Besides the proper structural resistance to the wind, but also to other predictable loads (snow, crops which are trained to hang, auxiliary equipment), the greenhouse must be built in such a way that the plastic film will remain well fastened, airtight, and without wrinkles, to avoid breaks caused by the wind. It must, as well, be easy to change the film. For this, the fastening system must be simple and efficient. The increasing costs of mounting the film and the plastic materials have favoured the use of special films with several years' durability. For longer durability, if possible, the structural elements susceptible to heating up by solar radiation which are in contact with the plastic film must be insulated, because excess temperatures contribute to shortening the shelf life of the plastic film.

When arcs or metal frames are used, the separation between them will depend on the predictable loads (wind, snow), normally does not exceed 3 m.

The greenhouse must be airtight, to prevent night cooling in those climates in which low night temperatures are expected, as well as to prevent undesirable leakage of CO_2. A proper ventilation system is needed, with airtight vents. The entrance of water from rainfall must be avoided.

Its volume must be large enough, not only to obtain a higher thermal inertia, but also to allow for crops that are trained to grow up high supports, and proper movement of the inside air necessary for natural ventilation. The unitary volume of the greenhouse is the quotient between the greenhouse inner volume (m^3) and the area that it covers (m^2), being equivalent to the average height.

Collection of rainfall water by means of gutters, for its later storage and use for irrigation, is not only of interest in areas of low rainfall, but also because the excellent quality of rain water makes it especially valuable for soiless cultivation, a technique for fast growth. The gutter must be 4 cm larger than the diameter of the drainpipe and must have a slope of 1% to avoid overflows (the minimum slope must be higher than 0.2% in any case). The drainpipes must have a cross-section of 7 cm^2 for each 10 m^2 of cover area that is to be drained, which caters for rainfall intensities of up to 75 mm h^{-1} (Aldrich and Bartok, 1994).

To avoid water dripping over the crops from condensation on the inner surface of the cover, it is important to build the greenhouse with roof angles greater than 26° (such angles also allow snow to run off the cover), and have an appropriate collection system, or to use anti-dripping plastic film. In unheated greenhouses, where climate control is quite limited, the slope of the roof becomes relevant to avoiding condensed water dripping from the roof cover; roofs with ogive shape might be of interest (Fig. 4.9).

Likewise, as a general rule, the greenhouse must maximize solar radiation transmission, at least in winter (when it is lower), for which proper roof geometry and orientation are fundamental.

4.9.3 Design criteria in areas with a Mediterranean climate

The most limiting climate conditions for greenhouse cultivation in Mediterranean climates are: (i) low night temperatures in winter; (ii) high daytime temperatures; (iii) high ambient humidity at night and low values during the day; and (iv) CO_2 depletion during the day (Zabeltitz, 1999; Von Elsner *et al.*, 2000a, b).

Therefore, it is especially necessary to achieve efficient ventilation, which allows for alleviation of the thermal excesses and extreme humidities, and prevents CO_2 deficits. Depending on the type of greenhouse and climate conditions it is advisable that the ventilation area is up to 30% of the ground area of the greenhouse. The increasing use of insect-proof screens in the vents, to avoid or limit the entrance of insects, decreases the efficiency of ventilation. Collection of rainfall water must also be a priority. In the low-cost type greenhouses, the general problem of condensed water dripping is aggravated in flat-roof greenhouses, inducing serious plant protection problems as it facilitates the development of diseases.

Thermal losses must be limited by choosing a suitable cladding material and making it as airtight as possible. Night heating may be necessary for the crop, during the critical winter months but its economic profitability is questionable in many cases.

4.9.4 Design criteria in humid tropical climates

The high rainfall during the whole year or during the rainfall season (which induces high RH), the stability of the temperatures (high during both the day and the night) throughout the year, and the solar radiation (which may be excessive in some cases), are the most outstanding characteristics of humid tropical climates (Loveless, 1983; Zabeltitz, 1999; Von Elsner *et al.*, 2000a, b).

As a consequence, in these greenhouses protection against the rainfall must prevail (the greenhouse umbrella effect) and there should be efficient permanent ventilation (with vents frequently equipped with screens to prevent the entrance of insects), as well as a good height and sufficient resistance to withstand strong hurricane winds which are usual in such climates. Figure 4.17 shows some of the solutions for humid tropical climates. Achieving a compromise between these requirements, at a low cost, is not easy.

4.9.5 Greenhouses for other climate conditions

In dry desert climates, the extreme temperature values are more acute than those

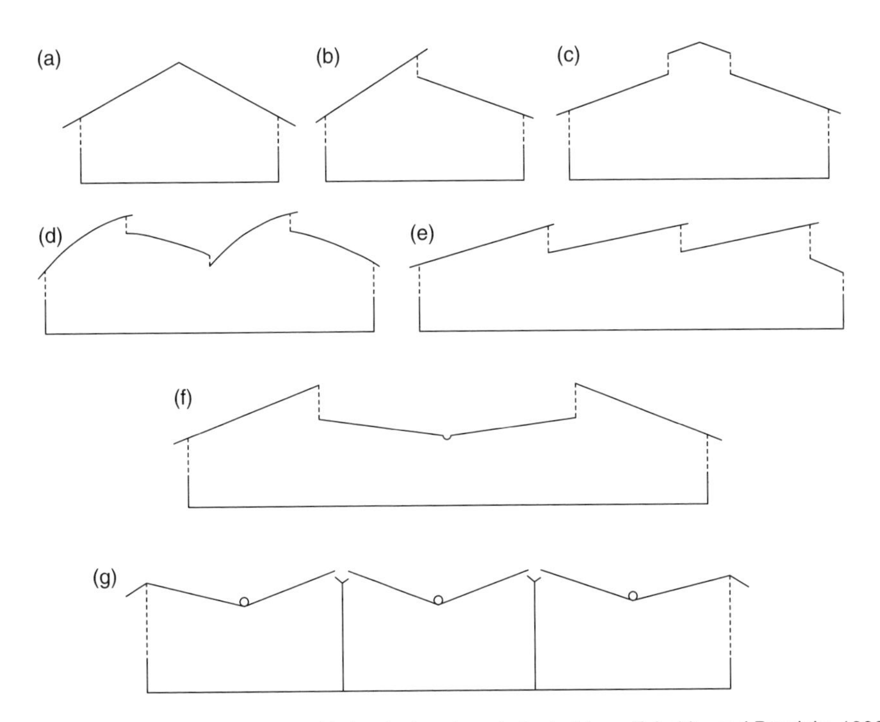

Fig. 4.17. Greenhouse structures used in tropical regions (adapted from Zabeltitz and Baudoin, 1999).

experienced in Mediterranean climates, and the ambient humidity is notably lower, the winds being frequently loaded with sand and with a very low water content (Zabeltitz, 1990, 1999; Von Elsner *et al.*, 2000a, b).

In these conditions, high ventilation capacity and efficiency is a priority (with the possibility of tightly closing the vents), and there is possibly a need for humidification systems (if the evapotranspiration of water is insufficient) to decrease the temperature and increase the RH (oasis effect). Preventing thermal losses at night is necessary (so choice of a proper cladding material and enough sealing are important) to avoid the need for night heating. The structural resistance to the wind is fundamental and the collection of rainfall water for irrigation is normally desirable.

Under cold climate conditions, the greenhouse effect must be enhanced and, normally, the maximum solar energy collection (interception) should be reached with proper roof geometry and cladding material as well as optimized greenhouse orientation. Limiting thermal losses is always desirable (using proper cladding material, thermal screens and being as airtight as possible; see Chapter 7).

Frequently, the insulation measures to reduce thermal losses imply a decrease in available solar radiation (the double wall decreases the transmission, the thermal screens generate shadows even when folded) so it is not easy to obtain a compromise solution which must be based on profitability criteria in each specific case. In these cold climates, the obvious choice between multi-span and single-span greenhouses is clearly for the first type. Heating is a must, not just during the winter months, and ventilation is necessary during the season of high radiation.

In some cases, greenhouse cladding with a screen (permeable to air and water) aims at achieving a windbreak effect, a shading effect, or plant protection (limiting the access of pests), when the natural thermal conditions are adequate for crop growth and, therefore, a greenhouse effect is not pursued.

4.10 Maximizing the Radiation Inside the Greenhouse

4.10.1 Introduction

In principle, except for some special cases (such as for crops with low light requirements), the objective of maximizing solar radiation inside the greenhouse must be pursued, especially during the months in which radiation is a limiting factor for production, as long as the costs do not hinder the primary goal of achieving good profitability for the farm. The increasingly clear and well-documented relationship between radiation and yield makes it a priority to maximize solar radiation. Artificial lighting is seldom used because it is of little economic interest (although recently this is changing in very sophisticated greenhouses) except for crops of high added value, or when it is used to modify the photoperiod.

4.10.2 Factors determining the available solar radiation

There are several factors that determine the quantity of available solar radiation. The sun's position in the sky in the different seasons of the year, the location of the greenhouse and the cloudiness influence the amount of available solar energy.

The latitude and time of year determine the angle of incidence of the solar rays over the Earth's surface as well as the daily number of sunlight hours. The angle influences the amount of radiation, reaching a maximum at the summer's solstice (21 June in the northern hemisphere) and a minimum during the winter's solstice (21 December). When the sun is very low in winter less energy impacts on the Earth's surface, because the Sun's rays have to cross a thicker atmospheric layer, and therefore, the Earth's atmosphere absorbs a higher proportion of energy.

Altitude and local climate conditions also modify the amount of solar radiation available for the plants.

The Sun's position in the horizon varies through each day, from sunrise to sunset, reaching a maximum value at noon, or shortly afterwards. In addition, the quality of the light varies through the day. As the atmosphere absorbs more of the UV and short blue wavelengths than IR and long red wavelengths, and the atmospheric layers crossed by the solar light are smaller at noon than at dawn or dusk, the light during the initial and final hours of the day have a higher proportion of red and IR radiation.

In the absence of clouds, during most of the day, the higher proportion of solar radiation corresponds to direct radiation (directly from the Sun) whereas during the first and last hours of the day diffuse radiation (of a non-directional nature) prevails over direct radiation. During the winter, when there are normally more clouds, the proportion of diffuse radiation is much higher.

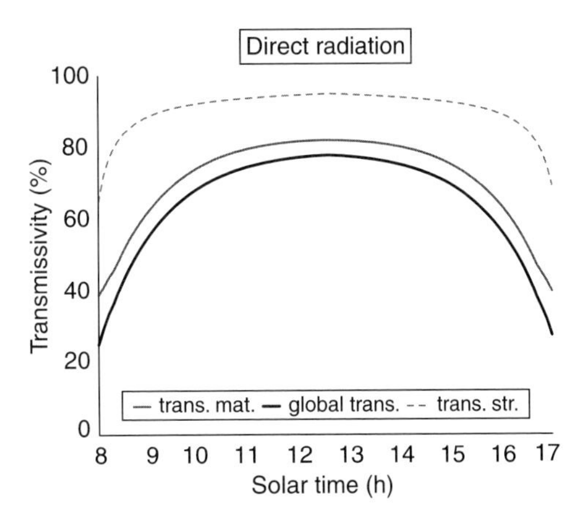

Fig. 4.18. Evolution of global transmissivity on 21 December in a low-cost greenhouse (thermal PE), with a symmetrical gable roof, with a 30° roof angle, oriented east–west, on a clear day. The global transmissivity is the product of the transmissivity of the material (trans. mat.) without the structure and the transmissivity of the structure (trans. str.) without the material.

In this sense, the ratio 'cladding area (roof plus sidewalls):greenhouse soil area' notably influences the heat losses and, as a consequence, the temperature level and the heating costs, thus influencing the economic viability.

In addition, as it is necessary to minimize the shadows caused by the structural elements, the geometry of the roof must also be considered accordingly.

4.10.3 Solar radiation inside the greenhouse

The main factors affecting the radiation transmitted inside a greenhouse are: (i) the type of structure; (ii) the shape and slope of the greenhouse roof; (iii) its orientation with respect to the Sun; (iv) the location of the greenhouse equipment (due to the shadows they generate); and (v) the characteristics of the cladding material (glass, plastic film, rigid panel). See Chapter 3 and Fig. 4.18 and Plate 9.

The transmission of solar radiation through the plastic or glass of a greenhouse will depend on the angle of incidence (Figs 3.3 and 3.4). The architectonical shapes of the roof (see Chapter 3) must tend to optimize the angle of incidence (Fig. 4.19), without losing sight of other relevant aspects of the design (cladding surface and its influence on the energy balance, resistance to the wind, volume and greenhouse dimensions, ventilation area) which may limit its economic viability.

4.10.4 Greenhouse orientation

The orientation of a greenhouse (direction of its longitudinal axis) is of great influence on the transmission of radiation inside the greenhouse during the winter months, when radiation is lower. When diffuse radiation prevails over direct radiation (cloudy days) orientation has less impact on transmission. The orientation is dependent on the geometry and slope of the roof, the latitude and the season of the year. In practice, the shape and topography of the plot, as well as the direction of the dominant winds, determine the orientation of the greenhouse (see Chapter 3).

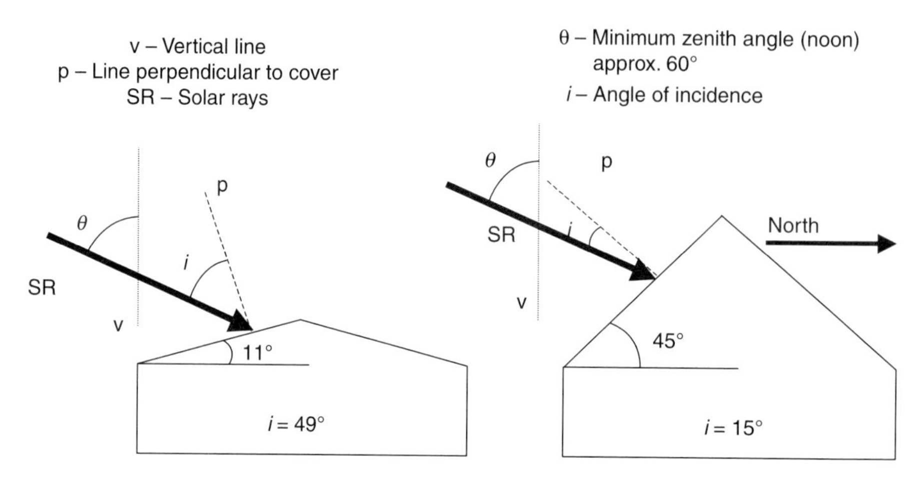

Fig. 4.19. Angle of incidence (*i*) of direct solar radiation in greenhouses oriented east–west, of low and high roof slopes, in the south of Spain (latitude 37°N) in the winter solstice at noon.

The objective is to capture the maximum solar radiation in winter, in the case of the single-span greenhouse, for latitudes above 30°N (in the northern hemisphere), as long as the roof slope is sufficient. It will be desirable, as a general rule, to orient the greenhouse east–west, to maximize the capture of light in winter (Fig. 4.20 and Plate 10). The north–south orientation produces a more uniform distribution of radiation at different points in the greenhouse, than the east–west orientation (Fig. 4.21), especially if the roof slope is low, and in a more marked way if direct radiation prevails over the diffuse (see Chapter 3).

When analysing multi-span greenhouses the problem is more complex, due to the shadows that each span projects over the adjacent one (Fig. 4.22).

It is desirable, for each case to study the specific problem before installing the greenhouse and to consider this together with other aspects, as previously indicated (shape of the roof, slopes, priority growing season), because the design is usually more important than the orientation.

Small deviations from the optimal orientation (of the order of 15°) have very little influence on the transmissivity, but if they reach 30° they start to be significant.

An important aspect to consider in single-span greenhouses, once the orientation has been chosen, is the separation between them to avoid shadows (Fig. 4.23). In latitudes of the Mediterranean Basin a minimum separation of up to 8 m is advisable (if built at the same elevation).

Nowadays, solar radiation transmissivity models have been developed for different types of design of greenhouse structure in the Mediterranean, and these are an efficient tool for the designer (Soriano *et al.*, 2004b).

Summarizing, to maximize radiation, losses due to structural elements and equipment must be minimized. The site selection, orientation, shape and slope of the roof and cladding material used are primary aspects to capture the maximum possible radiation inside the greenhouse, bearing in mind that the best technical solution (Fig. 4.24) is not always the most economically suitable (Fig. 4.25).

To maintain high transmission proper management of the greenhouse is required. Some of the measures that may help to achieve the goal of maximizing the solar radiation available for the crop in the greenhouse are: (i) cleaning the cover; (ii) limiting condensation on the inner surface of the cover (as it reduces transmissivity) by means of good climate control (which limits high

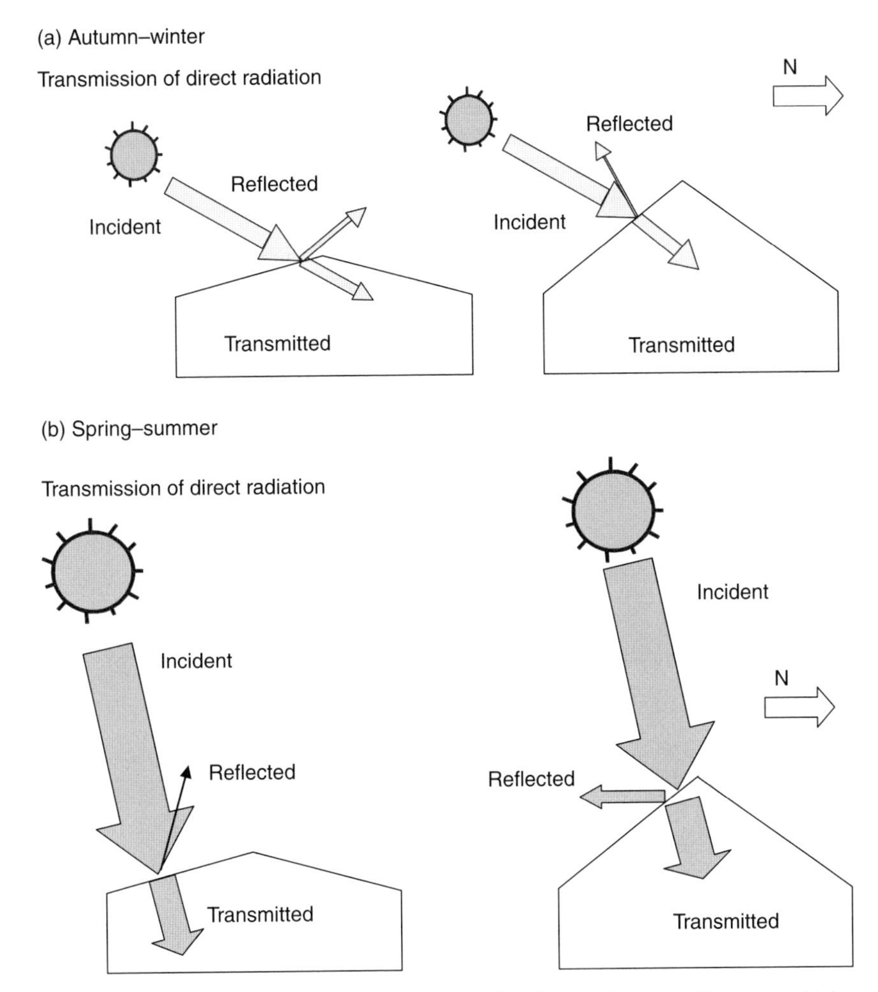

Fig. 4.20. Transmission and reflection of direct solar radiation in greenhouses with low and high roof slopes, in the south of Spain (37°N) in autumn–winter (a) and spring–summer (b).

humidity in the air) or using drip irrigation and/or mulching (which reduces water evaporation from the soil); (iii) orienting the crop rows north–south; and (iv) painting the structural elements in white and using white mulch to reflect the radiation.

4.11 Normalization of Greenhouse Structures

Regulation UNE-EN-13031 is the Spanish version of European regulation EN-13031-1 (*Greenhouses: design and construction. Part 1: Greenhouses for commercial cultivation*). This regulation specifies the general principles and requirements of mechanical resistance and stability, state of use and durability for the design and construction of commercial greenhouse structures (including the foundations) for the cultivation of plants and crops.

Regulation UNE-EN-1303212 has replaced the experimental regulation UNE-76208.

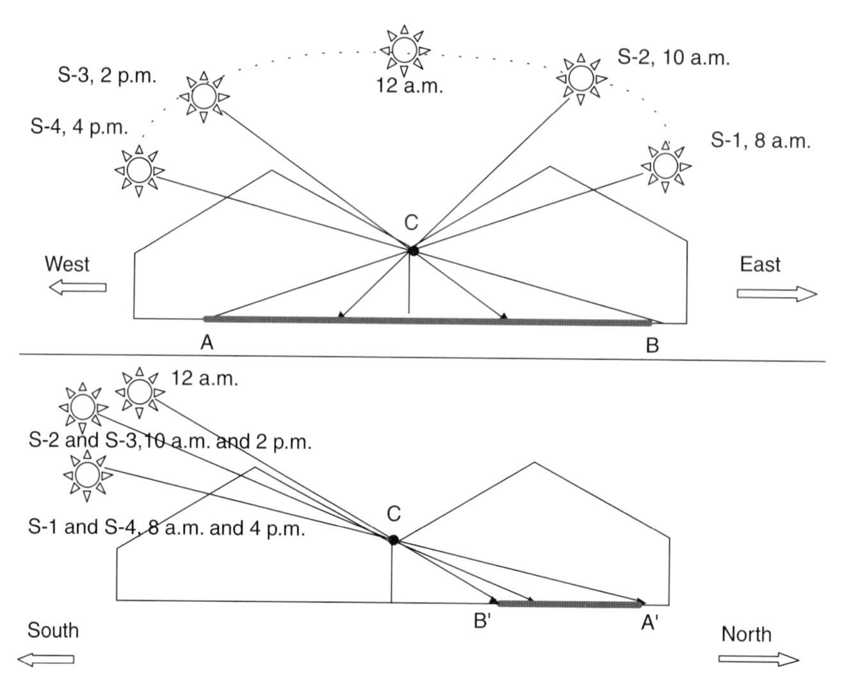

Fig. 4.21. Distribution of shadows created by the gutter (C) through the day in two greenhouses with east–west orientation (bottom, A'B') and north–south (top, AB). The sun's elevation is shown between 8 a.m. and 4 p.m. (solar time) on the 21 December (latitude 37°N). The shadows are distributed more uniformly in the north–south oriented greenhouse.

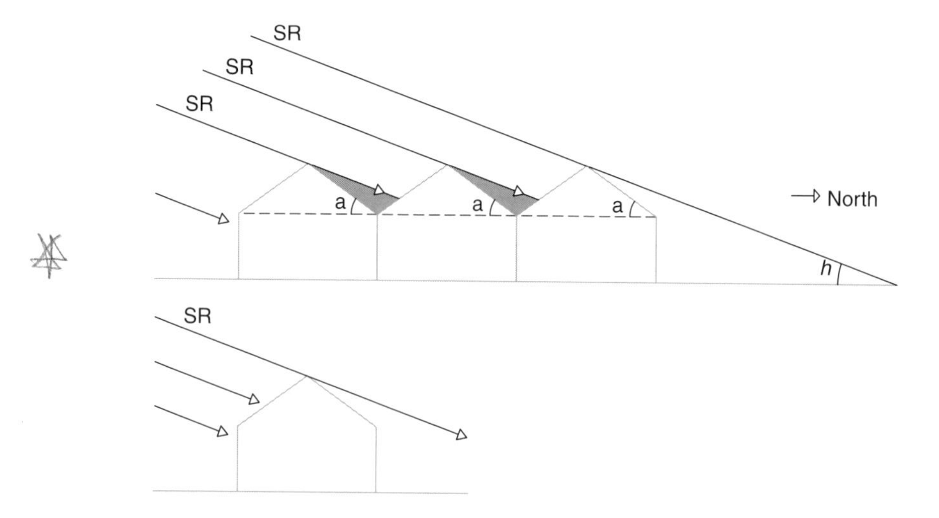

Fig. 4.22. In a multi-span greenhouse oriented east–west, if the roof angle of the north side (a) is greater than the angle of the elevation of the Sun (*h*) shadows are produced (shown as grey shading) as one span casts a shadow over the adjacent span. In a single-span greenhouse these shadows are not produced. SR, Solar rays.

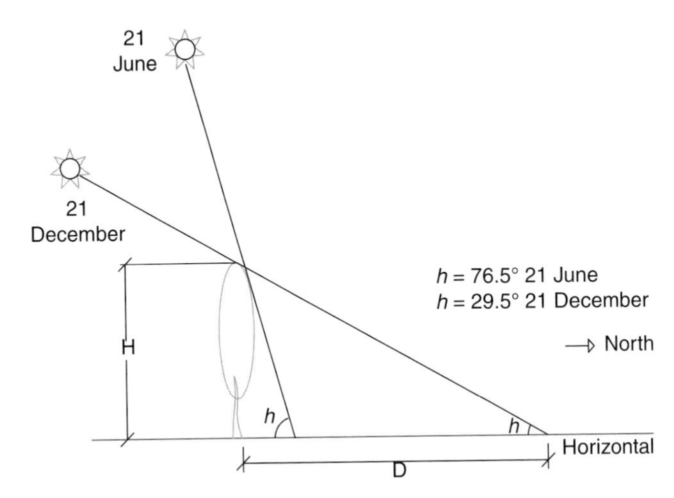

Fig. 4.23. Shadows created by an object (a tree) at noon during the summer and winter solstices. D, Length of the shadow; H, height of the tree; h, elevation angle of the Sun (see Fig. 2.8). Latitude: 37°N. (D = H cotangent h).

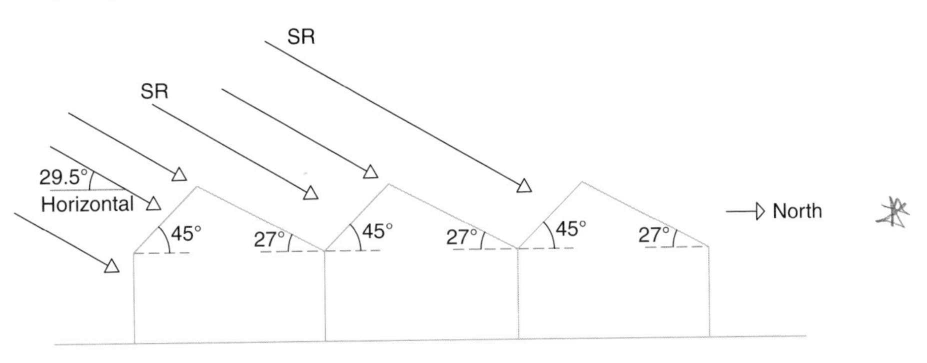

Fig. 4.24. A greenhouse prototype with an asymmetric roof, oriented east–west, with angles of 45° on the south side of the roof and 27° on the north side, does not generate shadows between spans at noon in the south of Spain (37°N), in the winter solstice. This is because the elevation angle of the Sun (h) is approximately 29.5°. The higher angle of the south-facing roof (45°), where most of the radiation is captured on these dates, induces a high transmissivity (see Fig. 4.19). SR, Solar rays (Castilla *et al.*, 2001).

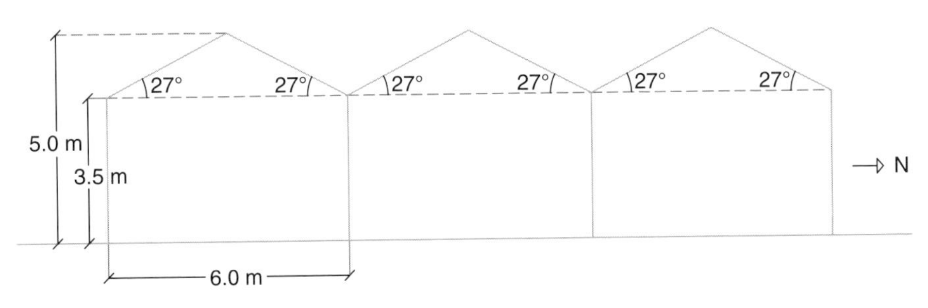

Fig. 4.25. An efficient solution to achieve a good transmissivity in autumn–winter, at an affordable cost, in low-cost multi-span greenhouses for the south of Spain is to build symmetrical spans, with roof slopes 1:2 (equivalent to angles of approximately 27°), oriented east–west. Although this greenhouse is less transmissive in the winter solstice than the greenhouse with an asymmetric 45°/27° roof shown in Fig. 4.24, it is easier to build, cheaper and offers less resistance to the wind (Soriano *et al.*, 2004a, b).

4.12 Summary

- Nowadays, the previously considered main function of a greenhouse (increase of the temperature in relation to the open field, as a consequence of the 'greenhouse effect') in some cases remains secondary, while the 'shading effect' or the 'windbreak effect' of the greenhouse under specific climate conditions are considered of equal or even greater importance.
- The suitability of a location for greenhouse cultivation is determined by its climatic conditions (mainly temperature and radiation) as well as other factors of a socio-economic nature.
- In the last few decades flexible plastic films, with their low weight in relation to the materials previously used in greenhouses (glass), have resulted in a considerable reduction in supporting structures and their cost, and have allowed for a massive expansion in the use of greenhouses all over the world.
- The most common plastic materials used in the form of flexible films as greenhouse covering materials are LDPEs (in their normal, long-life and thermal variants), EVA copolymer and plasticized PVC, although this last material is not used very extensively except in Asia.
- Incorporation of different additives improves the characteristics of plastic films. The multilayer films (formed by coextrusion of several layers of different materials) enable several desirable characteristics to be combined in a single film, which is not possible with a single material.
- A good plastic film must have high transmission to solar radiation, a limited transmission to long-wave IR radiation (Earth's radiation), as well as durability in line with its thickness, formulation and cost. The longevity of a plastic film will depend on the type of solar radiation received, as UV rays coming from the sun are responsible for ageing plastic materials.
- Rigid plastic materials are used in the form of simple or alveolar panels and normally need a more expensive structure than that used with flexible films. The most commonly used rigid materials are polyester, PC, PVC and PMMA.
- The rigidity characteristics of the covering materials determine, to a great extent, the shape of the greenhouse and the covering materials. Conventionally, glass panels have not been used in curved roofs, but recently curved glass panels have become available on the market.
- The main aspects to consider in the choice of a greenhouse must be: (i) transmission to PAR, which determines the production potential; (ii) the solidity and longevity; (iii) the functionality and ease of maintenance; and (iv) the economics of the energy required and price.
- The most common greenhouse types are the artisan greenhouse and the industrial greenhouse. Artisan greenhouses use plastic films and are cheap. In Spain, the most common type is the artisan low-cost parral greenhouse. Among the industrial greenhouses, the most common are the Venlo type, which uses glass as cladding, and multi-span greenhouses, which usually have a curved-roof shape or are multi-tunnel greenhouses with flexible plastic-film covering although they also allow for the use of semi-rigid panels.
- Among the criteria that should be considered for the design and construction of plastic-film greenhouses (which are heavily influenced by the climate and latitude of the location) the maximization of the light is the most important, as well as providing proper insulation and sufficient ventilation. In addition, such greenhouses should be structurally sound (against wind, snow, etc.) and the film should be easy to assemble so that it remains

tight and well fixed. They should have a large inner air volume, be reasonably airtight, have systems for the collection of rain water, and should minimize the amount of water that drips over the crop from condensation on the inner surface of the cover. Prioritization of these criteria will vary in each case.

- The greenhouse equipment (heating, humidification, ventilation) will depend on local conditions.
- The greenhouse orientation and the roof geometry are fundamental to maximize the radiation inside the greenhouse, especially in autumn and winter in Mediterranean latitudes, when sunny days are predominant.

5

Greenhouse Heat Exchanges

5.1 Heat Transfer

Two bodies at different temperatures exchange energy in the form of heat, which flows from the hotter body to the colder body. There are three main modes of heat exchange: (i) conduction; (ii) radiation; and (iii) convection (with or without change of state).

The air renewal of the greenhouse involves a mass transfer, notoriously affecting the greenhouse energy balance.

5.1.1 Conduction

Conduction is the energy transport from molecule to molecule inside a body, solid or fluid. By conduction the energy is transported through a media at rest, which is not flowing in the direction of the energy transport (Fig. 5.1).

Conduction is the only mode of heat propagation inside a solid, or between two solids in physical contact, at different temperatures. The energy flows from the higher to the lower temperature. The rate of temperature change along the distance (d) is called the thermal gradient.

The main heat changes by conduction in a greenhouse take place between the soil surface and its deep layers, and the thermal losses through the greenhouse structure. See Appendix section A.4.1.

5.1.2 Convection

Convection without phase change

Convection is the energy transport by a fluid in the same direction as the flow or between a static surface and a fluid. The thermal exchanges by convection involve the displacement of matter. They take place mainly in fluids and induce their movement.

Convection is of the 'forced' type when it is provoked by an external mechanical action (e.g. air fan). 'Natural' convection is driven by density differences, derived from temperature differences, which generate the fluid movement (for instance, the warm air weighs less than the cold air and rises up).

Convection exchanges have great importance in the greenhouse. The air in contact with the heating pipes is heated by conduction and, once heated, moves by convection, heating the rest of the greenhouse air. This warm air contacts the plants and then heats the plants by conduction.

When it is windy, the greenhouse cover loses a lot of heat because of active (forced) convection. See Appendix section A.4.2.

Convection with phase change

EVAPORATION. Water evaporation consumes a lot of energy. Evaporation involves a change of phase, from liquid to gas. The energy required to make water pass from the liquid state to the gas state is called the 'latent heat of vaporization', and for water it is 2445 kJ kg^{-1}, at 20°C. The partial water vapour pressure rises, and so does its energy content, that is, its enthalpy increases. Evaporation is only possible when the water vapour pressure is lower than the saturation vapour pressure at a given temperature.

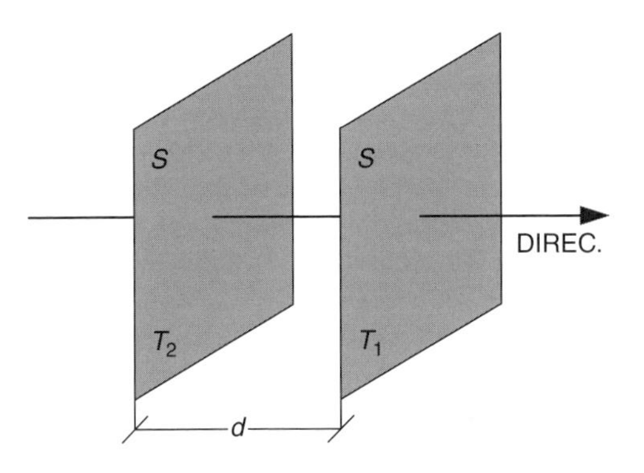

Fig. 5.1. Heat transfer by conduction between two plane, parallel surfaces in a direction perpendicular to both surfaces (DIREC) (see text).

If there is no change of state (from liquid to vapour) the heat required to increase the temperature of water by 1°C (specific heat of water) is relatively low, 4.2 kJ kg^{-1} of water.

Transpiration is a specific form of evaporation that happens in plants. As the transpiration consumes energy, it allows for the cooling of the transpiring organs, and it is the main mechanism by which the plants decrease their temperature. Water fogging or misting enables the greenhouse to be cooled (Photo 5.1; see Chapter 8).

CONDENSATION. Condensation is the inverse phenomenon to evaporation. Water passes from the gaseous state (water vapour) to the liquid state, releasing energy. It involves a decrease of the partial pressure of the water vapour in the air, and an increase in the temperature of the surfaces where water condenses.

Condensation happens only when the air water vapour partial pressure reaches or exceeds the saturation vapour pressure point at a given temperature. The dew point is the temperature below which condensation takes place.

From the phytosanitary point of view, condensation is of great importance as the

Photo 5.1. Water fogging or misting and proper ventilation enable the greenhouse to be cooled.

presence of water over the leaves favours the development of many fungal diseases (e.g. *Botrytis*). Besides, the formation of condensation droplets on the internal surface of the plastic films cladding the greenhouse involves important reductions in light transmission, and eventually dripping over the crops with harmful effects on their health (Photo 5.2).

Effects of condensation on transmissivity and the thermal balance

In plastic greenhouses, when the water condenses in the form of water drops, there is a decrease in the radiation transmissivity, as the droplets reflect part of the radiation. However, if the condensation occurs in a continuous film, not only is there no reduction in transmissivity, but in the case of special plastic films treated with special surfactants there can even be an increase (Fahnrich *et al.*, 1989).

Moreover, water condensation on the plastic cover improves the thermal balance of plastic greenhouses, and by avoiding long-wave radiation escaping from the inside, it decreases their global heat transfer coefficient (section 5.4).

The 'anti-dripping' plastic films do not avoid condensation, but avoid the formation of droplets. A proper roof slope and a corresponding system of collection channels at the greenhouse gutter allows for the effective removal of the condensation water (see Chapter 4).

5.1.3 Radiation

All bodies with temperatures above –273°C (0 K) emit energy through their surface, in the form of electromagnetic radiation following the Stefan–Bolzman law (see Appendix 1 section A.4.4). This radiant energy is transformed into thermal energy if it finds a body to absorb it.

The characteristics of the radiation (wavelength) which a warm surface emits depend on its temperature, following Wien's law (see Appendix 1, section A.1.4). The amount of energy of a certain radiation decreases as the wavelength increases.

Photo 5.2. Condensation of water vapour on the internal face of the greenhouse cover is of great importance (see text).

When two bodies are at different temperatures, separated by a permeable medium, there is a net heat transfer from the warm body to the cold body in the form of radiation (Fig. 5.2). The energy received by the surface of the cold body is divided in three parts: (i) a fraction is reflected; (ii) a second part is transmitted (crosses the body without heating it); and (iii) a third fraction is absorbed by the body and increases its temperature (Fig. 3.3).

Bodies have reflection, transmission and absorption properties for radiation which vary with the received wavelength. For each material the reflection coefficient or reflectivity (ρ), the transmission coefficient or transmissivity (τ) and the absorption coefficient or absorptivity (α) can be defined.

For a certain wavelength the sum of the three coefficients is equal to 1, or 100%, depending on how the coefficients are expressed (per unit or as a percentage), due to the energy conservation principle.

A grey body is any body in which the absorption coefficient is independent of the wavelength of the incoming radiation. In practice many materials can be treated as grey bodies. A black body is any surface in which the absorption coefficient is 1, whatever the wavelength (Rosenberg *et al.*, 1983); it is an ideal body.

The transmissivity and the reflectivity, are quite dependent on the angle of incidence of the radiation over the material's surface (see Chapter 4). The absorptivity is less dependent on such an angle and is more linked to the type of material and its thickness (Seeman, 1974).

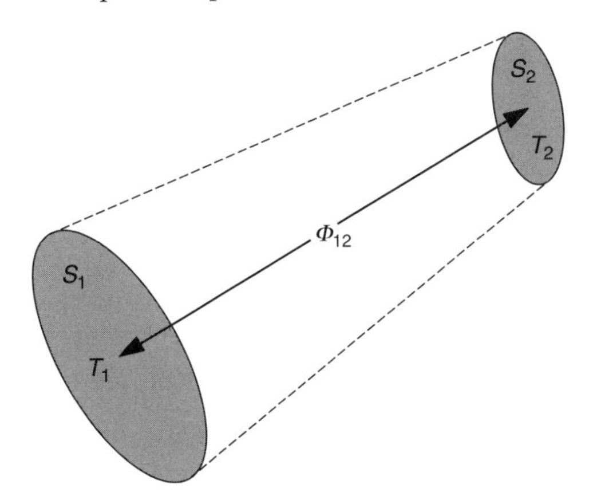

Fig. 5.2. Heat transmission by radiation between two bodies at different temperatures ($T1 > T2$) (see text).

When the angle of incidence is not known and we talk, generically, of these coefficients, the data usually refer to transmissivity and reflectivity for zero angle of incidence (rays perpendicular to the surface), which correspond to the maximum transmissivity and minimum reflectivity (Figs 3.3 and Plate 5).

When the radiation is diffuse, it does not have an angle of incidence as it comes from all possible directions and this transmissivity of diffuse radiation is different from the transmissivity of direct radiation.

Emissivity is the proportion (per unit) of the total radiation emitted by a body at a given temperature with respect to the one emitted by a black body of the same surface under the same conditions (see Appendix 1 section A.4.4). The absorption coefficient of a material to radiation of a certain wavelength is equal to its emissivity in this same wavelength, according to Kirchoff's law (Rosenberg *et al.*, 1983).

The emissivity and the absorptivity, in the same range of wavelength, have equal values, but if a body receives solar radiation and emits radiation in the IR band the absorptivity for the solar radiation is different from the emissivity for the IR radiation (Table 5.1).

For non-metallic bodies, such as plant leaves and white paint, the emissivity is high (from 0.7 to 1) at normal crop temperatures, whereas for metals, especially if they have been polished, it is low (0.05–0.3) (Bot and Van de Braak, 1995).

It is essential to know the spectral distribution characteristics of transmissivity of a covering material if the plants which are grown in the greenhouse have colours generated by anthocyanins (pigment with colourations between red and violet), because if the material is not transmissive within a certain range of UV (290–360 nm) it may prevent such colouration (Takakura, 1989), as may happen in aubergine. Indeed the lack of UV radiation, to which the eyes of

Table 5.1. Characteristics of absorptivity to solar radiation and emissivity of several surfaces (at 13°C), both expressed per unit (adapted from Aldrich and Bartok, 1994).

Surface	Absorptivity to solar radiation	Emissivity (at 13°C)
Concrete	0.60	0.88
Red brick	0.55	0.92
Glass	0.03	0.90
White paint	0.35	0.95
Dry soil	0.78	0.90
Wet soil	0.90	0.95
Aluminium		
Commercial	0.32	0.10
Painted in white	0.20	0.91
Painted in black	0.96	0.88
Galvanized steel		
Commercial	0.80	0.28
Painted in white	0.34	0.90

bees and bumblebees are sensitive, may induce problems in their mobility, affecting flower pollination (see Chapter 4).

The heat exchanges by radiation are essential in greenhouses. The surfaces of a greenhouse exchange heat by radiation between them. The greenhouses are heated absorbing an important part of solar radiation and get cooler radiating energy towards the sky.

The heating pipes, besides heating the plants directly by convection, also do it directly by radiation.

5.2 Heat Exchanges by Air Renewal in the Greenhouse

The interior air of the greenhouse is usually warmer and more humid than the outside air. The renewal of the interior air with external air involves a decrease of its energy content (enthalpy; see Appendix 1).

5.3 Heat Exchanges in the Greenhouse and Energy Balance

The calculation of the energy balance of a greenhouse is useful, especially when calculating the capacity of the heating system to be installed.

To analyse the energy balance of the 'greenhouse complex' it is possible to divide it into different subsets, for instance: the soil, the crop, the interior air volume and the cover. Then the energy balances of each one of them can be analysed independently, which is easier, integrating them later.

In practice, simplifications are used which do not consider some elements of the energy balances that have less influence overall; this allows for sufficient approximation of the energy balance.

The soil surface absorbs part of the solar energy, exchanges energy by IR radiation with the crop canopy, with the heating pipes and with the walls, cover and other elements of the greenhouse, and by convection with the greenhouse air. The soil surface is cooled by water evaporation and by exchanging energy with the deeper layers of the soil, cooling or heating itself depending on the season.

The vegetation absorbs an important part of the solar energy that it receives and exchanges energy by IR radiation with the soil surface, with the heating pipes, with the wall, cover and other elements of the greenhouse, and by convection, with the greenhouse air. The vegetation, in addition, loses energy by transpiration and may, eventually, gain energy by condensation.

The heating pipes absorb some solar energy and, if the boiler is on, they may receive energy from the hot water. The pipes exchange heat mainly by convection with the air and by radiation with the vegetation, the soil and the cover.

The interior air exchanges energy mostly by convection with all the greenhouse surfaces: soil, plants, heating pipes and cover. The renewal of the interior air by external air, normally drier and cooler, produces a decrease in its enthalpy (energy content).

The cover absorbs a small amount of the received solar radiation, exchanging energy by IR radiation towards the interior of the greenhouse and towards the exterior. In addition, it exchanges energy by convection with the external air, through its external surface, and with the interior air, through its internal surface. It may collect energy when water vapour condenses on the cover and will cool when the condensed water evaporates.

A relevant fact to consider is the thermal inertia of the greenhouse, which will depend on what its components are made of. Therefore, the relationship between the thermal capacities of the air/cover/plants/ and soil (up to a depth of 20 cm) of the greenhouse is of the order of 1/3/10/100 (Day and Bailey, 1999), which means that the air thermal inertia of the greenhouse is minimal and therefore its temperature responds quickly to energy balance changes (as the air is heated), whereas the response of the soil temperature is slow, since its thermal inertia is much larger.

5.4 Simplified Greenhouse Energy Balances

If we consider all the greenhouse heat exchanges overall, by radiation, conduction and convection, through the cladding, we may quantify their amounts (per time unit) as:

$$Q = K(Ti - Te)\ Sc \quad (5.1)$$

Ti = Interior temperature (°C)
Te = External temperature (°C)
Q = Amount of heat exchanged between the interior and the exterior (W)
Sc = Cladding surface (m^2)
K = Global heat transfer coefficient of a greenhouse covering material, characteristic of each covering material (W m^{-2} °C^{-1}) (see Table 5.2 and Appendix 1 section A.4.9).

To simplify, it could be assumed that the solar energy penetrating the greenhouse is responsible for heating the greenhouse and for evapotranspiration, neglecting the energy used for photosynthesis, among other simplifications, the instantaneous energy balance would be approximately:

Solar radiation – Evapotranspiration
+ Heating = Overall losses
+ Air renewal

Table 5.2. 'Global heat transfer coefficient' (K in W m^{-2} °C^{-1}) for some greenhouse covering materials, measured under normalized conditions (temperatures: exterior: –10°C, interior: +20°C, wind: 4 m s^{-1}). (Source: Nisen and Deltour, 1986.)

Cover		Clear sky	Overcast sky
Single	PE	8.8–9.0	7.1–7.2
	EVA	7.8	6.6
	PVC	7.6	6.4
	Polyester	7.2	6.2
	Glass (4 mm)	6.1	5.5
Double	PE + PE	6.4	4.2
	PC (6 mm)	3.5	3.2
	Glass + glass	3.1	2.8

(for calculations on air renewal see Appendix 1 section A.4.5)

For the approximate calculation of the heating in the night when the requirements are higher, the following simplified energy balance equation can be applied (solar radiation being nil) (Montero *et al.*, 1998):

Heating = Overall losses + Air renewals (see Appendix 1)

$$Qc = K(Ti - Te)\ Sc + m\ Cp\ (Ti - Te) \quad (5.2)$$

Where:
Qc = Heating requirements (W)
m = Air mass renewed per unit time (kg s^{-1})
Cp = Air specific heat (J kg^{-1} °C^{-1})

Some authors estimate the heating requirements under conditions of closed ventilators (i.e. when air renewals are only by infiltration, and this renewal represents only 10% of the heating requirement) (Boodley, 1996), as follows:

$$Qc = 1.1\ K(Ti - Te)\ Sc \quad (5.3)$$

5.5 Summary

- There are three fundamental modes of energy exchange in the form of heat: (i) conduction; (ii) radiation; and (iii) convection (with or without change of state).
- Energy is transported through a medium at rest by means of conduction. In greenhouses the heat exchanges by

conduction between the soil surface and its deep layers are important.

- We call convection the energy transport by a fluid in the direction of flow, or between a static surface and a fluid. It is of crucial importance in a greenhouse. The air in contact with the heating pipes is heated by conduction and, once heated, moves by convection, heating the rest of the greenhouse air. This warm air contacts the plants and then heats the plants by conduction. When it is windy, the greenhouse cover loses a lot of heat because of active (forced) convection.
- Water evaporation consumes a lot of energy, as water goes from the liquid state to a gas. We call the latent heat of vapourization the energy needed to evaporate 1 kg of water, at 20°C, and its value is 2445 kJ kg^{-1} of evaporated water. If there is no change of state (from liquid to vapour) the heat required to increase its temperature by 1°C (specific heat of water) is low, of the order of 4.2 kJ kg^{-1} of water.
- The condensation of the air water vapour is the inverse phenomenon to evaporation and releases a large amount of energy. Condensation only occurs if the partial pressure of water vapour reaches a value known as 'saturation pressure'.
- The dew point is the temperature below which condensation occurs.
- Water transpiration by the plants is a particular form of evaporation, which allows plants to cool down and decrease their temperature.
- Water condensation in a greenhouse starts at the colder spots, normally on the greenhouse cover. Condensation in the internal surface of the cover affects the light transmission and improves the insulation conditions of the cover.
- All bodies with temperatures above −273°C emit energy from their surface in the form of electromagnetic radiation. This radiant energy is transformed into thermal energy if it impacts a body that absorbs it.
- The radiation received by a body can be reflected, absorbed or transmitted through it. The greenhouse covering materials must be transmissive to solar radiation.
- The heat exchanges by air renewal in the greenhouse are very important for the energy balance. The humid air contains more energy than dry air at the same temperature, as it incorporates the energy used for the evaporation of the water. The renewal of the internal air, which is usually warmer and more humid, by external air (cooler and drier) involves a great loss of energy from the greenhouse.

6

Crop Physiology: Photosynthesis, Growth, Development and Productivity

6.1 Introduction

The relationship between the different organs of a plant are represented schematically in Fig. 6.1. The leaves receive solar energy for its conversion into biomass (vegetable matter) by means of photosynthesis. The required gas exchange for photosynthesis takes place through the stomata of the leaves, as does transpiration (transfer of water from the leaves to the surrounding air). The roots, besides anchoring the plant to the soil or the substrate, absorb water and mineral elements and may also serve as reserve storage organs. The stem and the branches, besides accomplishing a support function, contain the vessels through which the ascending and descending sap flows between the different organs of the plant. The apical meristems are responsible, as their cells multiply, for the formation of new organs. Other meristems are responsible for branching and for other growth functions. The meristems are centres of intense biochemical activity, especially in the synthesis of hormones which regulate the harmonic growth of the plant (Berninger, 1989).

The plant can be conceived as a set of 'sources' providing compounds necessary for the activities of its parts and a set of 'sinks' or destinations where these compounds are moved and consumed. The roots are a source of water and mineral elements, whereas the leaves are a source of carbohydrates. The main 'sinks' are the growing parts (young leaves, flower buds, fruits) and the reserve storage organs (tubers, bulbs). To operate appropriately the plant requires a balance between 'sources' and 'sinks'.

Physiologists distinguish between two aspects of the vegetative activity of the plant: (i) growth; and (ii) development. Both are conditioned, directly or indirectly, by environmental factors. Growth is a quantitative notion corresponding to the variation in size and weight of the different plant organs (Berninger, 1989). Development is a qualitative notion related to the changes of stage: (i) germination; (ii) leaves and internodes succession; (iii) bud differentiation; and (iv) flowering.

The yield, in a broad sense, includes not only the quantitative aspects of production but also the qualitative ones. These are more complex to measure, due to the difficulties in evaluating quality, whose impact on the economic return of high added value vegetables for fresh consumption is large.

Good product quality is determined, normally, by proper crop growth and development conditions. The amount produced, quantified in vegetables by the weight of

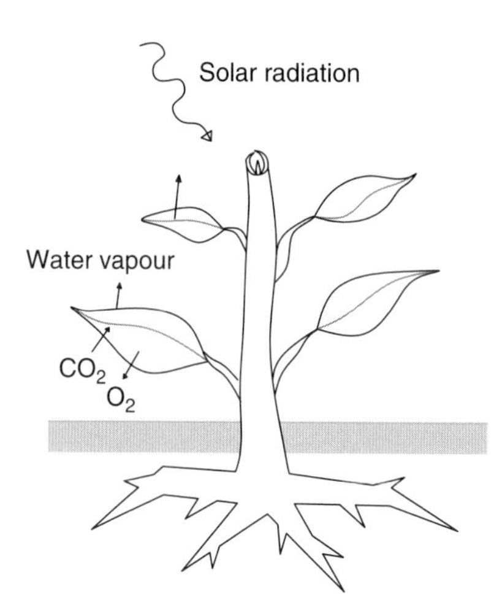

	Sources of:	Sinks of:
Apex	Hormones	Assimilates, water
Young leaves	Hormones	Assimilates, water
Adult leaves	Carbohydrates	Water, assimilates
Stems and roots	(Reserves)	(Reserves)
Roots	Water, hormones, mineral elements	Assimilates

Fig. 6.1. Schematic representation of the organization and relationship between the various organs in a plant (adapted from Berninger, 1989).

the product in different grades and referenced to the surface unit, relies on the efficient use of incident solar radiation by the crop, which in itself, requires appropriate plant density and properly arranged and managed leaf canopies to achieve an optimal economic profit.

6.2 Physiological Functions and Growth

The main physiological functions involved in the growth of a plant are: (i) water and mineral element absorption by the roots; (ii) water vapour transpiration through the leaves; (iii) photosynthesis; and (iv) respiration (Berninger, 1989).

Water absorption is influenced by climate conditions (radiation, temperature), plant conditions (water stress) and soil conditions (water availability, aeration). To optimize water and mineral element absorption it will be necessary to have: (i) an appropriate soil or substrate (well drained, able to store the required amounts of water and oxygen to ensure a sufficient availability between two consecutive irrigation episodes); (ii) a balanced soil solution (with an appropriate concentration of nutrients for the crop, and adequate salinity and pH); and (iii) water and nutrient supply which cover the crop requirements, performed with proper periodicity.

When water and mineral supply is achieved by drip irrigation, the root system develops less than in the case of surface irrigation, and therefore the aerial part:root ratio is higher. Consequently, drip-irrigated plants can allocate more assimilates to the upper organs which are the parts of interest in the common vegetable crops, but these plants are more sensitive to an accidental water deficit.

Transpiration is the evaporation of water by the plant, mainly through the stomata of the leaves. The energy required to evaporate the water basically comes from the solar radiation; therefore, transpiration is directly related to solar radiation. In heated greenhouses, the energy supplied may also contribute to transpiration. In Mediterranean greenhouses, the irrigation water requirements to cover the demand of the vegetables crops range from 0.5–1 mm day^{-1} in winter to 4.0–5.0 mm day^{-1} in unshaded greenhouses in the summer (Castilla, 1995). With intermittent surface irrigation (when a single irrigation could apply more than 10 mm of water), the use of cold water must be avoided, as it may induce undesirable cooling of the rhizosphere (Berninger, 1989).

Photosynthesis allows plants to convert different inorganic compounds into vegetable organic matter (or biomass), using energy from the Sun. Respiration supplies the plants with the energy required for them to function, consuming part of the biomass generated in the photosynthesis. When photosynthesis is greater than respiration, the surplus biomass is used by the plant to 'fuel' its growth and development. The climate control in the greenhouse is aimed at optimizing this balance (photosynthesis/respiration), to achieve the desired growth and productivity.

Of the whole fresh weight of the greenhouse plants, approximately 90% is usually water and the remaining 10% is organic matter (Levanon *et al.*, 1986). In order for the plant to grow, the difference between photosynthesis (carbon absorption in the form of CO_2 to be converted into biomass) and respiration (energy and CO_2 release) must be positive; in other words, the 'carbon balance' (or, otherwise, the net photosynthesis) must be positive.

The productive process is complex, with short- and long-term responses. The short-term responses (minutes, hours) are the water and assimilate status, processes that supply energy, construction materials and water for tissue growth, whereas in the long term the productive process may be characterized by the accumulation of dry matter and the development and distribution of such dry matter, not to forget the quality of the product (Challa *et al.*, 1995).

6.3 Photosynthesis

6.3.1 Introduction

Photosynthesis is the process by which plants, using the Sun's energy, synthesize organic compounds from inorganic substances.

Every existing living being needs energy for its growth and conservation. In the vegetable kingdom the energy used comes from the Sun, whereas animals, being unable to directly use the Sun's energy, use energy stored in plants or in other animals on which they feed. Therefore, the primary source of all the metabolized energy used on our planet is the Sun and photosynthesis is fundamental to the preservation of living beings.

Fossil fuels (coal, oil) are decomposition products of animals and land and sea plants and the energy that they store was captured by living organisms millions of years ago, coming initially from the Sun's radiation.

In photosynthesis, the most important step, chemically, is the conversion of carbon dioxide (CO_2) and water into carbohydrates and oxygen. The reaction, schematically, is:

$$6CO_2 + 6H_2O \rightarrow C_6H_{12}O_6 + 6O_2 + \text{Energy} \quad (6.1)$$

Photosynthesis may be described as the process in which solar energy is converted into chemical energy by plant tissues in the presence of chlorophyll. This chemical energy is stored in the form of different compounds (carbohydrates, mainly, ATP and NADPH). By means of this process carbon is fixed into carbohydrate molecules, and oxygen (O_2) along with highly energetic compounds (ATP and NADPH) are released to be later used by the plant in the synthesis of amino acids, organic acids and other more complex organic compounds. All these compounds are transported to the growing parts, to become part of the plant's structures, contributing to the generation of biomass.

The majority of plants cultivated in greenhouses are of the C3 type (metabolism in C3), so called because of the type of chemical reactions in their photosynthetic process. Other plants, called C4, are less responsive to atmospheric CO_2 content; they are usually plants from tropical areas. From the cultivated species, C4 plants are species such as sugarcane, maize, millet and sorghum. A third type of photosynthesis is CAM metabolism, undertaken by succulent plants, which are characterized by their ability to pre-fix CO_2 in the dark, during the night. Therefore, the stomata do not have to open during the day, avoiding the loss of precious water supplies.

Photosynthesis depends on a series of external and internal factors. The internal factors are the characteristics of the leaf (structure, chlorophyll content), the accumulation of products assimilated in the chloroplasts of the leaves, the availability of water, mineral nutrients and enzymes, among others (Hall and Rao, 1977). The most relevant external factors are the radiation incident on the leaves (quantity and quality), temperature, the ambient humidity and the concentration of CO_2 and oxygen in the surrounding air (Hall and Rao, 1977).

Liebig's 'law of minimums' and Blackman's 'principle of the limiting factors' explain the interactions between the several factors that simultaneously influence photosynthesis and the speed of the photosynthetic process (Mastalerz, 1977). The Blackman principle enunciates: 'when a process is conditioned by different factors, the velocity of the process is limited by the velocity of the slowest factor' (i.e. 'the limiting factor imposes a limit which prevents the effect of other factors beyond this limit') (Mastalerz, 1977). When the limiting factor ceases to be, the remaining factors may express their corresponding effect beyond the previous limit.

Nevertheless, there is certain short-lived compensation between climatic factors at certain developmental stages, as is the case with light and temperature (Berninger, 1989), so that poor light conditions may be compensated for by select thermal conditions and vice versa.

6.3.2 The stomata

The surface of vascular plants has some openings, called stomata, through which the gas exchanges between the plant and the environment take place. The stomata are located in the under surface of the leaves. By opening the stomata the plants take CO_2 from the air, but lose water vapour to the exterior, which induces cooling of the leaf as the water evaporates as well as a water flux towards the leaves which allows for the transport of water and nutrients from the roots to the foliage.

Every stoma is surrounded by two guard (occlusive) cells and may have other associated auxiliary cells. The guard cells control the size of the opening (or pore called the ostiole) of the stoma (Fig. 6.2). Depending on the turgor of the guard cells, which is determined by their water content, the size of the opening is regulated.

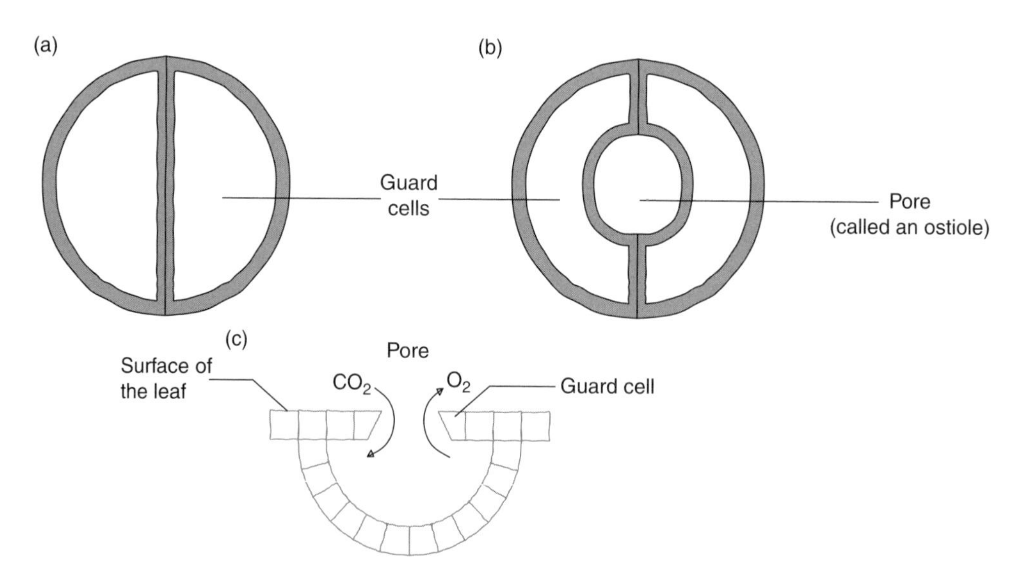

Fig. 6.2. A stoma, in closed (a) and open (b) positions. (c) Cross-section through a stoma.

The turgor changes come as a response to different external stimuli, such as light, CO_2 content, presence of potassium ions (K^+) and water supply. Under normal conditions of water supply, most of the higher plants open their stomata during the day, as a response to light, and close them at night. If the water conditions are less favourable, which affects their turgor, the stomata may close partially or totally. The CO_2 content in the intercellular spaces also affects the stoma opening. The temperature also affects the movement of the stomata, influencing the speed of response, which is slower at lower temperatures.

Regulating the opening, the stomata maintain a balance between photosynthesis and water transpiration, to achieve the higher levels of photosynthetic assimilation while avoiding desiccation.

In plants with CAM metabolism, the stomata open at night and close during the day, to preserve precious water supplies in extremely dry climates.

6.3.3 Internal factors affecting photosynthesis

To access the chloroplasts of the leaf tissues (where photosynthesis takes place) the CO_2 must diffuse from the external air to the stomatal cavity (Fig. 6.2). Access to the stomata by CO_2 is hindered by the stability of the air layers, which surround the leaf (boundary layer) and the stomatal cavity. The CO_2 must overcome these two barriers, which are quantified by their resistance: (i) the boundary layer resistance; and (ii) the stomatal resistance (Gijzen, 1995a). Both barriers affect water vapour, CO_2 and O_2 fluxes, influencing photosynthesis as well as transpiration and respiration.

All the factors which induce stomatal closure decrease photosynthesis. Lack of air movement also reduces photosynthesis as the thickness of the boundary layer increases, and the resistance to the diffusion of CO_2 molecules increases (Nobel, 1974a, b) (Fig. 6.3). The boundary layer is the layer of motionless air that surrounds the leaves where gaseous exchange takes place by molecular diffusion. In the greenhouse, the absence of wind compared with open field cultivation generates thick boundary layers. The resistance to gaseous diffusion of the boundary layer of the leaves may be notably higher than that of stomata when air movements are very weak, such is the case in closed greenhouses (Urban, 1997a). In practice, the grower must maintain a certain air movement in the greenhouse for efficient photosynthesis and proper production.

The accumulation of photosynthetically assimilated products in the leaves may have a depressant effect on their own photosynthesis. These assimilates must be transported to other organs of the plant. This transport process is regulated by several factors. For instance, high temperature, as well as the presence of nitrates, favours this translocation of assimilates. The lack of nitrogen involves accumulation of starch in the leaves, decreasing photosynthesis (Acock *et al.*, 1990).

The availability of proteins is fundamental in photosynthesis. A reduction of nitrogen involves a decrease in the rate of photosynthesis (Urban, 1997a).

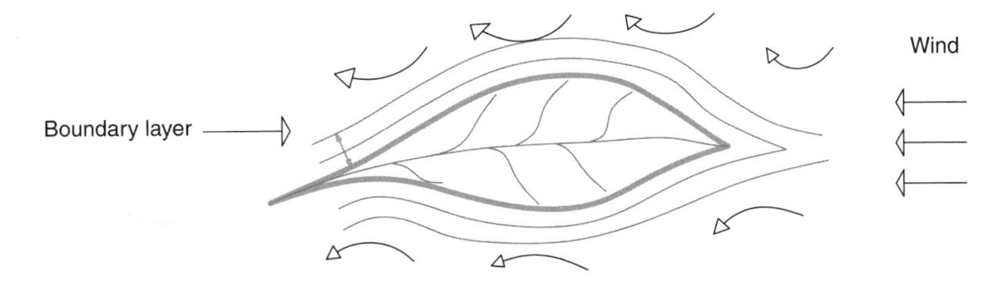

Fig. 6.3. The leaf and its boundary layer (layer of motionless air which surrounds the leaf).

Photosynthesis per unit leaf area reaches a maximum approximately when the leaves reach their maximum size and decreases as the leaves age, as they lose functionality from a photosynthetic point of view.

6.3.4 External factors influencing photosynthesis

Radiation

The quantity of photosynthesis carried out by a plant is influenced by three properties of the light: (i) the 'quality' of the light; (ii) its intensity; and (iii) its duration.

Only a fraction of the global solar radiation is used in photosynthesis. This fraction is known as photosynthetically active radiation (PAR). Within the PAR range (400–700 nm) not all the photons of different wavelengths have the same photosynthetic efficiency. Within the range of 500–600 nm (green colour) the radiation is not well absorbed by the chlorophyll (reflecting part of it), giving the plants their typical green colour (Monteith and Unsworth, 1990).

The absorption spectrum of the photosynthetic pigments shows that green light is less efficient, the main absorption peaks (intervals of higher efficiency) being in the red and the blue areas, due to the light absorption by the carotenoids that accompany the chlorophyll in the chloroplast membranes (Plate 11) (Whatley and Whatley, 1984). The red light is more efficient than the blue light for photosynthesis (McCree, 1972). Within a canopy, the lower layers receive radiation with a higher proportion of green light than the higher levels, which filter the light.

There is a curvilinear response of photosynthesis of individual leaves to the absorbed PAR, if other factors such as CO_2 and temperature are not limiting (Fig. 6.4) (Urban, 1997a). At low radiation, the photosynthesis may be lower than the respiration losses. Gains and losses become equal at the radiation compensation point (Fig. 6.4). A plant cannot subsist for a long time below the compensation point. With high radiation the increase in photosynthesis reaches a maximum and is no longer proportional to the increase in radiation; this point is the light saturation point (Fig. 6.4). When the radiation is very low artificial light is more

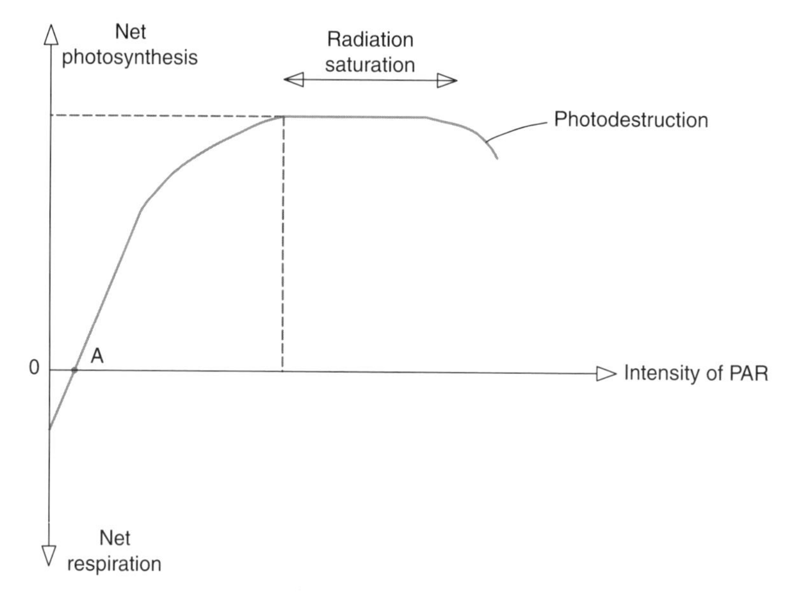

Fig. 6.4. Response of net photosynthesis to photosynthetically active radiation (PAR). A, Radiation compensation point (adapted from Hall and Rao, 1977).

efficient in increasing photosynthesis than when the radiation is high (Fig. 6.4).

In general, if the photoperiod is reduced or extended and the radiation intensity is varied in such a way that the accumulated radiation is the same, the growth rates will be similar (Langhams and Tibbitts, 1997).

It must be expected that the longer the duration of the light period, the more photosynthesis will take place. But it may be that, due to incapacity to store all the starch (produced by photosynthesis) the leaf stops its assimilating activity.

In practice, at the level of an isolated leaf it is possible to reach this saturation point, but in a fully developed leaf canopy (where the lower leaves receive little radiation as they are shaded) it is nearly impossible to reach saturation to radiation by a commercial greenhouse crop (Plate 12) (Hanan, 1998).

Within closed canopies it is very unlikely that saturation radiation levels can be reached with global radiation intensities below 1000 W m^{-2} (Bakker, 1995). A well-developed crop (high leaf area index, LAI) obviously will photosynthesize more than another one with lower LAI (Fig. 6.5).

This radiation saturation level also depends on other factors, for example temperature and of the CO_2 content of the air (Fig. 6.6) (Urban, 1997a). The excess of radiation may damage the photosynthetic complex, for example destroying chlorophyll, making the foliage chlorotic and eventually reducing the photosynthetic productivity of the plants.

Radiation controls photosynthesis not only by its intensity (Fig. 6.5), but also through the available wavelengths and its duration. The proper radiation level to saturate the photosynthetic system of many plants (C3 type) is around 400 μmol m^{-2} s^{-1}, when supplied for 16 h day^{-1}, whereas other plants (C4 type) require levels of 500 μmol m^{-2} s^{-1} or higher to maximize their growth (Langhams and Tibbitts, 1997). By contrast, some ornamental plants develop well with levels ranging from 10 to 50 μmol m^{-2} s^{-1} for a period of 8 h (Langhams and Tibbitts, 1997). We distinguish between shadow plants and plants with high light requirements, depending on their response to radiation. In shadow plants, the saturation point is reached with low radiation levels, which is not the case for radiation-demanding plants, in which the compensation point is higher. The edible horticultural species usually have high compensation points (Urban, 1997a).

In the majority of the plants low levels of radiation intensity induce smaller leaves (with higher length:width ratio), longer

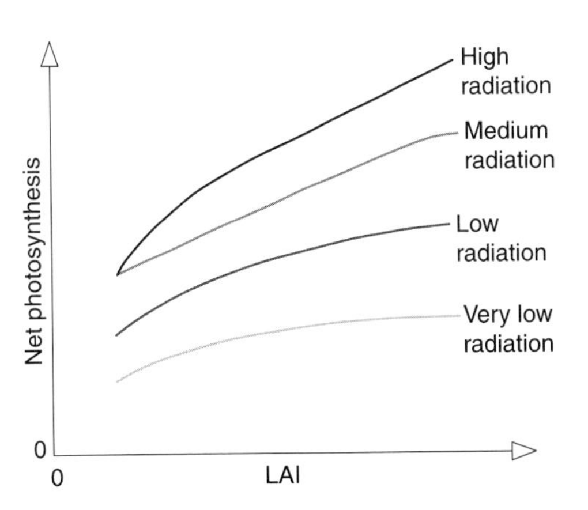

Fig. 6.5. Net photosynthesis as a function of the leaf area index (LAI) and the radiation intensity (adapted from Urban, 1997a).

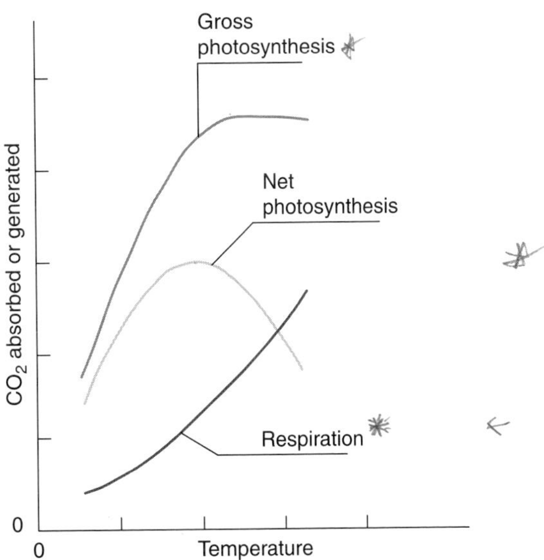

Fig. 6.6. Effect of temperature on respiration and gross and net photosynthesis of a C3 plant (adapted from Urban, 1997a).

internodes, lower chlorophyll concentration and lower dry weight. High radiation levels induce the stimulation of branching, the proliferation of growing points, possible photodestruction of chlorophyll (known as bleaching), and, in extreme cases, generation of symptoms of stress attributable to radiation excesses in some ranges of the spectrum, as the increase in the production of anthocyanins (Langhams and Tibbitts, 1997). High radiation levels may also induce, due to their higher energy supply, heating of the leaves increasing water use, and causing desiccation in extreme cases.

In nature, under low light conditions, in an adaptation process of the plants that involves a long-term mechanism of natural selection, the leaves tend to place themselves horizontally, to intercept the maximum radiation. When the light intensity is high, the leaves tend to adopt a more vertical position.

The efficiency in the use of radiation by photosynthesis (CO_2 fixed in relation to absorbed PAR) varies little among C3 plant species (Ehleringer and Pearcy, 1983), but varies more if we refer to incident radiation rather than to absorbed radiation.

At low latitudes the predictable maximum values of global radiation are slightly higher than 1000 W m^{-2}, in open field, elevated locations, with a semi-arid climate and low air turbidity (Hanan, 1998), whereas at sea level the maximum global radiation is of the order of 900 W m^{-2} (Salisbury, 1985). On the Spanish Mediterranean coast, the maximum values of global radiation intensity are close to 1000 W m^{-2} at the summer solstice. The average daily transmissivity, in low-cost type commercial greenhouses with a shallow roof slope, oscillates during this season around 61.5% as an average value (Morales *et al.*, 1998), which is very influenced by the dirtiness of the plastic cover, since washing the plastic results in an increase of transmissivity of about 14% (Montero *et al.*, 1985; Morales *et al.*, 1998). At the winter solstice, in low-roof-slope commercial greenhouses, the average transmissivity ranges from 57% for north–south orientation, to 63% for east–west orientation (Morales *et al.*, 2000), with less influence from dirt than during the summer months, because in the winter the rain usually washes the plastic cover. In low-cost parral-type greenhouses, increasing the slope of the south side of each span, with east–west orientation, allows increases in transmissivity of up to 73% at the winter solstice (Castilla *et al.*, 2001).

Temperature

Photosynthetic activity has a clear response to temperature; it is at a minimum at about 5°C, reaching an optimum at temperatures from 25 to 35°C in the majority of horticultural species and it decreases at higher temperatures (Urban, 1997a).

The optimum temperature increases with the radiation and CO_2 levels (Acock *et al.*, 1990). In practice, it is of no interest to maintain high temperatures with low radiation (not much heating on days with little light). Under high temperatures, crops grow better with high radiation; therefore, shadows must be avoided in usual horticultural crops (which are radiation demanding).

CO_2

At relatively low CO_2 levels in the air, if radiation and temperature are high enough not to become limiting factors, the photosynthetic rate is almost proportional to the air CO_2 content (Urban, 1997a) (Fig. 6.7). The critical CO_2 threshold below which the carbon balance is negative (respiration is higher than photosynthesis) is, normally, lower than 200 ppm (Gijzen, 1995a).

Higher CO_2 contents induce a higher value of the CO_2:O_2 ratio, increasing the activity of the enzymes which favour photosynthesis (ribulose-1,5-bisphosphate carboxylase/oxygenase, commonly known by the shorter name RuBisCO) and limiting photorespiration, improving the carbon balance (Urban, 1997a).

The CO_2 levels in the air have increased during the last century, from values of 280 ppm (Nederhoff, 1995) to levels of 360–370 ppm. It is forecasted that this increase will continue during the next years, due to human activity.

The atmospheric CO_2 values that maximize leaf photosynthesis are around

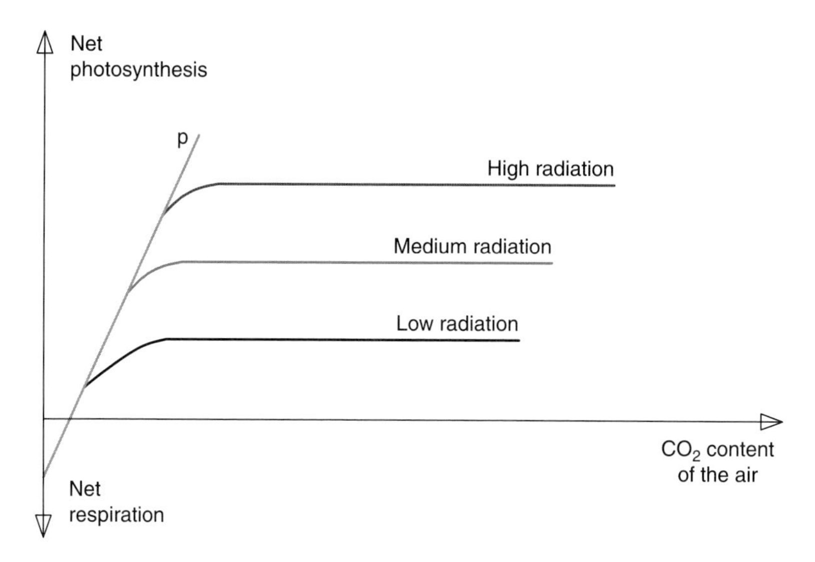

Fig. 6.7. Net photosynthesis as a function of the content of CO_2 in the air, for different radiation intensities. The graph portrays the 'limiting factors principle' applied to radiation (adapted from Urban, 1997a). P, Line of potential net photosynthesis.

1000 ppm (Kimball, 1986; Hicklenton, 1988). But this response at the leaf level does not guarantee a corresponding increase at a whole plant level in biomass production (Urban, 1997a). Besides, when plants are under high CO_2 levels for long periods, they become adapted to these levels and limit their response (Woodward, 1987).

On a practical level, values above 750 ppm have been recommended for tomato and cucumber crops, in Northern European countries (Urban, 1997a). The difficulty arises when ventilation is necessary due to excess temperature, which involves losing CO_2 to the atmosphere and higher costs. Therefore under Mediterranean conditions, where higher temperatures require frequent ventilation, the CO_2 fertilization strategy is different and it is recommended to maintain levels of 360 ppm of CO_2 in the internal air, while the vents are open, and 600–700 ppm when the vents are closed (Lorenzo *et al.*, 1997c, 2005).

Ambient humidity

The ambient humidity does not directly interfere with photosynthesis. Its role is indirect through its influence on stomatal opening. Under appropriate conditions of water supply (non-limiting irrigation) and in the absence of salinity problems, photosynthesis is not affected by a low environmental humidity (Urban, 1997a). However, if plants are under a very high evaporative demand caused by low humidity or if there are difficulties in the water supply from the roots, photosynthesis may be limited, because of stomatal closure due to the low water status of the leaves (Gijzen, 1995a, b). In fact, the positive effect of high humidities on photosynthesis is only observed when water absorption by the roots is limiting (Grange and Hand, 1987).

Therefore, under non-limiting irrigation conditions and in the absence of salinity problems, humidification of the atmosphere is not justified to improve photosynthesis.

Inhibition of photosynthesis

When carbohydrate demand is lower than the supply, for instance after the harvesting of fruit (sinks), there is an increase in the carbohydrate content (starch) in the leaves which may induce a photosynthesis reduction as a 'feedback' effect (Stitt, 1991).

The presence of highly polluting gases (SO_2, CO, NO_x) may also reduce photosynthesis if high levels are reached.

6.4 Photomorphogenesis

6.4.1 Introduction

Plants use solar radiation as a supply of energy and as a source of information (Hart, 1988). Photomorphogenesis is the effect of radiation on plant development. The mere presence of light, above a certain minimum, generates several responses in flowering, germination or phototropism.

The majority of photomorphogenic reactions are induced by wavelengths within the blue region (400–500 nm) or in the red or far red region (600–700 nm and 700–800 nm, respectively) and controlled by the pigment 'phytochrome' (Challa *et al.*, 1995). The most relevant wavelengths are around 660 nm (in the red region) and 725 nm (in the far red).

There are three main pigment groups associated with the relevant photo-responses of plants (Whatley and Whatley, 1984): (i) chlorophylls, involved in photosynthesis; (ii) phytochrome, involved in some morphogenic changes, in the perception of light duration and in the daily rhythms which affect some movements of the plants; and (iii) β-carotene or flavins, related to phototropism (Plate 11).

6.4.2 Vegetable pigments

Phytochrome

Phytochrome is located in the non-green (and etiolated) parts of the plants. Phytochrome seems to be involved with many different types of responses of plants. It is a very big and complex molecule that may adopt different forms depending on the type of radiation received.

Under the influence of red light (650 nm), phytochrome (P) adopts the form (P_{FR}), whereas if it is illuminated with far red light (725 nm), it adopts the form (P_R) (Whatley and Whatley, 1984) (Fig. 6.8).

The quantity of phytochrome present in the plant in the form of P_{FR} is expressed by the relation P_{FR}/P_{TOTAL}. To produce a certain morphogenic or biochemical effect a certain value of P_{FR}/P_{TOTAL} must be achieved. The P_{FR} coefficient with respect to the total amount of phytochrome (P_{TOTAL}) ranges between 0.1 when far red (FR) radiation prevails and 0.75–0.89 when red (R) radiation prevails (Langhams and Tibbitts, 1997).

The ratio of red/far red radiation (R/FR) is altered as the light is filtered by the leaves in the upper levels of a canopy. Therefore, different levels of the canopy receive light with different values of the R/FR ratio, altering its phytochrome. This will result in a different biochemical or morphogenic reaction.

The quality of the light (distribution of its spectrum) is, therefore, relevant in its action on phytochrome. Normally low light intensity is sufficient to obtain a response induced by phytochrome.

The phytochrome system may detect the duration of daily illumination (an environmental parameter that is constant for each location) which is relevant in such latitudes where there are large variations in daily illumination throughout the year.

The radiation intensity required for some photomorphogenic responses is only at the full moon level (0.01 μmol m^{-2} s^{-1}), but the majority of the responses are controlled by higher levels: 0.1–1.5 μmol m^{-2} s^{-1} (Langhams and Tibbitts, 1997).

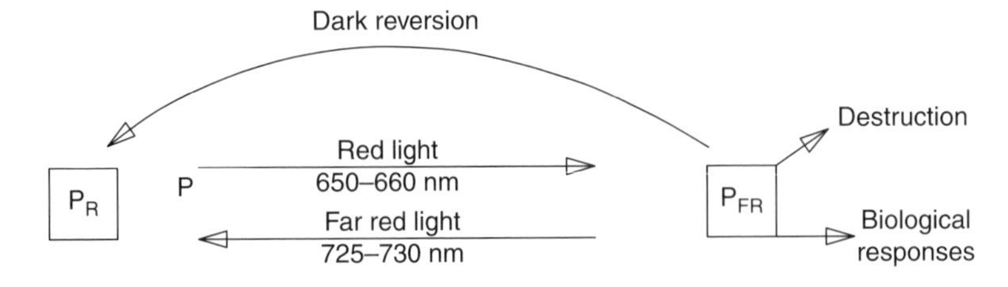

Fig. 6.8. Mode of action of phytochrome (according to Whatley and Whatley, 1984).

Other pigments

Not all plant movements are primarily controlled by phytochrome. In phototropism, the curvature of the stem of a plant towards the light due to lateral illumination, β-carotene or flavins are mainly responsible, although phytochrome also has some influence (Whatley and Whatley, 1984). The phototropic curvature is induced by blue and not red light.

The duration of the illumination to induce a response is low, at around 5 min (Whatley and Whatley, 1984). The relevance of the phototropic response is the fact that developing leaves search for the best illuminated position.

There are other photoreceptors (cryptochrome) but their effects are less known (Mohr, 1984).

6.4.3 Periodic rhythms in plants

In some plants phytochrome also regulates movements of the leaves from a horizontal position, in the morning, to a vertical ('sleep') position at night, following a certain daily rhythm. This regulation is achieved through the alteration in turgor (derived from their water content, influenced by the re-distribution of K^+ ions) of certain special cells in the petiole of the leaves (Whatley and Whatley, 1984).

The daily opening of some flowers is also regulated by light. The opening of the stomata starts at dawn, as the guard cells become turgid, due to the absorption of potassium which induces osmotic absorption of water (Fig. 6.2).

If the night darkness is interrupted, with artificial light, the stomata start to open but close when illumination ceases. The light, obviously, regulates the photosynthesis phase (during the day) and the corresponding translocation of assimilates from the leaves to the reserve organs (e.g. fruits and roots) during the night.

Some photo-nastic movements are initiated by radiation, such as opening and closing of flowers, movements of the leaves and stem turn.

If there is an excess of light in the blue wavelength range the length of the internodes is reduced. On the contrary, the internodes elongate in excess of far red. Therefore, equilibrium is necessary between blue and far red in the radiation spectrum for the normal development of some plants.

Other daily cycles in plants are the absorption of ions by the roots (influenced by transpiration), cell division, respiration and gutation or water expelled by glands in the edges of the leaves (Whatley and Whatley, 1984).

6.4.4 Photoperiodism

Photoperiodism is the control mechanism of plant development in response to a change in the period of illumination (photoperiod) to which plants are exposed each day (i.e. the duration of the day and the night in 24 h cycles). In this way, there are short-day plants that generally flower when the duration of the day is shorter than its critical photoperiod, normally less than 12 h, whereas the long-day plants flower when the duration of the day is longer than its critical photoperiod, usually more than 12.5 h (Langhams and Tibbitts, 1997). These photoperiod thresholds are not exact, being influenced by other factors such as the age of the plant or the climate conditions (temperature and radiation intensity). Those plants whose flowering is not dependent on the duration of the photoperiod are called day-neutral plants (Vince-Prue, 1986). The majority of vegetables grown in greenhouse are day neutral. These 24 h rhythms are known as circadian rhythms and have similarities to those existing in animals, among them man (Vince-Prue, 1986). Many of the daily activities of plants are controlled by this endogenous circadian rhythm, with which light interacts in different ways (Hart, 1988).

The existing luminosity (with photomorphogenic response) before dawn and after dusk, plus moments of appearance and disappearance of the Sun in relation to the horizon, means that the duration of the photoperiodic day, in our latitudes, corresponds to the duration of the astronomic day increased by a period of 40–60 min (Berninger, 1989).

The seeds of some species require radiation to germinate, which they get if their seeds are on or near to the soil surface (less than 5 mm deep) and receive red radiation (Langhams and Tibbits, 1997).

The seasonal control of plant development by means of photoperiodism allows them to be in synchrony with the climate conditions and other organisms (Hart, 1988).

6.5 Respiration

Respiration is the essential process of energy release, which is necessary for the processes of life. It involves absorption of O_2 and the release of CO_2, with consumption of carbohydrates in a reaction that we may consider as the reverse of photosynthesis.

Respiration consumes carbohydrates produced by photosynthesis. It has two basic components: (i) maintenance respiration, which is proportional to the dry weight of the plant or of its active organs (excluding reserves); and (ii) growth respiration, which is proportional to the products of photosynthesis. There is a third form of respiration, photorespiration, relevant in C3 plants, which only exists in the presence of light, which fulfils a defence function against the toxic effect of some ions (Berninger, 1989).

Growth respiration is less sensitive to temperature than maintenance respiration, which doubles for every 7–10°C increase in temperature. Therefore it is desirable to limit high temperatures, especially at night, to improve the overall carbon balance. Growth respiration consumes, approximately 20–30% of the photosynthesized carbohydrates (Berninger, 1989).

6.6 Distribution of Assimilates and Sink–Source Relations

6.6.1 Introduction

The translocation or distribution of assimilates is the transport of these from the production sites (sources) to the places in which they are used (sinks).

Normally the harvestable product of a crop is only a part of the total produced biomass. A good agronomic management must ensure that the distribution of assimilates is mainly destined for the harvestable organs of the plant.

The mineral elements are mainly transported through the xylem. The organic elements are transported through the phloem.

In many horticultural species the most important differences in harvest between cultivars are a result of the differences in the distribution of assimilates (Challa *et al.*, 1995).

This distribution of assimilates is regulated, mainly, by the 'sink strength' of individual organs, which is the capacity of a sink to accumulate assimilates and is related to its growing potential (Marcelis and De Koning, 1995). Climate factors influence the distribution of assimilates in the short term, affecting the sink strength of different organs and, in the long term, altering their number (Marcelis, 1989; Marcelis and De Koning, 1995).

6.6.2 Distribution of assimilates between organs

Different organs compete for assimilates but this is regulated by hormones (Russel, 1977). A short supply of water and nutrients increases the distribution of assimilates to the root, to favour its growth, and in this way, reduce these deficiencies (Brouwer, 1981).

Generally assimilates produced in a certain point are transported to the closest sink (Wardlaw, 1968). To maximize photosynthesis, it is necessary for the sinks to have enough capacity to consume the available assimilates, otherwise they would induce a reduction in photosynthesis, which would re-adjust to the actual assimilate demand (Giménez, 1992); or, the assimilate distribution to other organs would be prioritized, such as in crops of undetermined growth, when the scarcity of fruits (due to pruning or harvest) induces a higher vegetative growth (stem and leaves).

During the initial stage of vegetative growth, roots, stems and leaves compete for the assimilates produced by the leaves. The young leaves initially need to import assimilates until they are self-sufficient, normally before reaching their final size (Giménez, 1992).

From the time of flowering, the fruits are the main sinks, attracting the available assimilates and limiting the translocation to the vegetative organs. Then, if the crop is of determined growth, the growth of stems and leaves slows down until it stops. But if the crop is of undetermined growth, there is a coexistence of growth of fruits and vegetative organs, whose balance must be controlled by cultural practices (removal of stems, leaves or fruit; but, also, appropriate nutritional and environmental control).

During the senescence of the leaves, and when the demand for assimilates by the sinks is not satisfied by production in the active sources, the remobilization of carbohydrates, nitrogen compounds and other mobile compounds from the senescent leaves to other active sinks of the plant offers another option (Giménez, 1992). In a similar way, remobilization occurs if the plant has reserve organs (e.g. tubers).

6.6.3 Management of the assimilate distribution

In fruit vegetables (horticultural crops that are grown for their fruit), it is essential to achieve rapid leaf development, to ensure optimal development of the future sources of assimilates (leaves) that will meet the future high demand for assimilates by the fruits. So it is wise to maintain high temperatures at the beginning of the cropping cycle, as well as pruning the first fruits to avoid competition with leaf development. Later, with a developed crop, the quantity of fruit is regulated by means of pruning (and the harvest itself if they are staggered), adapting it according to plant density. The vegetative growth adapts to the fruit load by means of pruning and elimination of stems.

In plants of undetermined growth, the pattern of distribution of assimilates between fruits and vegetative organs is not constant through the cycle (Marcelis and De Koning, 1995). Low temperatures, in general, limit translocation. At the end of a day in which the rate of photosynthesis has been high, the assimilates accumulate in the leaves and they can, if they are not transported, limit the photosynthesis of the following day. For this reason, it is recommended to maintain a greenhouse night temperature that is high enough to transport these assimilates from the leaves to other organs. By contrast, at the end of a day in which the rate of photosynthesis has been low (for instance, a day with low radiation) it is not necessary to transport so many assimilates, so the night temperature does not have to be as high (Calvert and Slack, 1974). Other environmental factors, such as light and CO_2, have only an indirect influence in the distribution of assimilates by affecting the rate of photosynthesis and, as a consequence, the availability of assimilates (Marcelis and De Koning, 1995).

Temperature can alter the distribution of assimilates by creating new sinks. In general, high temperature enlarges the internodes and decreases branching (Challa *et al.*, 1995).

Humidity affects the size of the leaves. For instance, high humidities induce larger leaves in cucumber, which does not occur in tomato (Bakker, 1991).

High CO_2 levels increase the aerial part:root ratio, inducing thicker leaves and favouring lateral branching, decreasing the apical dominance (Enoch, 1990).

The priority in the demand for translocation of assimilates towards a particular sink organ depends on: (i) the nature of the sink; (ii) its stage of development; (iii) its age; and (iv) its position within the plant. Normally, sink organs located in lower positions promote the translocation of assimilates towards them, with respect to ones in higher locations (Urban, 1997a), at least when the distance is not excessive.

Competition between organs which act as sinks, obviously, influences the growth and development of the plant and, therefore, its yield. The management of pruning, of the nitrogen supply, of the carbonic fertilization and temperature allow translocation of assimilates to be manipulated to the benefit of the organs desired by the grower (Urban, 1997a).

In young plants, the aim of controlling the climate must be to get rapid formation of leaf area to increase the potential for radiation interception. In plants of determined growth, the aim of climate control must be to get the maximum amount of assimilates going to the harvestable organs, whereas in plants of undetermined growth climate control must aim for a balance in the distribution of assimilates between harvestable organs and the rest of the plant (Marcelis and De Koning, 1995).

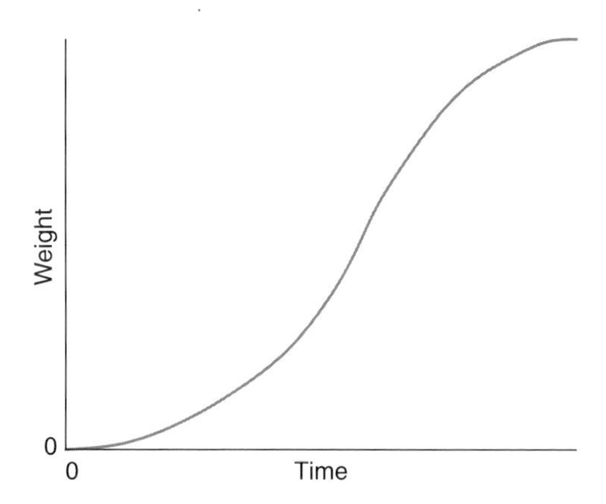

Fig. 6.9. Growth of a plant's organ (sigmoid curve).

6.7 Growth

6.7.1 Introduction

There are several ways of measuring growth: at an elementary level by variation in the dimensions of a leaf or of an internode; or at a more complex level by measuring a stem or whole plant, as well as measurements taken over different time scales.

Measurements of weight usually consider dry matter (which does not have the level of variability of fresh matter) requiring the destruction of the samples. So, representative samplings must be taken, with the aim of having accurate and comparable measurements.

Crop growth, in general, follows the pattern of a sigmoid curve (Fig. 6.9). Initially, when plants are young and the limiting factor is the interception of radiation (low leaf area), growth is exponential. When leaf area increases, the interception of radiation is less dependent on the leaf area, and growth is approximately linear. Finally, as senescence approaches, growth slows down.

In complex organs, such as a stem composed of many internodes, each internode has its own kinetic, but the sum of the elemental growth keeps the sigmoid pattern (Berninger, 1989).

In the short term (at hour scale), plant growth mainly depends on photosynthesis and respiration (carbon balance) and on the water status of the plant. In the long term, the productive process is determined by the accumulation of dry matter, by the development stage of the plant, by the distribution of this dry matter to the harvestable organs and by the quality of the produce (Challa *et al.*, 1995).

6.7.2 Influence of the microclimate on growth

All the climate factors interact with the crop. The greatest effect of radiation is on photosynthesis. The influence of temperature depends on the age of the plant; in young plants, its influence on leaf expansion, necessary to maximize the interception of radiation, is essential. For this reason, in the first stages of development it is necessary to optimize the temperature, to achieve rapid leaf development. Later, with well-developed crops the main role of temperature is on respiration.

The positive effect of CO_2 on growth has been widely documented in C3 type plants (Kimball, 1986) with an average increase in yield of 30% at 1000 ppm of CO_2; levels of CO_2 supplementation greater than 1000 ppm is not recommended because of cost considerations but also to avoid the incidence of polluting gases (Hand, 1990).

The ambient humidity has little effect on crop growth and development if certain values of VPD are not exceeded. In conditions in Northern Europe this level is 1 kPa (Grange and Hand, 1987), whereas in Mediterranean conditions the VPD may reach 3 kPa as there is a certain adaptation to unfavourable conditions (Lorenzo *et al.*, 1997b). High humidity values favour leaf expansion, but may induce a deficit of calcium in the leaf. In addition, it may promote the development of diseases; if water condenses, fungal spores may germinate in the water droplets formed. The large influence of humidity on transpiration allows for the manipulation of transpiration to avoid deficits in nutrients, when transpiration is low (Challa *et al.*, 1995).

6.7.3 Growth analysis

The dry weight (or biomass) of a plant or of a particular organ of a plant (and its evolution over time) is the parameter commonly used to quantify growth.

The crop growth rate (CGR, expressed in g m^{-2} s^{-1}) or net CGR or the accumulation of dry matter quantifies biomass production per ground area in a given unit of time. Sometimes, it is called the dry matter or biomass accumulation rate. In unheated Mediterranean greenhouses, average values of dry matter accumulation of 6.5 g m^{-2} s^{-1} in tomato, which may be as high as 9.5 g m^{-2} s^{-1} (values close to those obtained in climatized greenhouses) when climate conditions are favourable, have been documented (Castilla and Fereres, 1990). In cucumber production without heating, in an autumn–winter cycle, the biomass accumulation rates are of the order of 5.0 g m^{-2} s^{-1}, whereas values obtained in greenhouses with climate control in the spring cycle are around 9.9 g m^{-2} s^{-1} (Castilla *et al.*, 1991). In a winter melon crop without heating average values of 9.7 g m^{-2} s^{-1} have been obtained, in a late cycle (Castilla *et al.*, 1996).

Other indicators used to characterize growth are the growth rate of the main stem and the leaf appearance rate. In unheated Mediterranean greenhouses, in autumn–winter cycles leaf appearance rates of 0.22 and 0.4 leaves day^{-1}, for tomato and cucumber, respectively, have been measured (Castilla and Fereres, 1990; Castilla *et al.*, 1991), lower than those obtained under optimal conditions. In a similar way, the growth rate of the main stem reaches average values of 4.5 cm day^{-1} in a trained cucumber crop, for an autumn–winter cycle, without heating, a value lower than 7.4 cm day^{-1} reported for optimal conditions (Castilla *et al.*, 1991).

Another index widely used to characterize growth is the LAI (leaf area index) that quantifies the surface of leaves per unit ground area (Photo 6.1).

Other growth indexes are detailed in Appendix 1.

Photo 6.1. A good leaf development allows for a better interception of solar radiation. The plant density must be adapted for this objective.

6.8 Development

6.8.1 Introduction

Development is a qualitative notion of the stage of the plant. It is the ordered change towards a higher or more complex stage of the plant (Challa *et al.*, 1995). The time intervals between different developmental stages constitute the development phases. The development of a crop follows a basic pattern according to its genetic make-up, which may be modified, although not changed, by the environment (Challa *et al.*, 1995).

Flower differentiation is a development phase and takes place when a series of conditions are met. Some of these conditions are internal, for instance, that the plant reaches a certain maturity (corresponding to an age or a certain number of leaves) that allows the meristems to differentiate from vegetative to flowering. Other conditions are external, for instance, the existence of certain conditions of photoperiod or temperature. One of the most used development indicators, at the whole plant level, is the number of leaves.

When the photoperiod and nutrition conditions are favourable, the rate of development depends primarily on temperature, its response being linear from the lower thermal threshold (known as vegetative zero, characteristic of each species, below which there is no growth) to the top limit (characteristic of each species) from which the development rate decreases. The vegetative zero of the majority of the horticultural species ranges from 0 to 6°C.

The thermal integral received from a certain moment determines, in many cases, the beginning of a certain developmental stage. In a similar way, the photo-thermal integral (thermal integral corrected according to the received radiation) determines, when reaching a certain value, the beginning of a developmental stage, for instance flowering (Berninger, 1989).

When considering how a crop responds to temperature, in terms of growth, development and production, the daily average temperature is considered, such responses being independent of the thermal regime (day/night temperature), within certain limits. This capacity of integrating the thermal fluctuations in time (responding to the average thermal values) is not limited to periods of 24 h (being possibly higher), but is observed only in developed crops with closed canopies, which completely cover the ground (Challa *et al.*, 1995). This can optimize short- and long-term responses: for instance, using the daytime temperature to optimize photosynthesis and managing the night temperature to obtain the desired average thermal values (Challa *et al.*, 1995).

6.8.2 Development stages in greenhouse crops

The most important development stages in greenhouse crops are: (i) the germination and sprouting (of bulbs and corms); (ii) flowering; and (iii) the formation of reserve organs.

The essential factors for good germination are humidity and temperature, although some seeds need a pre-treatment that interrupts their latency. After germination the seedlings need light to expand their first leaves.

Many species don't need flower induction, as they flower when they reach adult stage: such is the case with the majority of vegetables. Other species need exogenous signals to flower under natural environmental conditions, such as photoperiodic signals, alone or associated with temperature or radiation. The direct control of temperature on flower induction occurs, for instance, in Chinese cabbage whose flowers are induced by low temperatures, commercially depreciating the product (Hernández, 1996).

In the first stages of development, many plants do not flower, even if they receive the right stimuli. However, with greenhouse vegetables such as tomato and pepper, transplanting takes place at an advanced developmental stage with flowers clearly developed, to maximize the utilization of expensive production inputs.

The formation of reserve organs is influenced by growth regulators. This is of no interest to the usual greenhouse vegetables, although it is important in ornamental horticulture.

The shape and size of the product (fruit, leaf) are very important from a qualitative point of view. Although the effects of radiation on tomato fruit size are due to the availability of assimilates (Cockshull, 1992), other aspects of radiation, such as its quality, influence morphogenesis which affects the size and shape of leaves, flowers and fruits (Challa *et al.*, 1995).

6.9 Bioproductivity

6.9.1 Bioproductivity and harvest index (HI)

In any crop there are four factors which determine its net productivity (P_n): (i) the amount of incident PAR; (ii) the efficiency of interception of this radiation by the green organs of the plant (ε_i); (iii) the efficiency in photosynthetic conversion of the PAR into biomass (ε_b); and (iv) the biomass losses due to respiration (R). Such factors are related (Coombs *et al.*, 1985):

$$P_n = \text{PAR} \times (\varepsilon_i \times \varepsilon_b) - R \qquad (6.2)$$

P_n = Net productivity or net gain of biomass (g m^{-2}) or net photosynthesis, result of deducting the respiration losses from the gross photosynthesis
ε_i = Efficiency of light interception (PAR) by the crop, expressed per unit
ε_b = Efficiency of light (PAR) conversion into biomass (g MJ^{-1})
PAR = accumulated PAR (MJ m^{-2})
R = Biomass losses due to respiration

The product ($\varepsilon_i \times \varepsilon_b$) is called the efficiency in the use of radiation (light) by the crop (Baille, 1995).

The economic performance of a crop is the amount of this productivity (P_n) which is destined to the harvestable organs (the fruit in the case of tomato or pepper, or the leaves in the case of lettuce). The proportion of the total biomass represented by the harvestable organs of the crop is known as the 'harvest index' (HI) (Coombs *et al.*, 1985).

Improvement in the productivity of the crops can be achieved by minimizing the respiration losses and maximizing the PAR, the efficiency in light interception (ε_i) and the efficiency in biomass conversion (ε_b).

In practice, the improvement in the productivity of many crops has been achieved by a better light interception, derived from suitable fertilization and proper cultural practices (Coombs *et al.*, 1985). The improvement in the efficiency of the conversion of light into biomass in greenhouses is feasible, mainly by means of CO_2 enrichment, which also decreases photorespiration (Coombs *et al.*, 1985) and by avoiding suboptimal climate conditions. In plastic greenhouses, the use of light-diffusing covering materials improves the efficiency of radiation conversion (ε_b), as the proportion of diffuse radiation increases (Baille, 1995), and its use is positive if it does not significantly reduce the PAR transmissivity.

The improvement in the HI has been possible by breeding, as well as improving the cultural practices (fertilization, protection against pests).

In vegetables, the HI (referred to as dry weight) varies depending on the cultural practices and the cultivar used. In Mediterranean greenhouses, HIs expressed as a decimal per unit or a percentage, have been estimated at 0.3 for tomato grown in a winter cycle (Castilla and Fereres, 1990), very influenced by the suboptimal climate conditions and the low dry matter content in the fruit. In pepper, values of HI from 0.36 to 0.46 are usual in unheated greenhouses (Martínez-Raya and Castilla, 1993), whereas in cucumber grown in an autumn cycle it reaches a value of 0.59 (Castilla *et al.*, 1991). The climate conditions notably affect the HI values. For instance, in melon cultivated in an unheated greenhouse in an early cycle (colder), the HI is higher than in a later cycle due to the lower vegetative growth with suboptimal thermal conditions (Castilla *et al.*, 1996). Also, a notable influence on the HI values has been the length of the cycle as well as the pruning and elimination of stems, which must be done correctly and on time (suppressing newly formed shoots, without allowing their growth). In sophisticated greenhouses, therefore, HI values are expected to be higher than those obtained in unheated greenhouses during the cold season (see Table 6.2 in Appendix 1 section A.5.4).

6.9.2 Interception of radiation by the crop

Leaf area index (LAI) and crop growth rate

The interception of solar radiation by the leaves is essential to convert the solar energy into vegetable matter (biomass). At the beginning of a crop growing cycle, when the plants are small, a large part of the radiation is not intercepted, impinging on the ground and not being profited by the crop.

The basic parameter that relates the radiation intercepted by a crop and the incident solar radiation is the LAI (Watson, 1958). This quantifies the surface of leaves of a crop per unit soil ground area: LAI = leaves' surface (m^2)/soil surface (m^2).

At the beginning of the crop growing cycle, leaf development is slow and the LAI increases slowly. At this stage, a large part of the radiation is not intercepted by the crop. Later, the LAI increases exponentially if there are no limiting factors for growth (lack of water, inappropriate temperatures) until it reaches its maximum values (Fig. 6.10). Later changes in LAI depend of the type of growth of the crop. In plants of determined growth, after reaching the maximum values of LAI, this parameter decreases when senescence starts. In crops of undetermined growth, high values of LAI will be maintained during a great part of the cycle, as the senescent leaf area is compensated for by the production of new leaves.

Temperature has a great influence on the growth and development of the leaves; so, temperature needs to be managed accordingly to achieve the maximum LAI in the minimum possible time.

A LAI index between 3 and 4 is considered necessary, so that radiation interception reaches 95%, with the usual planting

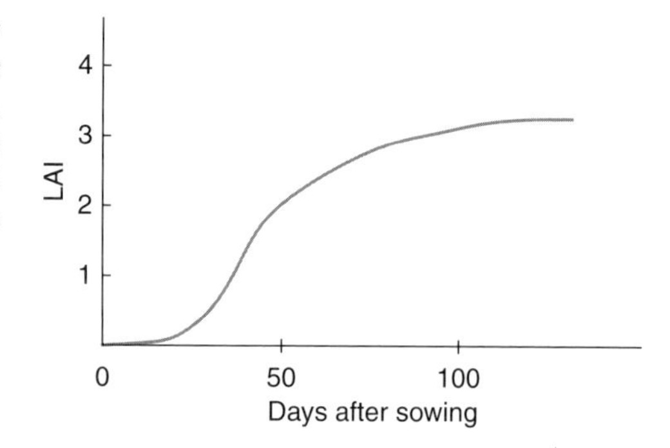

Fig. 6.10. Evolution of the leaf area index (LAI) of a cucumber crop, along its cropping cycle (autumn–winter), in an unheated plastic greenhouse (Mediterranean area).

densities for herbaceous crops (Giménez, 1992). Crops with more vertical leaves (garlic, onion, gladiolus, cereals) may reach higher values with LAI of 5–10 for maximum interception. In greenhouses, as the crops are grown in rows (in paired lines in many occasions) the situation is more complex. Some authors (Baille, 1995) estimate, as an approximation, that interception is 100% (ε_i = 1, in Eqn 6.2) if the LAI is equal or higher than 3. Until the plant does reach a LAI of 4, photosynthesis rates increase in parallel to the LAI (Challa and Schapendok, 1984).

The 'critical LAI' (Broughman, 1956) is the value above which there are no more increases in the crop's growth rate, which usually corresponds to an interception of 95% (Giménez, 1992), that is, ε_i = 0.95.

Later increases in LAI, when the interception is virtually total, there may be a decrease in the growth rate in some cases, when the photosynthesis rate of the shadowed leaves does not compensate for the respiration losses. In other cases, an acclimation of the shadowed lower leaves occurs, adapting their respiration rates to the photosynthesis rates, without changing their growth rate (Giménez, 1992).

The influence of cultural practices (manual defoliation, plant density, training, etc.) on the LAI is relevant. In Mediterranean greenhouses, values of LAI of 3.5 for cucumber in autumn–winter have been established, whereas for undetermined growth green-bean values of 6.2 have been recorded

(Castilla and Lopez-Galvez, 1994). In pepper, a high plant density allows for the achievement of LAI values of 5.0, versus an index of 3.2 with low plant density (Lorenzo and Castilla, 1995). In the same way, under good temperature conditions, the LAI is higher. For a greenhouse melon crop, the value established for an early cycle (colder) was 2.5, versus 4.6 for a late cycle (Castilla *et al.*, 1996).

The radiation absorbed by a crop, and which is used for photosynthesis, is much more difficult to calculate than the intercepted radiation (see Appendix 1 sections A.5.1 and A.5.2).

Light penetration in the crop

A fraction of the radiation incident on top of a canopy penetrates into the vegetation, depending on the amount of leaves (that is, the LAI) and their disposition in the canopy.

The extinction coefficient (see Appendix 1 section A.5.1) represents the efficiency of the canopy to intercept radiation in its different layers (Giménez, 1992).

A crop with horizontal leaves requires less leaf area to intercept the same proportion of radiation than another crop with more erect leaves (which let more light pass to the lowers layers) (Fig. 6.11).

Diffuse radiation, being non-directional, has greater penetration in the vegetation as it is more efficient than direct radiation which causes more shadows in the crop's lower layers.

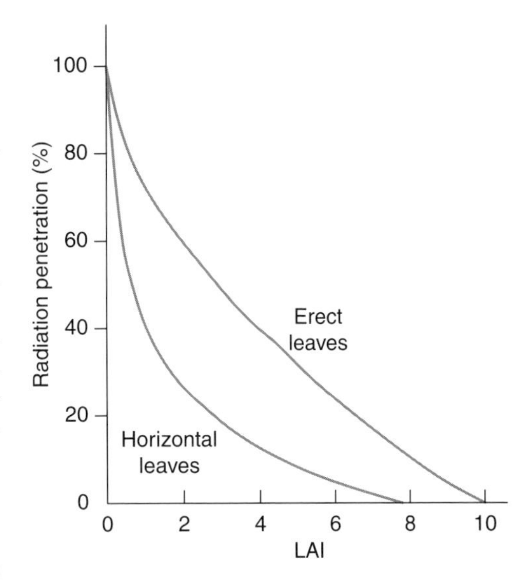

Fig. 6.11. Penetration of solar radiation in a canopy as a function of the leaf area index (LAI) for two types of crops: one with erect leaves and one with horizontal leaves (according to Giménez, 1992).

6.9.3 Efficiency in the use of solar radiation

There is a linear relationship between the accumulation of biomass (dry matter) during a certain period and the accumulated PAR during that period, which was intercepted by the crop (section 6.9.1), provided there are no other limiting factors present.

The efficiency in the conversion of the intercepted radiation into dry matter, under non-limiting crop growth conditions, depends on the type of the plant: C4 plants, CAM plants, C3 legume plants, C3 non-legume plants (Varlet-Grancher *et al.*, 1982).

C4 plants are less efficient converters of light into biomass than C3 plants, but their C4 mechanism allows them to limit photorespiration losses (in relation to C3 plants), so they are more productive (Whatley and Whatley, 1984).

In the absence of other limiting factors, for an adult horticultural crop which covers the ground well, and under normal conditions, the average efficiency of conversion (in grams of biomass per megajoule of global radiation) is of the order of 1 g MJ^{-1} (Baille, 1995). The maximum value of the efficiency of conversion of radiation into dry matter is of the order of 2.5 g MJ^{-1} of accumulated global radiation or 5.0 g MJ^{-1} of accumulated PAR (Russell *et al.*, 1989).

The efficiency of the conversion will depend on the radiation conditions. If all radiation is diffuse, the efficiency will be higher than if direct radiation prevails, with values ranging between 0.8 and 1.4 g MJ^{-1} (Challa *et al.*, 1995). For this reason, light-diffusing plastics are so interesting in greenhouse production, provided they do not limit transmissivity. With CO_2 enrichment, these indices may be increased by up to 20% (Baille, 1995) or 30% (Challa *et al.*, 1995) if a concentration of 1000 ppm of CO_2 is maintained. It has been reported that a reduction of 1% of radiation caused a decrease of 1% in the yield of cucumber and tomato (De Visser and Vesseur, 1982; Cockshull *et al.*, 1992).

On a fresh weight basis, assuming a HI of 0.7 in tomato, a dry matter content in the fruit of 5%, and, an efficiency in the use of radiation of 1 g of dry matter MJ^{-1} of global radiation, the tomato productivity would be 14 g of fresh fruit MJ^{-1} for a developed crop without other limiting factors under normal conditions (Baille, 1995).

The potential greenhouse production may be roughly estimated for average meteorological conditions and depending on the greenhouse characteristics (Challa and Bakker, 1999). In practice, it is estimated that the proportion of absorbed PAR, which is used for dry matter production, ranges from 4% to a maximum of 10%, in the best conditions (Baille, 1999).

The use of radiation in a greenhouse is one of the most efficient of all agricultural ecosystems, although radiation is reduced compared with that of an open field. However, in terms of total energy use, modern heated greenhouses are the most intensive of all the agricultural ecosystems (Baille, 1999) (Fig. 6.12; see Appendix 1 section A.5.2).

6.9.4 Strategies to maximize the use of radiation

Crop management must pursue optimization of the photosynthetic process, to maximize the yield and the quality. Early sowings, when the climate conditions are good for plant growth or planting with sufficiently developed transplants, allow for a good and early interception of radiation. Achieving a fast leaf development by the use of cultivars adapted to the local climate conditions allow for an improvement of radiation interception and photosynthesis. Cultural practices, such as optimum management of fertilization and irrigation, also affect the final yield, as photosynthesis is improved.

The management of plant density is one of the ways used to achieve efficient interception of radiation (Giménez, 1992; Papadopoulos and Pararajasinghma, 1997). However, in fruit vegetables, besides adapting the plant density to the climate conditions (radiation, mainly), it must be taken into account that a high density may affect the fruit size (Castilla, 1995). For a certain density, pruning and training of the plants must pursue the optimization of photosynthesis and promote the distribution of assimilate towards the plant organs which are required (Papadopoulos and Ormrod, 1988, 1991).

Diffuse radiation represents an important fraction of solar radiation entering

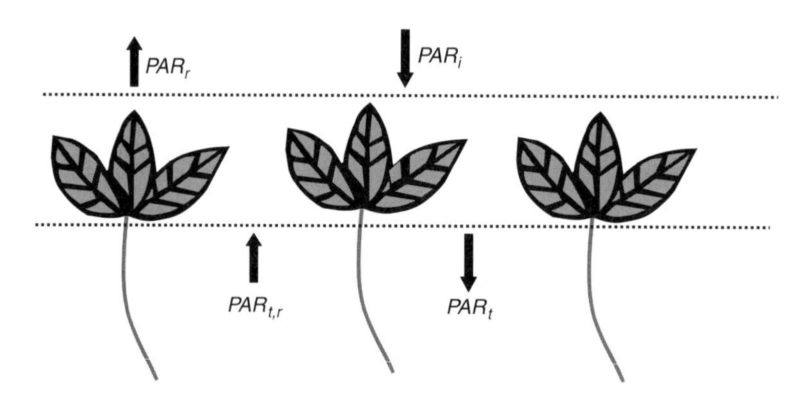

Fig. 6.12. The different components of the PAR for the calculation of the radiation absorbed by a crop. PAR_i, PAR incident on the crop; PAR_r, PAR reflected by the crop or albedo; PAR_t, PAR transmitted at ground level; $PAR_{t,r}$, PAR reflected from the ground (see Appendix 1 section A.5.2, adapted from Baille, 1999).

greenhouses (Baille and Tchamitchian, 1993). Increasing the relative fraction of diffuse radiation in greenhouses contributes to higher radiation uniformity within the greenhouse (Kurata, 1992) and to yield increase, due to higher radiation efficiency (Baille, 1999). The increasing use of highly diffusing cover materials (mainly PE films) in the Mediterranean area contributes to less direct solar radiation inside the greenhouse, and thus more diffuse radiation (Cabrera *et al.*, 2009); this increase in diffuse radiation usually results in higher yield (Magán *et al.*, 2011).

Crops such as fruit vegetables with a high plant canopy utilize diffuse radiation better than direct radiation, as diffuse radiation penetrates the middle and lower layers of a high-canopy crop and results in a better horizontal radiation distribution in the greenhouse (Hemming *et al.*, 2008). For this reason, as stated by Cabrera *et al.* (2009), starting with the pioneering work of Deltour and Nissen (1970), laboratory studies aimed at characterizing the diffusive properties of greenhouse cladding materials have become of paramount interest (Pearson *et al.*, 1995; Wang and Deltour, 1999; Montero *et al.*, 2001; Pollet *et al.*, 2005).

6.10 Production Quality

6.10.1 Introduction

Besides the fresh weight of the harvestable products, their quality determines the yield of greenhouse vegetables. Quality is a combination of attributes, properties or characteristics which give each product a value, depending on its use (Kader, 2000). Quality may be defined as the group of characteristics by which the product and the production mode satisfy the demand of the buyers, traders and distributors, and the expectations of the consumers (Vonk-Noordegraaf and Welles, 1995). The quality criteria, obviously, are not the same throughout the distribution chain, from the grower to the consumer, varying depending on the product and the way it is consumed (Kader, 2000).

We may distinguish between external quality, which includes those visible attributes (shape, colour) and internal quality (flavour, shelf life) which cannot be evaluated at a glance (Kader, 2000). Several aspects of quality are measurable (analytical quality) whereas some others are subjective (emotional quality). This emotional quality is, sometimes, related to the mode of production, such as integrated or organic production (Vonk-Noordegraaf and Welles, 1995).

Sometimes, in tomato, a certain decrease in production may be compensated for by an increase in the organoleptic quality of the product (Adams and Ho, 1995; Schnitzler and Gruda, 2003), maintaining or even increasing the economic return.

For the consumer, the appearance is the most important qualitative criterion when buying (at least, until the product is consumed), so the size, the shape and the uniformity, the colour and the absence of visible defects are the aspects most usually considered as qualitative elements when choosing the product (Urban, 1997a). Although frequently their contribution to the decision-making power in the distribution chain is limited, at least in the short term, in relation to other agents of the chain, such as the purchasing managers.

For more information about quality see Chapters 15 and 16.

6.10.2 Effects of climate factors on quality

A high rate of photosynthesis affects the production of sugars and acids, which are very important compounds in the flavour of fruit vegetables (Vonk-Noordegraaf and Welles, 1995). High radiation favours the sugar content and decreases the acid content (Janse, 1984) while it limits the harmful accumulation of nitrates in leafy vegetables.

The external quality is also affected by the light, at first through photosynthesis, since with higher photosynthesis levels (under proper conditions of competition between organs) larger fruit sizes are expected (Cockshull *et al.*, 1992). In addition, in crops like cucumber, low radiation levels give rise to fruit of light green colour that soon turn yellowish, something that is associated with low quality (Vonk-Noordegraaf and Welles, 1995). Excessive radiation impinging on cucumber fruit induces a green colour that is too intense and a skin that is too thick. In tomato, in order to achieve a proper colour, growers prune leaves to favour the penetration of light to the fruits, in low radiation periods, or provide various forms of shading to protect the fruit from intense radiation and overheating. On the other hand, greenhouse light-diffusing covering materials limit intense direct radiation on the fruits, contributing to improve their quality.

Unfavourable climatic conditions for fruit set affect fruit quality. Low temperatures limit fruit set of several fruit vegetables. In tomato, for instance, low temperatures induce the formation of irregular fruits, with bad colour and that are slow to ripen (Castilla, 1995). Low night temperatures increase the number of malformed fruits in pepper.

With high radiation, the temperature of fruits exposed to the Sun's rays may exceed the temperature of the air by up to 10°C, as fruits have very low transpiration, and exposed fruits may be sunburned. Therefore, in fruits like tomato it might be necessary to shade them during periods of high radiation.

The firmness of the fruit decreases when radiation is low, ambient humidity is limited and temperatures are extreme: lower than 13°C or higher than 25°C (Zuang, 1984). These extreme temperatures, same as excessive day–night thermal differences (greater than 10°C) negatively influence the colour of tomato fruits (Zuang, 1984).

It must be remembered that colour and firmness are frequently antagonistic. Harvesting before full colouration in tomato, with good firmness but with lower sugar content, imposed by the need to send the product to faraway markets involves suboptimal organoleptic characteristics for the consumer (Kader, 1996).

A high ambient humidity limits transpiration, which decreases the calcium content in the tissues, affecting the quality of the fruits. A high CO_2 level improves the quality and quantity of the produce indirectly, as photosynthesis is increased.

In general, climate factors which favour photosynthesis and, as a consequence, the synthesis of sugars, improve the organoleptic quality of the fruits, because the flavour, in fruit vegetables, depends primarily on the sugar and volatiles content, as well as on the acidity (Hobson, 1988). In melon, for instance, flavour improves with an increase in the dry matter content (sugars). In tomato, an increase in radiation favours the content of sugars and acids (Urban, 1997a).

6.10.3 Other factors affecting quality

Fruit size is very influenced by the quality of the fruit set, there being a linear relationship between fruit size and number of seeds, so an improvement in fruit set will induce an increase in fruit size (Castilla, 1995). The vibration of the flowers or the use of bumblebees or bees to improve pollination and fruit set is very beneficial.

The distribution of assimilates, obviously, is of primary importance in the size of the fruit. Therefore, cultural practices must limit the competition for assimilates: for instance, pruning the tomato trusses to allow a certain number of fruit which may achieve proper size or eliminating axillary shoots that compete with the fruits.

The distribution of assimilates also affects the shape and uniformity of cucumber fruit, because an excess of fruit induces a higher number of malformed fruit (Urban, 1997a).

An issue of increasing public interest during recent years is the absence of phytosanitaries and heavy metals in the commercial product, because the consumer is demanding safe, healthy and high quality food (Viaene *et al.*, 2000).

Water and nutrient supply under proper salinity conditions must be optimized to achieve a good quality, as they determine the quality of the harvestable product together with the genetic characteristics, the climate conditions and the crop management practices.

The postharvest storage conditions are more linked to the genetic characteristics of the cultivar and to the mineral nutrition than to the climatic conditions during the crop cycle (Urban, 1997a). Good postharvest management in all the links of the distribution chain is of utmost importance to ensure that the product has the proper quality when it gets to the consumer.

6.11 Summary

- Growth is a consequence of a positive carbon balance. That is, when the net photosynthesis (difference between gross photosynthesis and respiration) is positive.
- Gross photosynthesis is determined by the PAR, which is the main limiting factor of the productive process. The CO_2 concentration in the air is the main factor to optimize the efficiency in the use of the intercepted PAR. The influence of temperature in photosynthesis is limited, except when it reaches extreme values.
- The boundary layer (static air layer surrounding the leaf) is very thick inside a greenhouse due to the absence of wind, and may have a notable effect limiting photosynthesis, if there is not a minimum air movement inside the greenhouse.
- Suboptimal conditions for photosynthesis involve irreparable losses of potential yield.
- Photomorphogenesis is the effect of radiation on plant development. The majority of morphogenic responses are controlled by the pigment phytochrome. Among these responses is photoperiodism.
- Respiration is quantitatively important in the carbon balance. Primarily it depends on temperature, increasing with it, so its control is fundamental to optimize the net photosynthesis (carbon balance).
- The distribution of assimilates must be oriented towards the maximization of the biomass in the harvestable organs, by means of cultural practices and proper agronomic management (pruning, thinning, fertilization, climate control). Of all the climate factors, temperature is the main tool to manipulate the distribution of biomass. Light and CO_2 do not have direct effects in the partition of biomass, although their availability influences photosynthesis.
- Radiation is the main factor determining bioproductivity, together with the radiation intercepted by the canopy, the efficiency in the conversion of intercepted radiation into biomass and respiration losses.
- The crop characteristics (arrangement of the plants and the plant rows, leaf area) and of the greenhouse (transmissivity to radiation, radiation diffusion) affect the amount of intercepted radiation and its efficiency of use.
- In the short term, growth mainly depends on photosynthesis and respiration (carbon balance) and the water status of the plant.
- In the long term, the productive process is determined by the accumulation of dry matter and by plant development, by the distribution of dry matter to the organs which are going to be harvested and by the quality of the product.
- The greatest effect of radiation on growth is through photosynthesis. Temperature affects leaf growth, primordial development in young crops, and respiration, basic for the carbon balance.
- The positive effect of CO_2 on growth is important. The effect of ambient humidity is limited, except when extreme values occur.
- Given conditions of an appropriate photoperiod, water supply and nutrition, the development rate depends mainly on temperature. The thermal integral received from a certain moment determines, in many cases, the beginning of a certain stage of development.
- The capacity of the plants to integrate temperature fluctuations in periods larger than 1 day (24 h) may allow for more efficient management: for instance, using

the day temperature to optimize photosynthesis and the night temperature to achieve the desired average values.

- The quality criteria of a product are not the same throughout the distribution chain between grower and consumer, varying depending on the product and its final use.
- Appearance is the most important qualitative criterion for the consumer, at least until the product is consumed.
- The genetic characteristics, the climatic conditions, crop management, the water and nutrient supply and the absence of residues are the main determinants of quality, as well as the postharvest storage and management conditions.
- In essence, the production process must be dependent on the economic objectives of the grower: producing the quality required by the market at competitive prices.

7

Facilities and Active and Passive Climate Control Equipment: Low Temperature Management – Heating

7.1 Introduction

The greenhouse temperature depends on the energy balance. In order to avoid low temperatures heat losses must be reduced and heat supplies provided, taking into account that when one of the energy balance components is altered, other components, which should not be modified, may be also altered.

When the natural energy supply (for temperature increase) is not enough, we must resort to artificial supply, by means of heating.

7.2 Reduction of Heat Losses

This aspect must be considered when building the greenhouse. The reduction of heat losses is achieved, mainly, by reducing the heat exchange surfaces and the heat losses per unit surface, using insulation materials and windbreaks.

7.2.1 Reduction of the exchange surfaces

The reduction of greenhouse heat losses is achieved, in the first place, by limiting the heat exchange surfaces with the exterior of the greenhouse, building compact greenhouses, so that the proportion of sidewalls is reduced to a minimum (area of sidewalls versus greenhouse covered area) (Fig. 7.1).

Greenhouses with a high roof slope have a larger roof area than those with a low slope, which in turn involves a larger exchange area resulting in greater heat losses through the cover (Fig. 7.1). In a similar way, corrugated covering materials (for instance, rigid polyester panels) increase the exchange surface in relation to covering materials that are flat.

7.2.2 Reduction of heat losses per unit surface

Besides limiting the exchange surfaces, losses per unit of these surfaces must be reduced as well. Radiation losses could be limited using proper covering materials (low transmission to far IR radiation, such as thermal PE, for instance). Besides, radiative losses can be decreased using thermal screens deployed during the night. In cases where there is a heating system, a management regime that lowers the temperature of the cover will decrease radiation losses (because these losses increase with the temperature of the cover).

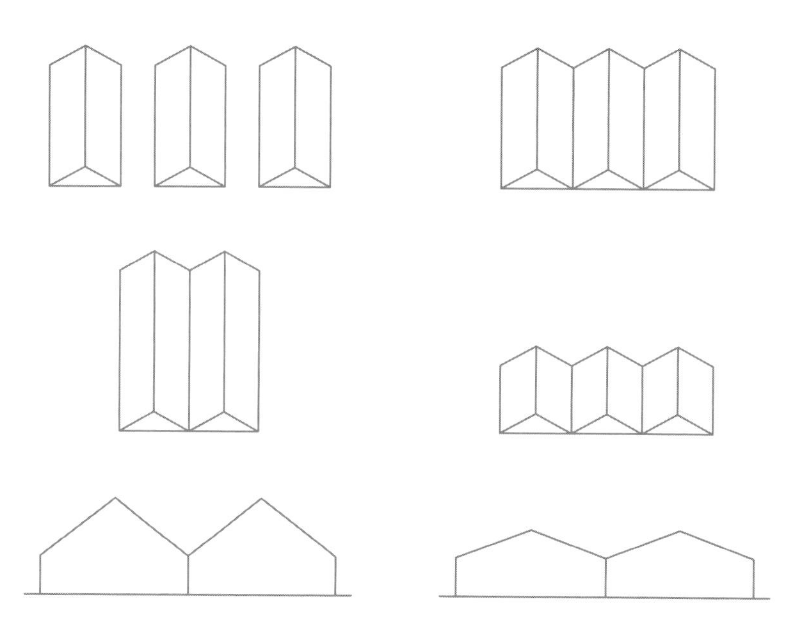

Fig. 7.1. Compact greenhouses have a lower proportion of sidewalls with respect to the greenhouse ground area, limiting thermal losses. Square-shaped greenhouses have less perimeter and less sidewall area than rectangular greenhouses of equal ground cover area. Equally, greenhouses with a high roof slope have a larger exchange surface and greater losses of heat through the cover.

Conduction and convection losses decrease by minimizing the effects of the exterior wind, for example protecting with windbreaks, and locating and orienting the greenhouse properly (Fig. 7.2). A reduction in the 'thermal bridges' (points through which heat, by conduction, escapes to the exterior of the greenhouse through the structural elements which are good heat conductors) also limit the heat losses from the structural elements to the exterior, but above all the most efficient means are the use of a double cover and thermal screens. For more related details, see section 7.3.

Losses due to air renewal are another type of heat losses. They can be limited by means of good insulation of walls and vents, to improve how airtight the greenhouse can be, and by minimizing the effects of the external wind. When a greenhouse is closed, it is not possible for it to be perfectly airtight as there is air infiltration through holes and slits. The amount of infiltration depends on the type of greenhouse and of the external wind (Table 7.1).

Night heat losses due to leakage in a closed low-cost greenhouse in Almeria are 10% of the total losses in the absence of wind, and with wind velocities of 4 m s^{-1} they are more than 30% of the total losses (López, 2003).

7.2.3 Total heat losses

Heat losses depend on the temperature differences between the greenhouse and the exterior. Radiation losses are predominant in unheated greenhouses. Conduction and convection losses are, proportionally, higher in heated greenhouses.

With the aim of integrating the set of losses through the walls, by conduction, convection and radiation, the 'global heat transfer coefficient' (K) is used (Table 5.2), which is evaluated under standard conditions of temperature, wind and cloudiness.

The amount of heat exchange between the greenhouse and the exterior (Q, excluding the heat lost by air renewal, due to night leakage) is detailed in Chapter 5 (section 5.4).

Some authors increase the value of Q in case of strong winds (Aldrich and Bartok, 1994), whereas others consider that the

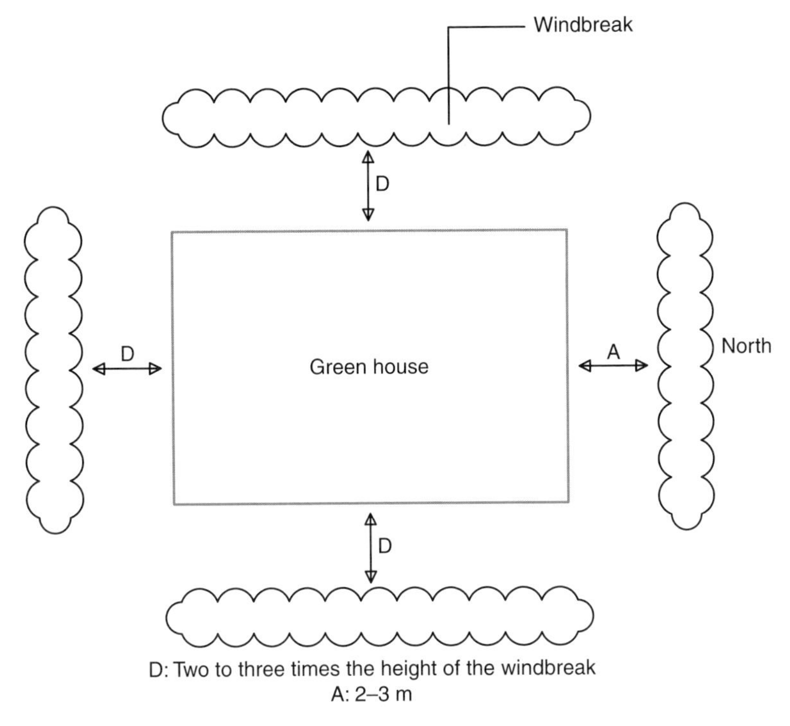

Fig. 7.2. The wind affects heat losses and leakage. Distances from windbreaks to the greenhouse (northern hemisphere, medium latitude) to avoid shadows (adapted from Urban, 1997a).

Table 7.1. Air infiltration in closed greenhouses (in greenhouse volumes per hour). The lowest values correspond to conditions with no wind. (Source: ASAE, 1984; Nelson, 1985; López *et al.*, 2001.)

Greenhouse type	Air volumes h^{-1}
PE: double layer	0.5–1.0
PE: multi-tunnel	1.0–1.5
PE: low-cost parral-type	1.3–6.0
Glass, new	0.75–1.5
Glass, old, good maintenance	1.0–2.0
Glass, old, bad maintenance	2.0–4.0

value of Q reported here is sufficient or even high (Hanan, 1998). For more details, see Chapter 5 (energy balances discussed in sections 5.3 and 5.4).

7.3 Insulation Devices

The main measures to decrease heat losses are shown in Fig. 7.3. They are: (i) the doubling the external sidewalls; (ii) the use of mobile thermal screens (especially in heated greenhouses); and (iii) the use of permanent double covers, which may be either inflatable (when flexible films are used) or cellular in the case of rigid panels.

Other measures are: (i) the use of windbreaks; (ii) specifically located insulation (e.g. the north wall, in the northern hemisphere); and (iii) providing double protection as a temporary measure.

7.3.1 Inflated double cover

The inflated double cover may reduce the energy losses in heated greenhouses by approximately 30%. Maintaining an insulating air chamber of several centimetres thickness, between two plastic films, improves the insulation properties of the cover, reducing the global heat transfer coefficient (Table 5.2). The disadvantage is an associated reduction in light. This reduction will depend on the transmissivity characteristics of the plastic film used. Its implementation requires the use of special fixing systems and the ventilation system must be consequently adapted (Photo 7.1).

To maintain the inflating pressure, low power turbines are used (250 W for 1000 m^2) to maintain a pressure in excess of 40–60 Pa. Care must be taken to avoid internal condensation in the insulating air chamber, which may limit the light transmission even more, so it is advisable to take the air for inflation from the outside.

It must be clarified that the reduced light transmittance of the double-inflated plastic greenhouse might not be necessarily a disadvantage. For example, it might very well be impossible to grow a crop in the winter time with the low amount of available ambient light even if all the light was transmitted by the greenhouse cover. In this case, a potentially additional 5–10% light available inside a glass-covered greenhouse, as compared with the double-inflated plastic house, would be of little practical benefit. However, the reduced light availability and the more diffuse nature of the transmitted light in a double-inflated plastic house, in the summer months, has resulted in a cooler and more gentle greenhouse environment for crops (especially suitable for sensitive crops such as cucumber) with significant benefits to producers who are interested in long-season cropping.

7.3.2 Mobile thermal screens

The function of mobile thermal screens is threefold: (i) they limit the volume to be

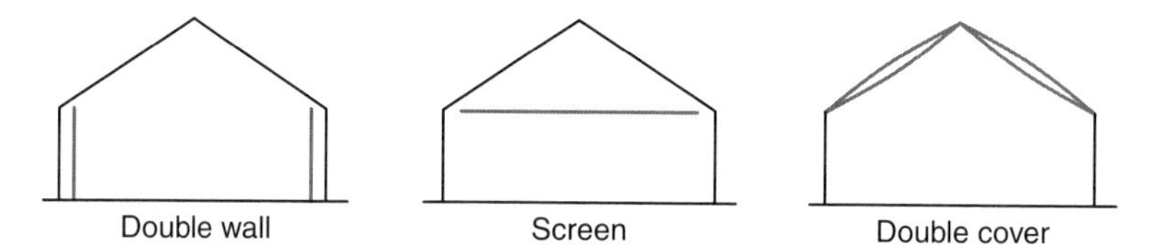

Fig. 7.3. Greenhouse insulation devices: double external walls, thermal screen and double roof cover.

Photo 7.1. Greenhouses made of inflated double plastic film improve the insulation conditions but require special fixing systems.

heated; (ii) decrease the radiation losses from the canopy and its surroundings; and (iii) limit the energy losses due to air leakage and convection, provided they are properly installed and they are not permeable, so that the enclosure is fairly airtight.

Thermal screens are used at night, in winter and in cold areas. The energy savings range from 5% with calm weather and overcast sky, to more than 60% with strong winds and clear sky (Urban, 1997a). The main problem they pose is the increase of the air humidity, if they are not permeable.

The screens can be made of linen or similar material, that may be woven, or plastic films, and aluminium strips are often present to improve efficiency (Photo 7.2). The strips have low emissivity of radiation, are highly reflective and do not transmit radiation. If they are used only at night, they do not need to be transmissive to solar radiation. Linens have very variable optical properties. The plastic films are very airtight. It is of fundamental importance to assemble screens properly in order to achieve maximum efficiency (for details see Chapter 4). They must be installed so that once they are folded shadows are minimized, because they limit light by at least 4% (Hanan, 1998).

The management of screens consists of extending them at dusk and folding them back at dawn. Folding must be progressive to avoid the sudden fall of the cold air mass (accumulated over the screen) over the plants. If they reduce radiation (due to being folded back late or extended early) they will limit photosynthesis, negatively affecting the yield. Their extension and retraction can be automated so that these occur at a prefixed time or when radiation reaches a certain threshold.

7.3.3 External double sidewalls

The insulation of the external sidewalls around the perimeter of the greenhouse is an efficient measure to limit heat losses. The thermal behaviour is similar to that of inflated walls. It is simple and cheap, resulting in little light reduction. The most commonly used material is a plastic film with air pockets.

Photo 7.2. Thermal screen.

7.3.4 Windbreaks

Windbreaks reduce the wind pressure and the risk of damage to the greenhouse. In addition, they limit heat losses by air leakage and by convection.

Windbreaks must be taller than the greenhouse and of a semi-permeable nature. Impermeable windbreaks generate turbulence and the protected distance is shorter (Fig. 7.4).

The distance protected by semi-permeable windbreaks (Fig. 7.4) ranges from 15 to 20 times their height (Van Eimern *et al.*, 1984). The orientation must be perpendicular to the dominant winds and the distance to the greenhouse must be from two to three times its height, in the east, west and south sides (in the northern hemisphere) to avoid shadows (Fig. 7.2). On the north side, it is enough to place them at a distance of 2 or 3 m (northern hemisphere).

7.3.5 Other insulation devices

In cold areas in China, the external north-oriented wall (in the northern hemisphere) of greenhouses is insulated to limit the heat losses (Fig. 7.5).

In China it is also common to use some kind of mats, which are extended at night over the greenhouse cover to decrease the heat losses, removing them during the daytime (Fig. 7.5), requiring a considerable amount of labour.

Among other peculiar insulation methods, it is worth mentioning the double-wall greenhouse prototype, whose air chamber is filled at night with a solid insulation material, which is removed during the day. This is very efficient from an insulation viewpoint but it has not been used to any great extent due to its cost (Short and Shoh, 1981).

Similarly, new technology has recently been developed and tested in Canada where the cavity inside the two layers of the double-inflated PE greenhouse was filled with foam. This had significant heat energy savings in the winter and provided greenhouse shading/cooling on demand in the summer months (Aberkani *et al.*., 2011).

Temporary double protection is a usual technique in Spanish greenhouses. The use of small tunnels or floating covers, inside the greenhouse, during the early stages of crop development of low crops (melon, watermelon) allows for higher temperatures and RH to be achieved. Contact between plants and the plastic must be avoided, to prevent scalds in case of high temperatures of the plastic.

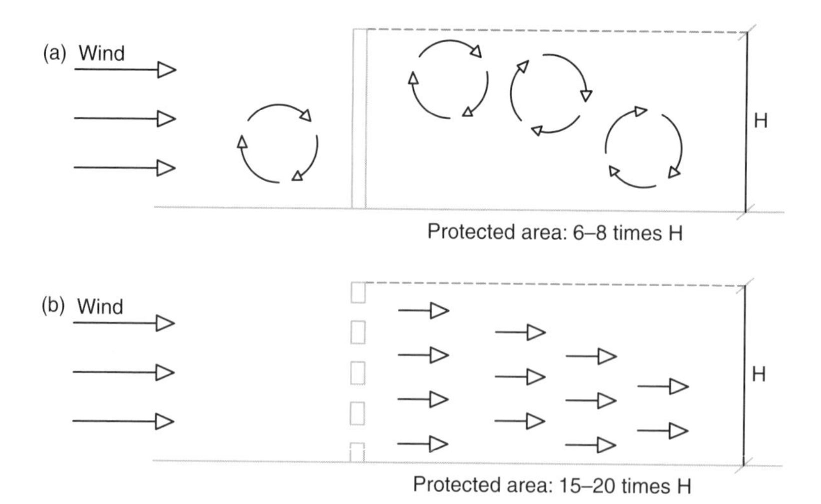

Fig. 7.4. Protection provided by windbreaks: (a) impermeable windbreak; (b) permeable windbreak. H, Height.

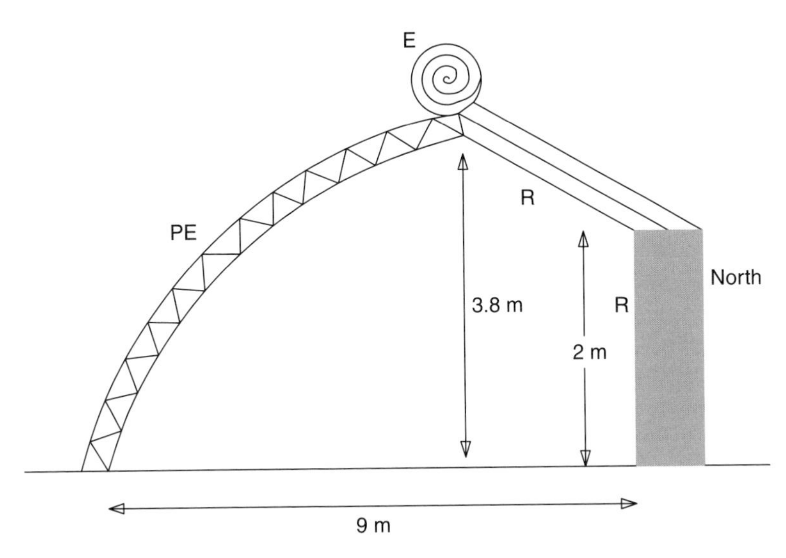

Fig. 7.5. A 'lean-to' solar greenhouse (improved version) built against a wall (on the north side) that is often used in cold areas of China. Rolling mats (E) on the upper part can be unrolled at night to decrease heat losses. R, Reflecting surface; PE, double PE plastic film.

When designing a greenhouse, placing the main transport corridor adjacent to the (colder) north wall of the greenhouse may result in better use of energy.

Double doors, or similar devices, are of great interest both from a thermal insulation point of view as well as to limit the entrance of insects. The limitation of 'thermal bridges' in greenhouse structures, by means of pertinent insulation, is a usual measure in cold areas.

The partitioning of the greenhouse by means of interior partitions, that may be fixed or mobile, allows for zones with independent microclimates to be created. They can separate different crops or different varieties, with different climatic requirements. Obviously, the partitions must be transparent to avoid light reduction. The ability to control the microclimate of each compartment independently in greenhouses with climate control must be planned in advance.

7.4 Heating

Heat inputs to raise the greenhouse temperature can be applied, depending on the grower's objectives, to the aerial part of the crop, or to the roots, or to both.

In the heating of the aerial organs the heat is transmitted by convection or radiation, whereas in heating the soil or substrate the heat is transmitted by conduction. What really matters is the temperature of the plants and not that of the air surrounding them.

The temperature of the plant surface results from heat exchanges such as convection from the air and radiation from all the surrounding surfaces, to which latent heat exchanges by transpiration must be incorporated, mainly during the day, and sometimes those of water vapour condensation in the form of droplets over the leaves.

7.4.1 Convective heating

The three usual air heating systems (Fig. 7.6) are: (i) fan coils; (ii) hot air generator; and (iii) heat pumps (water/air or air/air).

Fan coils transmit heat from a hot body to the air. Water fan coils are the most common. Hot water circulates through the metal pipes, which exchange heat with the air by convection. The contact surface is increased

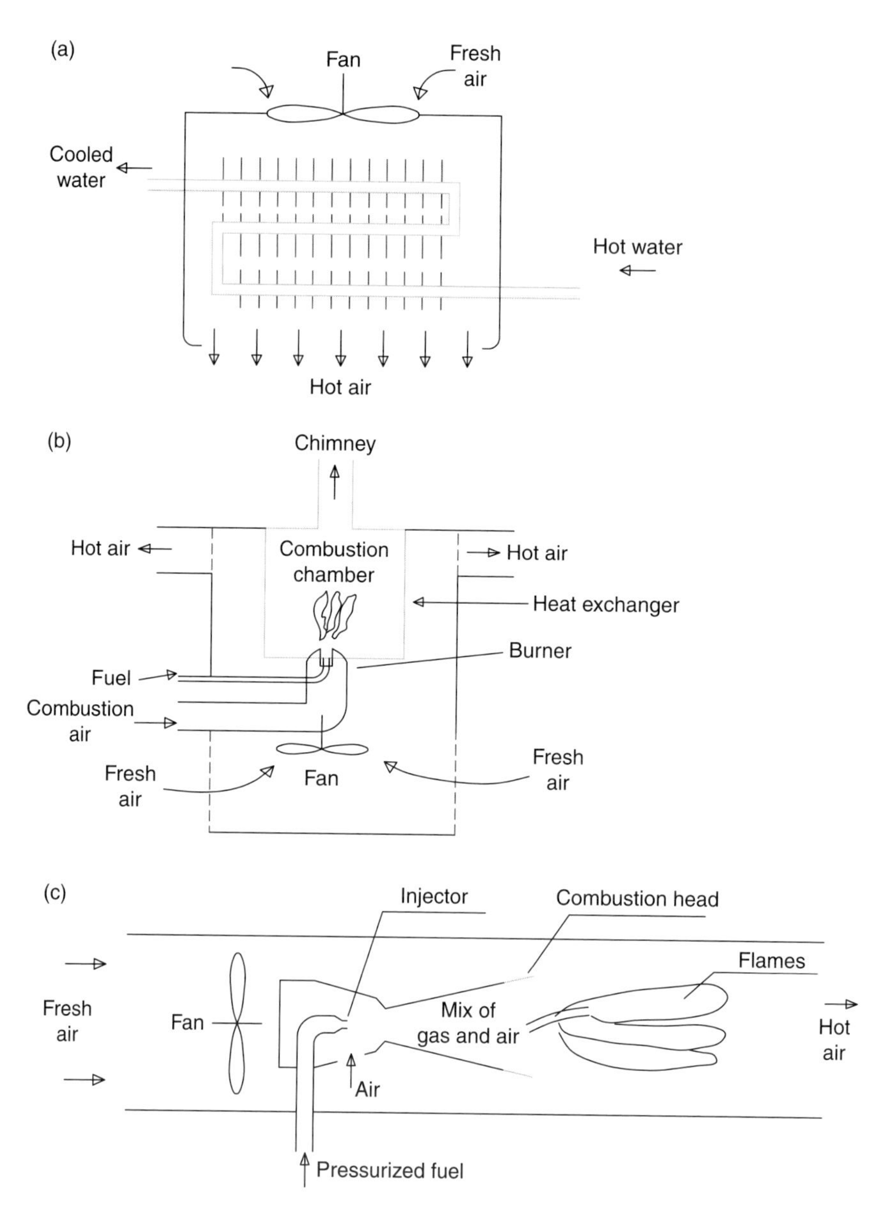

Fig. 7.6. Air heating systems: fan coil (a); hot air generator using indirect combustion (b) and direct combustion (c).

with fins and a fan improves the exchanges. Electric fan coils are expensive due to their power consumption.

Hot air generators burn fuel, normally a gas, which heats the air that is forced to circulate inside the greenhouse. They can use direct or indirect combustion. In those using direct combustion, the products of the combustion are sent to the greenhouse, so the fuel must be clean, with less than 0.03% sulfur content (Urban, 1997a). As they may generate other gases such as ethylene (Hanan, 1998), direct systems are normally used only as a defence against exceptionally low temperatures (of short duration) and as a support.

Indirect combustion hot air generators expel the combustion gases to the exterior. The heat distribution improves notably if the hot air is delivered through perforated flexible plastic pipes (located over the floor to avoid shadows) (Photo 7.3). It is recommended that tubes are designed that allow a flow from a third to a fourth of the volume of the greenhouse min^{-1}, with an air velocity at the start of the tube of 5–6 m s^{-1}. The total perforated area in the pipe must be, in total, from 150 to 200% of the section of the tube (ASAE, 2002). The length of the tube must not surpass 50 m; it being preferable to limit it to 30 m. The plant temperature is usually lower than the air, which involves the risk of water condensing on the plant surfaces and, as radiation heat is not supplied, the soil is not heated during the cold season (Bordes, 1992).

Heat pumps absorb heat from a cold to warm source (water or external air) and blow it inside the greenhouse (see section 7.4.4), but they are not very common.

Convective heating systems are not efficient. Besides, the warm air weighs less than the cold air and tends to rise, away from the plants. There may be temperature differences between the base and the top of the plants of 2–3°C (Urban, 1997a). The advantage is that the convective systems are cheap to install and respond quickly, but they have low thermal inertia. Their use is common in low-cost greenhouses, as a defence mechanism against very low temperatures. In this case, the set point temperature is usually maintained at around 5–7°C, to start running the system under extremely cool conditions (Urban, 1997a).

7.4.2 Radiative-convective heating

Hot water

In classic radiative heating systems using hot water at high temperature, the heating element is a tube or radiating surface which dissipates the heat by radiation and convection with the air (Bordes, 1992). Heat transmission by conduction only takes place inside the tube. The plant surface temperature is higher than that of the surrounding air, which limits water condensation over the plants and its influence in the development of diseases. These systems are more expensive than convective ones, but are more efficient and have higher thermal inertia.

Photo 7.3. Indirect combustion hot air generator and flexible plastic air distribution pipes.

These systems are also known as heating facilities by 'thermosyphon' (a method of passive heat exchange based on natural convection which circulates liquid without the necessity of a mechanical pump) because originally the water circulated naturally (propelled by the hot water being in a high position in a circuit and return of cold water in a low position). Nowadays, all systems are run by means of pumps which force the circulation. This improves their efficiency and allows for the use of smaller diameter pipes than the original natural circulation systems.

In conventional systems, water circulates at high temperature (50–80°C), at about 0.3–1.0 m above the ground (Fig. 7.7). Their thermal efficiency is not good, as they lose 50% of the heat by radiation towards the cover (Urban, 1997a). They must be used in association with thermal screens that limit these losses. The most commonly used pipes are circular in cross-section although there are other types, and they may suspended by chains above the crop so that they can be moved vertically as the crop grows (C in Fig. 7.8 and Fig. 7.9). They are usually built of steel but, also, of aluminium, or even plastic. Nowadays, the use of heating pipes in pairs is widespread; they are used as rails to transport trolleys and mechanized elements (Photo 7.4). To avoid condensation of water over the crop, sometimes additional pipes may be installed in the canopy (and managed at low temperature) (Kamp and Timmermen, 1996).

The use of water vapour as the fluid for distributing heat (still in use in very cold climates) has not received wide acceptance in Europe and is being displaced (in the few existing facilities) by hot-water heating systems.

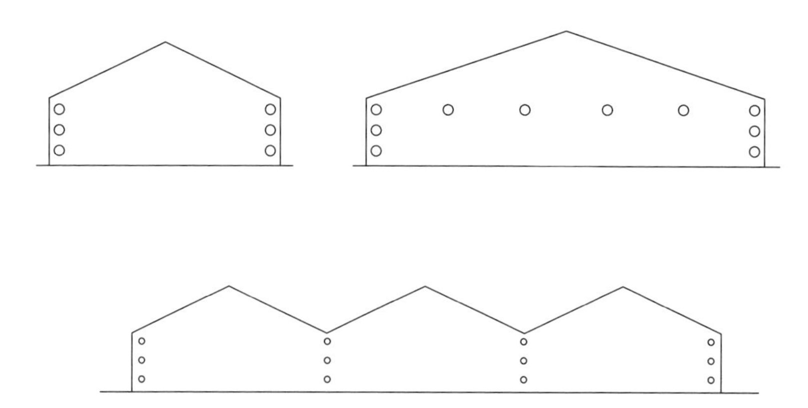

Fig. 7.7. Traditional location of water heating pipes (high temperature) in greenhouses, in narrow single-span greenhouses (top left), wide ones (top right) or multi-span (bottom).

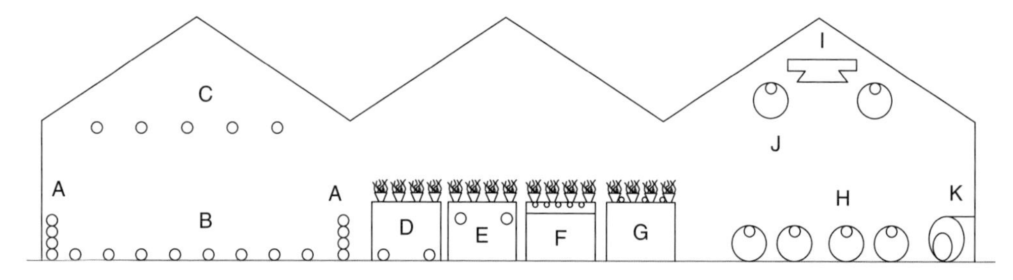

Fig. 7.8. Distribution scheme of the different elements for greenhouse heat distribution. A, Side radiant pipe; B, radiant pipe over the soil; C, aerial radiant pipe; D, radiant pipe over the soil for table cultivation; E, radiant pipe under the growing table; F, heating elements integrated in the growing table; G, heating elements over the growing table; H, PE hoses over the soil; I, fan coil with vertical discharge; J, aerial PE hoses; K, low hot air generator (adapted from Hernández, 2002).

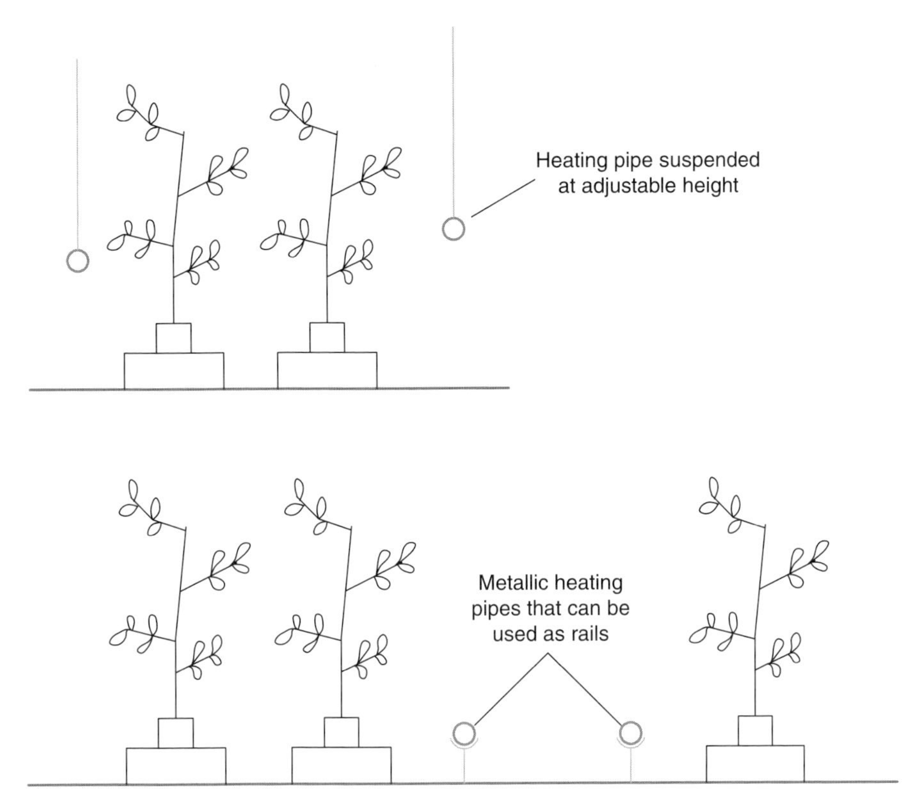

Fig. 7.9. Scheme of a water heating system (high temperature) with metallic pipes that are adjustable at different heights (top) and with metallic pipes, which can be used at the same time as rails for the movement of trolleys and machinery.

Photo 7.4. Metallic heating pipes that are also used as rails for the movement of trolleys and mechanized elements.

Heating with water at low temperature is a particular case (e.g. used is association with geothermal energy). It uses water at temperatures of 40–50°C. It is a localized heating system based on mobile or fixed pipes combined with the use of radiant sheets (radiant mulching) (Fig. 7.10). It is used as a base heating, or associated with a soil or substrate heating system, or as a support heating (Fig. 7.8).

The low-temperature heating system, as implemented with plastic materials in mild winter environments, is a cheap investment, easy to install and manage (Photo 7.5). It saves up to 30% of energy, in relation to the traditional system (Urban, 1997a). Its main disadvantages are that its response is slow, and that its power limited, due to the lower water temperatures. Its installation may require a large amount of pipes, which may be an obstacle. In fact, in order to maintain an air temperature of 20°C, it would be necessary to use four times more pipes, if the water circulates at 40°C, than if the water circulates at 80°C, to supply the same heat (Ellis, 1990). In cold

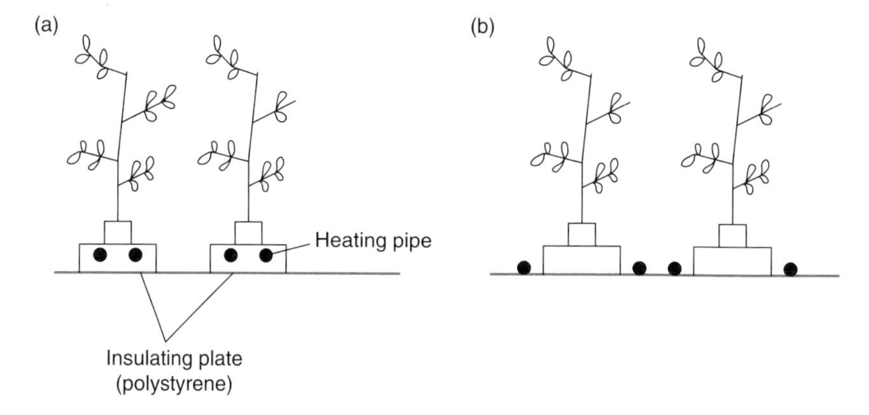

Fig. 7.10. Common hot-water heating systems (low temperature) of (a) soil or substrate and (b) soil–air. A soilless culture system is shown.

Photo 7.5. Heat distribution pipes of a low-temperature heating system.

areas it is recommended that the system is used in association with thermal screens and with a complementary heating system (hot air) which has a fast response.

The heat transfer from the heating pipes will depend on the type of pipe used (material and characteristics, such as diameter, with or without fins, etc.) and the temperature difference between the pipe and the greenhouse air. Table 7.2 summarizes some data in this respect.

Sometimes, low-temperature water heating systems are used to warm up the roots, besides heating the aerial part of the crop, in which case they heat by convection, radiation and conduction. The pipes may be made of PE or corrugated polypropylene. The water temperature is regulated by mixing hot water (from the boiler) with cold water (return) by means of three-way valves.

Low-temperature water heating systems are able to maintain higher values of soil or substrate temperature than hot-air heating systems (due to their location near the soil) if the same set temperatures of the greenhouse air are used (Lorenzo *et al.*, 1997a). This may be attributed to the higher thermal inertia of water heating systems that involves slightly higher energy consumption.

The response time of heating is of the order of 40–60 min in an efficient hot water system, whereas it decreases to 10 min in a hot air system (Day and Bailey, 1999).

Table 7.2. Heat transfer, in watts per linear metre of heating pipe, for several temperature differences between the pipe and the surrounding greenhouse air (adapted from Van de Braak, 1995).

	Steel pipe diameter (mm)			Plastic pipe diameter (mm)
Temperature difference (°C)	51	33.2	26.4	25
10	15	10	8	6
20	34	23	18	14
30	55	38	31	24
40	77	53	44	35
50	101	71	58	46
60	128	90	73	–
70	156	108	90	–
80	185	129	107	–

By IR radiation

In these systems, tubes heated to high temperatures (250–375°C) emit IR radiation which heats the plants (Van de Braak, 1995). The source of energy is usually propane or natural gas, for economic reasons, because electricity is very expensive. This system has a low thermal inertia and is not homogeneous, because the leaves that receive the radiation are much warmer than those in the shadows not exposed to this radiation, thus water condensates on these leaves (Hanan, 1998). Nowadays they are seldom used in greenhouses.

7.4.3 Soil or substrate heating

The heating of the roots is achieved by means of hoses or pipes buried in the soil, packed into concrete slabs or in contact with the crop substrate (Figs 7.8 and 7.10). Heat is transferred by conduction.

They require a conductive soil, which requires it to be wet. Insulation may be implemented at a certain depth to limit the losses. The water temperature must not exceed 45–50°C, to avoid drying of the roots and of the soil (Urban, 1997a), because dry soil is a bad heat conductor (Fig. 7.11). Polypropylene pipes of 20 mm diameter are commonly used, with water velocities of 0.6–0.9 m s^{-1}, providing up to 50 W m^{-2}, with pipes separated by 30–40 cm and avoiding tube lengths longer than 120 m (ASAE, 2002).

Heating pads are used in nurseries. They are composed of pads with a circulating water or electricity heating system. Careful irrigation must be performed, because the pads desiccate quickly, so it is necessary to install sprinklers or thin fogging.

In soilless growing systems it is frequent to heat the roots simply by locating the pipes of a conventional heating system near the substrate. Sometimes, a pipe is placed inside or below the substrate (with insulation below) (Fig. 7.10). Growing tables and crop benches may also be heated by radiant pipes or integrated heating elements (Fig. 7.8). These are used in nurseries and for cultivation of ornamental crops.

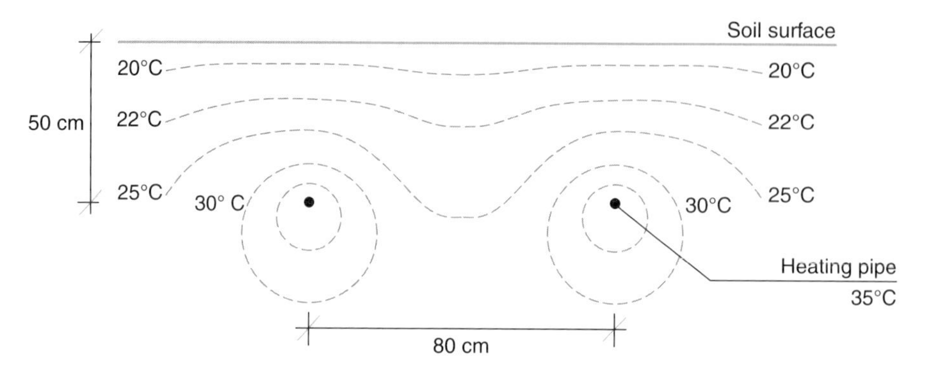

Fig. 7.11. Thermal profiles of a soil heated with pipes.

7.4.4 Heat production

Energy sources

When choosing the type of energy for heating it is necessary to know the main technical characteristics of the fuel, what is its state (solid, liquid or gas) at the usual conditions of temperature and pressure (an essential aspect for their use), its heating power and its content of impurities, mainly sulfur (Bordes, 1992).

Coal is cheap, but it is very polluting. It usually contains sulfur and its gases are corrosive. Coal boilers are more complex and expensive.

Wood and vegetable waste are quite voluminous. Their transport and storage is very expensive. They are also quite polluting.

Diesel is expensive. Fuel oil is cheaper and of lower performance. Natural gas is a fuel obtained from the purification of methane deposits. The disadvantage of natural gas is its high price, but its advantages are: (i) it contains no sulfur; (ii) it produces very little pollution; (iii) it is easy to use; and (iv) CO_2 can be recovered from its combustion gases.

The liquefied petroleum gases (LPGs) butane and propane are used. They have the same advantages as natural gas.

Gas boilers may use natural gas or LPG. Nowadays, the recovery of CO_2 from the combustion gases for carbon enrichment in these boilers is becoming widespread.

The boilers

Boilers are composed (Bordes, 1992) of: (i) a combustion chamber with a furnace (if solid fuels are used) or a burner (if the fuels are liquid, or powder or gases); (ii) a heat exchanger, where the combustion gases at high temperature transfer their heat to the fluid of the heating circuit (which can be water) through the conducting walls that separate the fluid and the air; (iii) a chimney to evacuate the smoke efficiently, passively with a good draught or actively with a fan; and (iv) safety devices and automatic equipment needed for their operation.

It is normal to recover the water vapour from the smoke to decrease the thermal losses (in the form of water vapour latent heat) and improve the performance of the boiler. The combustion must be done in the presence of an excess of air to achieve complete combustion and to optimize the performance of the boiler.

Heat distribution

When these systems were initially conceived circulation of hot water was spontaneous by 'thermosyphon'. The weight/pressure differences, due to water temperature differences, generated a circulation of water. Therefore, this type of heating was usually called 'thermosyphon' (Bordes, 1992). Nowadays these systems are provided with a circulation pump, to improve their efficiency and to distribute the heat without restrictions.

The distribution circuit's pumps must be insulated in the parts where heat transmission is not required.

In greenhouses of up to 9 m wide, the pipes fixed to the sidewalls may be all that are needed (Fig. 7.7), but in wider greenhouses it will be necessary to install heating pipes at one or more locations across the span (ASAE, 2002).

When the natural circulation of the air is not sufficient to achieve good uniformity of air temperature, it will be necessary to install fans to improve such air circulation (see Chapter 9).

Energy economies

In Europe, heating expenses represent from 15 to 35% of the expenses of greenhouse cultivation with a notable increase during the last decades (Chaux and Foury, 1994b). To decrease heating expenses without lowering the temperature, cheaper sources of energy are used. Also consumption is decreased while improving the efficiency and reducing the losses by radiation, convection, conduction and air leakage (Urban, 1997a).

During recent years, the efficiency of boilers has been greatly improved. Energy storage (as hot water) is used extensively in the CO_2 and heating supply systems, when the demands of both are not coincidental, as it is very cost effective (Van de Braak, 1995).

Radiation losses

The greenhouse cover exchanges heat by radiation with the sky (for calculations it is accepted that the temperature of the night sky is around 15°C lower than the outdoors air). The clouds act as a screen, decreasing radiation losses. In the south of France, with clear winter skies, radiation losses may represent 60% of the total losses in an airtight greenhouse (Urban, 1997a).

Losses by air leakage

These are more important the higher the humidity and temperature differences between the internal and the external air and the less airtight the greenhouse is. The wind increases heat losses by air leakage.

Losses by conduction/convection

Losses by conduction/convection are higher, the greater the temperature difference between inside and outside the greenhouse. They also increase with wind velocity. Therefore, the use of windbreaks may be of interest.

How airtight the greenhouse is, is fundamental. A very airtight greenhouse has a limited leakage rate (see Table 7.1). Junctions must be sealed. The 'thermal bridges' must be eliminated.

On the other hand, if the greenhouse is too airtight this may generate an excess of environmental contaminants and humidity and CO_2 depletion (if there is no CO_2 enrichment).

Alternative sources of energy

In the context of energy costs (year 2004), the majority of the alternative sources of energy are not economically competitive.

Heat pumps

Heat pumps extract heat from a low temperature source (cold source) to supply that heat to a heating circuit. In efficient heat pumps, temperatures above 55°C must not be expected (Bordes, 1992).

Heat pumps can be used to dehumidify the greenhouse environment but their use is expensive.

Geothermal and industrial hot waters

The geothermal gradient is approximately 1°C for every 30 m. Therefore, 1000–2000 m must be perforated to obtain waters of 40–70°C (Urban, 1997a). These waters usually have a high concentration of salts and are corrosive, being unavoidable-to-use heat exchangers.

The facilities are expensive, although the thermal energy is free, and the depreciation costs are high.

Cogeneration

The cogeneration of heat and electricity is becoming popular, when the greenhouse electricity consumption is high, because cogeneration usually supplies heat and electricity in a proportion of two to one (Van de Braak, 1995). It is also popular when it is possible to sell the electricity produced at attractive prices, as has happened in Europe in recent years.

Solar energy

Solar energy is free, available, excessive in summer but insufficient in winter, and requires high investments for its capture, and above all, its storage (Photo 7.6).

Biogas

Biogas has not been developed due to the low costs of conventional energy. The high proportion of CO_2 in biogas causes poor flame performance.

7.4.5 Sizing of the heating systems

See Chapter 5 Section 5.4 ('Simplified Greenhouse Energy Balances').

For the purpose of calculating the maximum heating requirements it is estimated that an internal greenhouse temperature (T_i) of 16°C (ASAE, 2002) covers the requirements of most plants and the external temperature (T_e) is the average of the minimum temperatures of the coldest month.

7.4.6 Heating and temperature management

In general, the optimum temperatures decrease with the age of the plant and vary depending on the process that we want to optimize (translocation, gross photosynthesis, root growth, harvest) (Hanan, 1998). With a higher temperature, of the order of 22–23°C, vegetative growth is activated in vegetables such as tomato, whereas lower values (18–19°C) favour fruit development (Kamp and Timmerman, 1996). The optimum temperature also depends on the radiation and CO_2 concentration in the air (see Chapter 6).

Although plants have a great capacity to integrate the temperatures in periods larger than 24 h, it is necessary not to overcome certain thermal limits, which usually range between 10 and 25°C for horticultural species. In general, if the temperatures are lower than the optimum there will be less

Photo 7.6. Low-cost solar panels for greenhouse heating (Experimental Station 'Las Palmerillas', Almeria).

high-quality yield (Hanan, 1998). It must not be forgotten that the economic objective must dictate the selection of the set point temperatures, if heating is used.

The thermal integral allows for quantification of the accumulated effects of temperature on the development processes of the plants; thus, the duration of one development stage depends, under normal conditions, on the thermal integral (see Chapter 3), even when the day and night temperatures are different (Slack and Hand, 1983).

In heated greenhouses, it has been recommended to adapt the night temperatures to the solar radiation of the previous day, to limit the heating costs without affecting growth (Gary, 1989). So, in a low radiation day higher night temperatures, as those required after a high radiation day, are not required, for a proper redistribution of the assimilates of the previous day. In a similar way, the use of decreasing temperatures throughout the night, in the form of cascade (split night temperature) will save costs without negatively affecting growth (Toki *et al.*, 1978). Relevant energy savings can be achieved adapting the heating conditions to the external environment, maintaining lower set point temperatures in the absence of wind (Spanomitsios, 2001).

The difference between day and night temperatures, which is known as DIF, influences different aspects of growth such as the elongation of the internodes and stem growth (Erwin and Heins, 1995). When DIF is positive (*T* during day higher than *T* during night) the internodes are longer and, on the contrary, when DIF is negative, plants are more compact, as internodes are shorter (Challa *et al.*, 1995). This DIF mechanism is related to phytochrome (Moe *et al.*, 1992) and is of particular interest in ornamental horticulture.

In daily management, the minimum temperature thresholds during the day must be from 6 to 11°C higher than those of the night on sunny days; a difference that must range from 3 to 6°C on cloudy days (ASAE, 2002). Other authors (Berninger, 1989) recommend limiting these differences in winter, maintaining the day temperature from 3 to 4°C higher than the night temperature on cloudy days and from 6 to 8°C higher on sunny days (Fig. 7.12).

The majority of horticultural species cultivated in greenhouses in the Mediterranean Basin experience a great reduction of their metabolic activity below 10–12°C (Nisen *et al.*, 1988), their optimum

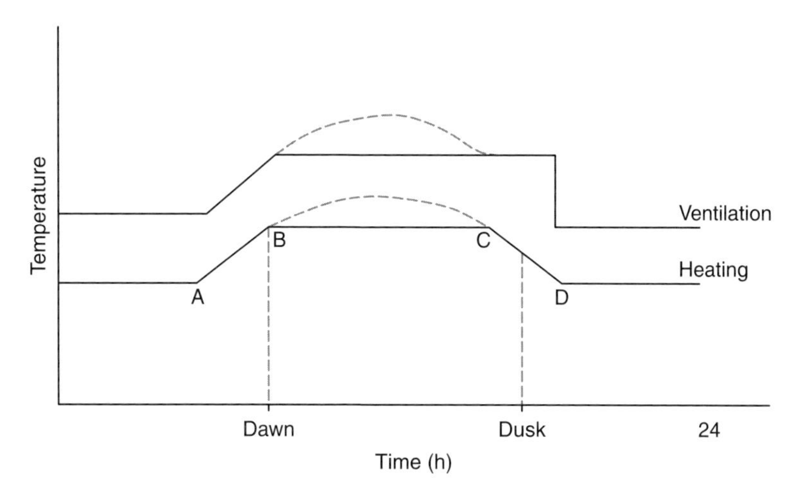

Fig. 7.12. Scheme of the set point temperatures for heating and ventilation management in a climatized greenhouse over 24 h. A, Starting point of the set point increase; B, sunrise (final point of the set point increase); C, starting point of the set point decrease (near sunset); D, final point of the set point decrease; dotted lines, possible high temperatures during daylight hours. The set points for heating and ventilation are calculated with regard to sunrise and sunset, when solar radiation begins and finishes, respectively (adapted from Bakker *et al.*, 1995).

temperatures varying over intervals from 15 to 20°C during the night and between 22 and 28°C during the day (Tesi, 2001). Table 7.3 summarizes optimum temperature intervals for the air and the substrate in different horticultural crops.

In sophisticated greenhouses, for optimum fruit quality, with well-developed crops, it is recommended, as a general rule, that the day/night temperature thresholds for heating, should be 19°C/17°C for tomato, 22°C/18°C for pepper, 21°C/19°C for cucumber and aubergine, whereas on less demanding crops, such as lettuce, temperatures should be limited to 12°C/6°C (Urban, 1997a). These values can be slightly decreased under low radiation conditions (Kamp and Timmerman, 1996). In any case, economic criteria must prevail when defining these thresholds for each specific case. In low technology greenhouses these temperature thresholds are usually lower, as the equipment and insulation levels are simpler.

During the winter, the management of set temperatures in Mediterranean greenhouses at 22°C (day)/18°C (night) in the beginning of the cycle, to continue at 20°C/16°C during the vegetative growth stage, and later at 18°C/14°C during the productive stage (Lorenzo *et al.*, 1997a) of a cucumber or a green bean crop induces high energy use, adversely affecting the economic viability of heating.

Table 7.3. Optimum air and substrate thermal levels for different horticultural crops. (Source: Tesi, 2001.)

Crop	Air (°C)		Substrate (°C)
	Day	Night	
Tomato	22–26	13–16	15–20
Cucumber	24–28	18–20	20–21
Melon	24–30	18–21	20–22
Green bean	21–28	16–18	15–20
Pepper	22–28	16–18	15–20
Aubergine	22–26	15–18	15–20
Lettuce	15–18	10–12	10–12
Strawberry	18–22	10–13	12–15
Carnation	18–21	10–12	10–15
Rose	20–25	14–16	15–18
Gerbera	20–24	13–15	18–20
Gladiolus	16–20	10–12	10–15

When using air heating in low-cost greenhouses that are not very airtight (Table 7.1), some authors recommend establishing a low set point temperature for heating. For instance, in a cucumber or green bean crop set points between 12 and 14°C are recommended, to limit fuel consumption (propane) below 5 kg m^{-2} (from November until the middle of March) (López *et al.*, 2000). Obviously, for each specific case, the economic conditions must determine the management of the heating system, with the aim of maintaining heating expenses which can lead to profitable results (Plate 13); energy consumption estimates in Mediterranean greenhouses grow exponentially with the set temperature (López, 2003; López *et al.*, 2003a, b).

The combined use of heating and CO_2 enrichment enhances the effect of both, allowing for yield increases (López *et al.*, 2000; Sánchez-Guerrero *et al.*, 2001), the profitability of which depends on the specific conditions of use.

When using heating in Mediterranean low-cost greenhouses (parral-type), it seems necessary to improve how airtight they are, in order to limit the thermal losses (López *et al.*, 2000), with the subsequent increase of the CO_2 fertilization efficiency.

In Mediterranean low technology greenhouses, the use of heating is usually limited to the winter months, in crops whose price is higher at this time of year, such as cucumber or green bean. Table 7.4 summarizes the results of a study of air heating in a climbing bean crop and demonstrates the interest, from an economic point of view, of maintaining low temperature set points when they do not limit the yield.

The use of fixed energy-saving screens allows for an increase in night temperatures but they decrease radiation so their use is of no interest (Castilla, 1994; López *et al.*, 2003a, b). However, some growers do use them, in order to limit the fall of water droplets (from condensed water vapour or from the rain in artisan low-cost greenhouses) over the crop. Mobile thermal screens improve the yield and adapt well to multi-tunnel greenhouses, being of more interest

Table 7.4. Yield results (early and total, in g m^{-2}) of a 126 day-cycle climbing bean crop, sown at the beginning of November, for different temperature set points (using hot air heating with propane, direct combustion). (Source: López *et al.*, 2003b.)

	Yield (g m^{-2})[b]		Estimated energy use
Treatment[a]	Early	Total	(MJ m^{-2})
T-14/T-12	1074a	2869a	180
T-14	954a	2863a	250
T 14-12	795a	2767a	180
T-12	231b	1952b	120
Control (no heating at all)	58b	1123c	–

[a]Treatments: T-12, minimum set point 12°C air temperature; T-14, minimum set point 14°C; T-14/T-12, set point 14°C/12°C, during the vegetative productive stages of the cycle; T 14-12 (split), set point 14°C during first half of the night, 12°C during the second half.
[b]Numbers in the same column followed by a different letter (a, b, c) indicate significant differences ($P = 0.95$).

in heated greenhouses than in unheated ones (Meca *et al.*, 2003). In low-cost parral-type greenhouses, their implementation is difficult, due to the large number of internal supports, and so their efficiency is lower. In this case, again consideration of profitability will determine whether their use is worthwhile.

It is necessary to generate information on the set point values that optimize climate control (heating, CO_2, humidity), from the economic point of view in the different crop growing conditions.

7.5 Summary

- To limit energy losses in the greenhouse, heat losses must be reduced and natural inputs favoured, and if the latter are insufficient use heating.
- The reduction of greenhouse heat losses is achieved, mainly, by decreasing the heat exchange surfaces (roof cover and sidewalls) with the exterior and the losses per unit area, using proper covering materials and insulation devices (thermal screens, double covers) and windbreaks.
- Greenhouse heating can be directed to the aerial part of the crop, to the roots (soil or substrate) or to both.
- Hot air heating is usually performed by means of fan coils or using hot air generators with direct or indirect combustion. The heat distribution takes place by convection. They have low thermal inertia and mainly heat the aerial parts of the crop.
- Hot-water heating systems distribute the heat through pipes by radiation and convection. They have more thermal inertia than hot air systems; they heat the air and also the soil or substrate. The systems that use water at high temperature (50–80°C) are more expensive than the low temperature systems (40–50°C).
- Soil or substrate heating transmits the heat by conduction and it normally uses hot water (at low temperature) or electrical resistances.
- The ideal fuel is natural gas, followed by LPG (butane and propane) and diesel, but their prices are high. Other fuels (fuel oil, coal, wood, vegetable waste) may be of interest, for economic reasons.
- Temperature management, when heating is applied, must be done using economic criteria, because usually the optimum temperatures for the plant's growth do not coincide with those at which the highest profitability for the grower is achieved. Therefore, temperature management must be adapted to the specific growing conditions of each area.

8

Management of High Temperatures: Cooling

8.1 Introduction

The battle against high temperatures inside the greenhouse is focused on decreasing the energy inputs and eliminating their excesses. If heating is used, the artificial energy input (by the heating system) is eliminated by turning it off. In low thermal inertia systems (air heating) the response is immediate, whereas in those with high inertia (water heating) there is a delay in the response.

The decrease in natural inputs is focused, in practice, on limiting solar radiation, by means of shading, inside or outside the greenhouse.

The increase in energy losses is achieved with ventilation, natural (or static) and forced (or mechanical), as a first step. Every ventilation system can only decrease the interior air temperature to the value of the outside air if the renewed air has the same humidity. This decrease of the interior air temperature by ventilation is, in many cases, enough to achieve acceptable thermal levels inside the greenhouse when the external air temperature is not excessive (i.e. ventilation counteracts the greenhouse effect). If the interior temperature must be further decreased, active cooling methods will have to be used; most commonly, by evaporating water.

Air cooling by conventional methods (refrigerator circuit) is not economical, under normal conditions.

At plant level, the first measure to limit high temperatures is to irrigate properly, so that the plants can transpire to the maximum and decrease their temperature, complemented by efficient air renewal through ventilation.

8.2 Function of Ventilation

Aeration, or ventilation, is the air exchange between the greenhouse and the exterior. This air exchange takes place through the greenhouse openings (vents and slits) (Photo 8.1). The air renewal allows the evacuation of the excess heat and a decrease in the air temperature, modifying the atmospheric humidity (exchanging the interior air with high water vapour due to plant transpiration), and modifying the gas composition of the atmosphere (especially the CO_2).

If the air leaving the greenhouse is dry, the energy evacuated is very limited due to the low specific heat of dry air (1 kJ kg^{-1} °C^{-1}, at 20°C). If the outgoing air is humid, the temperature decrease will be much higher, as the energy evacuated with the humid greenhouse air is much bigger (the energy to evaporate 1 kg of water, or the latent heat of water evaporation is 2445 kJ kg^{-1}).

Photo 8.1. The vents must be mechanized to facilitate their automatic opening and closing.

Therefore, the humidity difference between the interior and exterior is more important than the temperature difference, for greenhouse cooling purposes (Fig. 8.1).

The air renewal rate (R) is expressed as:

$$R = \frac{\text{volume of air exchanged}}{\text{greenhouse volume} \times \text{hour}} = \frac{\text{m}^3}{\text{m}^3 \times \text{h}} \quad (8.1)$$

It may also be expressed based on the greenhouse ground area. The ratio ventilation per square metre of ground (V) will be:

$$V = \frac{\text{volume of air exchaged}}{\text{greenhouse ground area} \times \text{hour}} = \frac{\text{m}^3}{\text{m}^2 \times \text{h}} \quad (8.2)$$

The relationship between both ratios is:

$$V = R \times H \quad (8.3)$$

where H = average greenhouse height (m)

The air exchanges between the interior and the exterior take place by leakage through slits and through the vents due to pressure differences between the interior and the exterior air. The wind and differences in air density between inside and

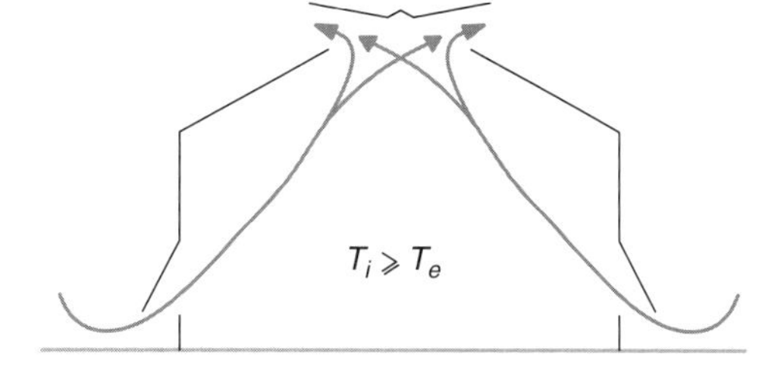

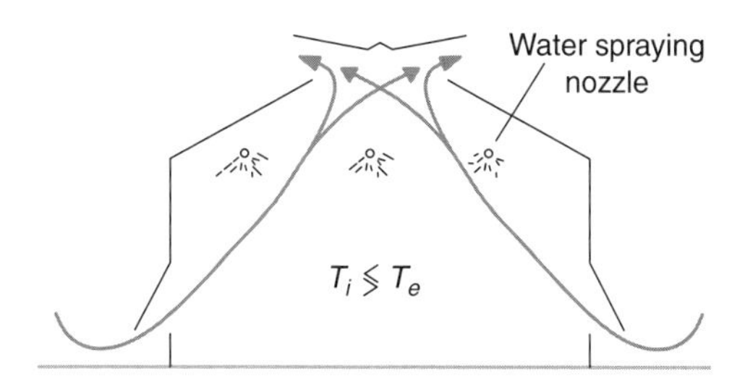

Fig. 8.1. Idealized scheme of natural ventilation in a greenhouse. (a) In a greenhouse with dry soil, with no irrigation or crop, the heat removal, as the air is renewed, is very inefficient because the dry air carries very little heat. (b) In a greenhouse with water fogging, the heat removal potential is high because water absorbs a lot of heat as it changes into vapour and then is removed from the greenhouse with ventilation. A greenhouse with a crop is also easier to cool because the transpired water absorbs heat as it changes phase from liquid to gas and then is removed through the vents. T_i, Interior air temperature; T_e, external air temperature.

outside the greenhouse generate these pressure gradients. The air densities are affected by the temperature and, to a lesser extent, by the air composition (the humidity especially).

As ventilation affects the conditions of the confined air, essential in the 'greenhouse effect', the knowledge of the ventilation flux is fundamental in the management of the greenhouse climate (Bot and Van de Braak, 1995).

8.3 How Airtight is the Greenhouse?

A greenhouse is not an airtight construction, suffering minor or major leakage losses. The importance of the exchange due to leakages depends on the quality of the construction and varies a lot with the wind velocity (see Table 7.1). When the wind is weak, the difference between the internal and the external temperature is the main influence. The losses increase with wind velocity.

How airtight a greenhouse is may be measured by calculating the air renewal coefficient with the help of a tracer gas or by creating a pressure differential between the inside of the greenhouse and the outside (i.e. lower or higher pressure inside the greenhouse compared with outside). It can also be quantified by means of the thermal balance.

The advantages of an airtight greenhouse are: (i) a decrease in the thermal losses (and therefore an increase in the amount of energy saved); and (ii) a decrease in the CO_2 leakages, if carbon enrichment is practised. The disadvantages of a greenhouse being too airtight are the build-up of air humidity and the higher risk of toxicity in the case of pollution or pesticide application.

8.4 Natural Ventilation

Natural ventilation allows for the renewal of the interior hot air by external fresh air. It is achieved by means of permanent or temporary openings in the roof, in the sidewalls or in the front walls. It is the cheapest and most commonly used system.

The efficiency of ventilation, quantified by the air exchange rate (R), depends on the climate conditions: (i) external wind velocity and direction; and (ii) temperature difference between inside and outside the greenhouse. These two effects, the wind effect and the thermal (buoyancy) effect, generate pressure differences which force the air to move (natural convection), from a high pressure area to a low pressure area. The ventilation efficiency also depends on the characteristics of the openings (area and position) and of the canopy (arrangement of the crop rows in relation to the sidewall vents).

8.4.1 The thermal effect

The existence of temperature gradients powers the convective movements, as the warm air rises and the cold air descends. When there is no wind, the air exchange rate depends on temperature difference alone between the interior and the exterior (Fig. 8.2).

A roof opening favours ventilation (chimney effect) (Fig. 8.2). The efficiency of the roof ventilation depends on the greenhouse height. Due to the chimney effect the taller greenhouses ventilate better, so it is advisable to build them at least 3 m high (Urban, 1997a).

The effect of the temperature gradient on ventilation is important with weak winds, high radiation and limited openings.

In Mediterranean greenhouses, the thermal (buoyancy) effect is of little importance in ventilation if the wind velocity exceeds 1–2 m s^{-1} (Muñoz, 1998; Pérez-Parra, 2002),

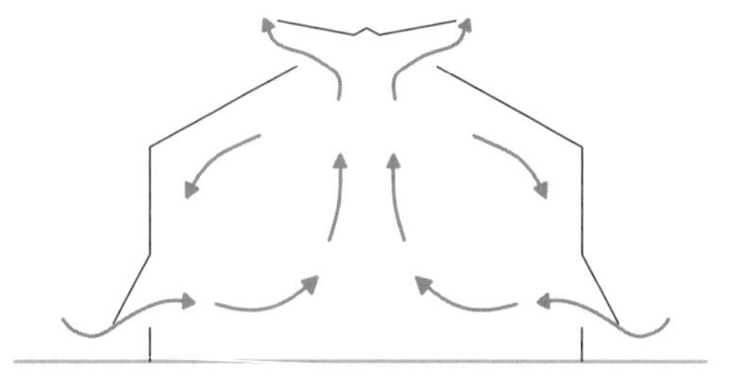

Fig. 8.2. Greenhouse ventilation fluxes when the wind velocity is zero and there is only a thermal effect.

values normally exceeded in the Mediterranean coastal area during daylight hours. However, with the use of low porosity anti-insect screens (nets), ventilation by the buoyancy effect is gaining importance.

In areas where the density of greenhouses is high, so the structures are very close, the wind effect is very restricted (see more details in the Appendix 1).

8.4.2 The wind effect

With low wind velocities (less than 2 m s^{-1}), ventilation depends mainly of the temperature differences between the greenhouse and the exterior. With wind velocities greater than 2 m s^{-1}, the number of exchanges of the air volume of the greenhouse is proportional to the velocity, varying with: (i) the number of spans of the greenhouse (Kozai and Sase, 1978); (ii) the dimension of the spans (dynamic pressure coefficient, Fig. 8.3); and (iii) the wind direction.

The wind loads on the greenhouse structure depend on the dynamic pressure of the wind, which varies depending on the effective height of the greenhouse and the cladding surface affected, mainly (see Appendix 1).

The pressures generated by the wind (Fig. 8.3) are positive on the side exposed to the wind (windward) and negative over the roof and in the side protected from the wind (leeward). This pressure distribution is altered when opening the vents.

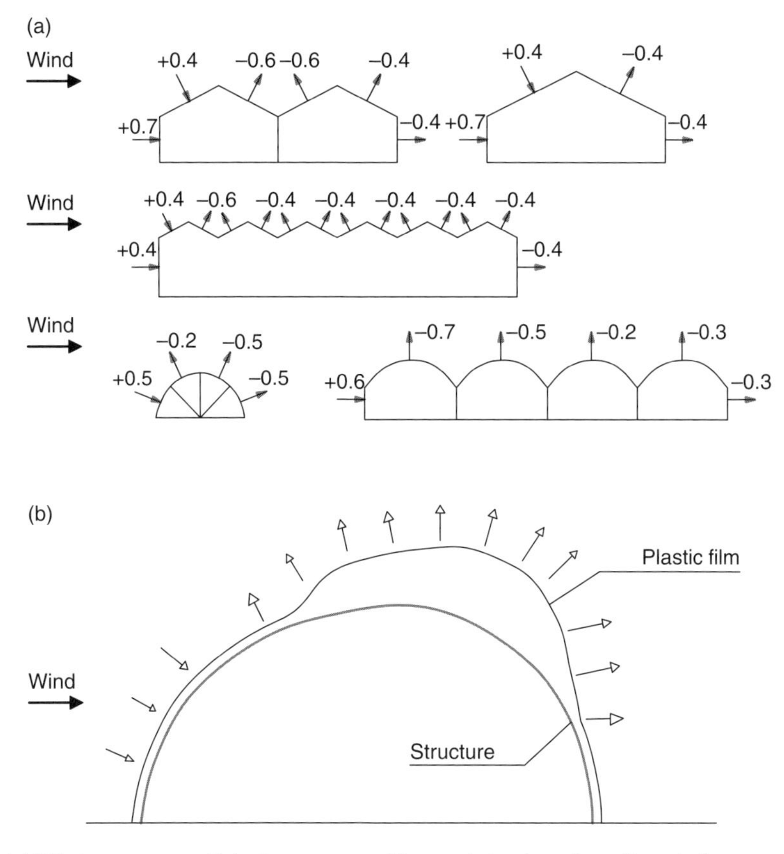

Fig. 8.3. (a) Wind pressure coefficients over several types of structure. A positive wind pressure coefficient shows high pressure, whereas a negative value shows suction (adapted from Zabeltitz, 1999). (b) Scheme of the effects of a perpendicular wind to the ridge on a tunnel covered with a plastic film, showing the effects of suction on the plastic film, which may break it.

The orientation of the vents which open facing the wind (windward) favour ventilation in relation to the vents which open towards the side sheltered from the wind (leeward) (Fig. 8.4), especially if they are of the flap type (Montero and Antón, 2000a, b).

When the wind grows in intensity, the roof vent that must be more open is the one opposite to the wind direction (Fig. 8.5); the suction created by the external wind forces the air out of the greenhouse, while it is quite risky to open the vent facing the wind (Wacquant, 2000). If there are strong winds, all the greenhouse vents must be closed to avoid them breaking.

The wind effect is very small if vents on the roof are not complemented with sidewall openings (Montero and Antón, 2000a) (see Fig. 8.6).

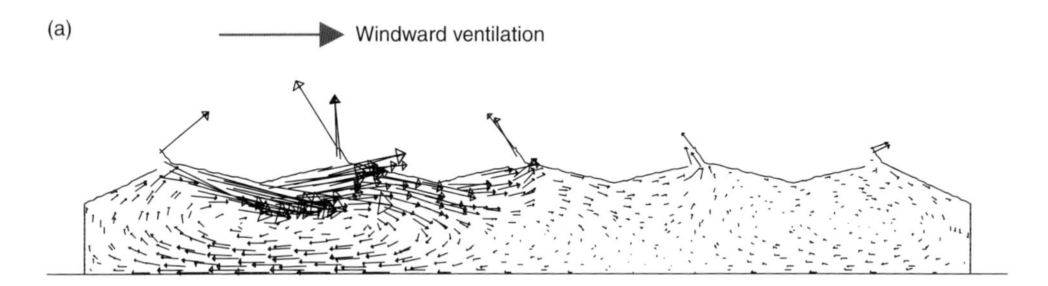

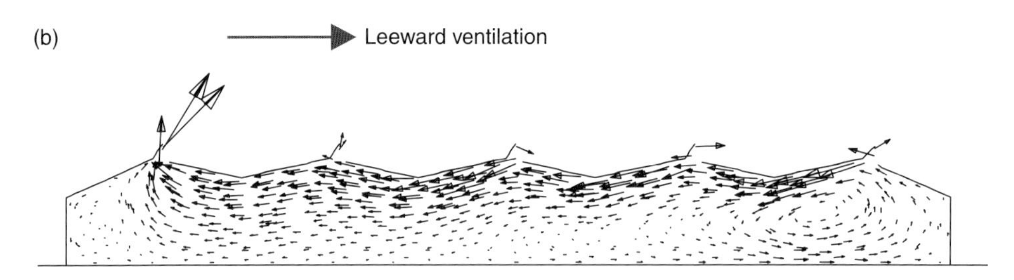

Fig. 8.4. Diagrams resulting from a ventilation study using flow visualization techniques, depending on the wind direction: (a) windward (facing the wind) and (b) leeward (side sheltered from the wind). The size of the arrows indicates the intensity of ventilation. Low-cost greenhouse, five spans, with hinged flapping-type vents and wind velocity of 4 m s^{-1} (data from 'Las Palmerillas' Experimental Station, Cajamar Foundation, Almeria, Spain).

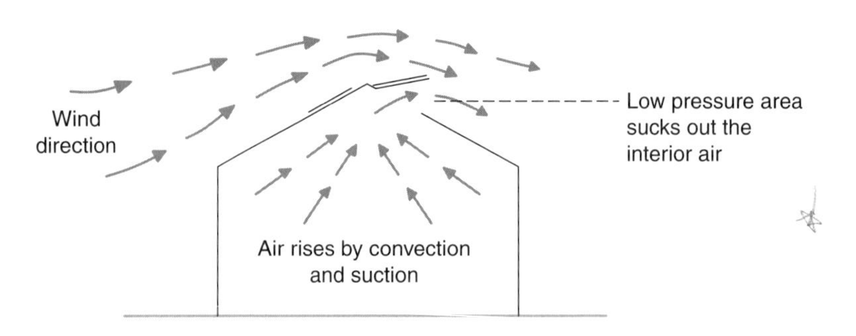

Fig. 8.5. The suction created by the wind on the leeward vent (protected from the wind) contributes to the extraction of the greenhouse air, if the wind is strong. If there are sidewall vents (for air entrance) the efficiency of ventilation is improved.

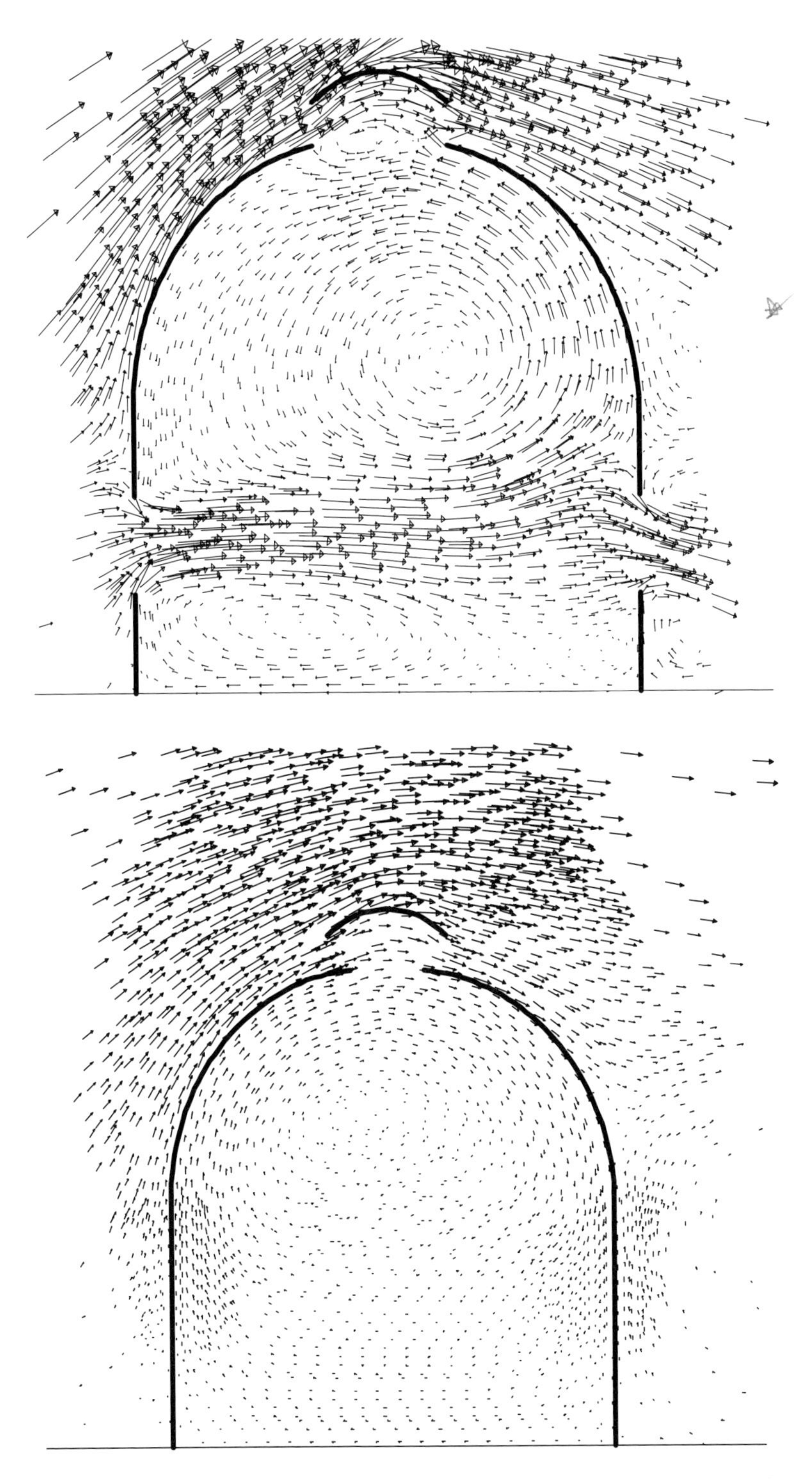

Fig. 8.6. Roof vents of the 'small hat' type are of low efficiency when used alone (bottom). Therefore, they must be associated with sidewall vents (top) to improve their performance. The size of the arrows indicates the intensity of ventilation. (Source: J.I. Montero and E. Baeza.)

Nowadays, ventilation can be automated. It is necessary to monitor conditions with an anemometer, vane and rain detector to close the vents in case of excessive winds (depending on their direction) or in case of rain (For more details, see Appendix 1).

8.4.3 Characteristics of the openings

The openings can be characterized by the opening area and by their positions. The opening area, in the case of hinged vents, is at most that of the frame (hole) (Fig. 8.7).

In the case of a long vent of continuous shutter, the maximum opening is achieved with an angle of 60° (Wacquent, 2000).

The opening area index, which relates the total ventilator opening area to the ground area of the greenhouse, expressed as a percentage allows for a comparison between different greenhouses.

The air exchange rate increases with the opening area ratio. There is an optimum value for this ratio, above which a complementary opening is less efficient (Fig. 8.8 and Plate 14). This optimum opening area ratio, in unscreened vents, ranges between 15 and 20% for tunnels (with well positioned vents) and between 25 and 33% for multi-span greenhouses (Wacquant, 2000).

In median latitudes, the vents are usually continuous, along the greenhouse, preferably on the ridge, with a recommended opening area index of 15–25% (ASAE, 1988) under high radiation conditions. The most efficient and versatile ventilation systems have vents on both sides of the ridge and in the sidewalls. When the vents have screens these open area ratios must be increased.

To maximize ventilation it is essential that the arrangement of the vents complements the convective movements inside the greenhouse with the pressure differences in the walls generated by the wind.

The sidewall ventilation is very important in small greenhouses, contributing equally or even more than the roof vents to the air exchange, but in wide greenhouses (width over 35 m) roof ventilation predominates (Pérez-Parra *et al.*, 2003b).

In single- or double-span tunnels, the most efficient ventilation is achieved combining ridge vents with sidewall vents (in the proportion 1.5 to 1.0). The chimney effect, using sidewall and roof openings, is of special interest if the wind is less than 1 m s^{-1}, multiplying by three the efficiency of a single opening (Fig. 8.8).With medium or strong wind, the roof ventilation is sufficient.

In multi-span greenhouses of large area the roof vents located in both sides of each

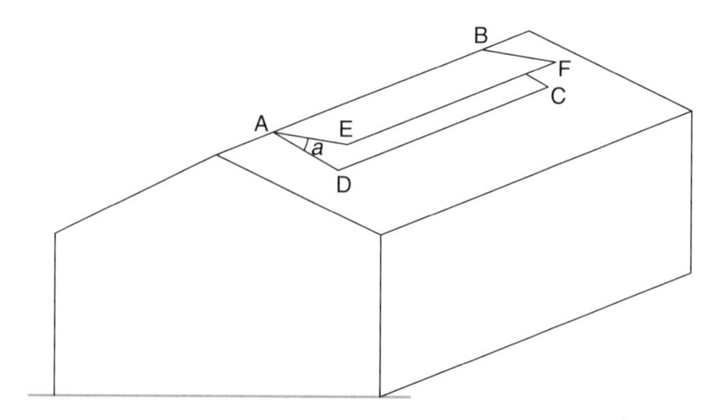

Fig. 8.7. The useful ventilation area is, at its maximum, the frame of the vent (ABCD in the figure). A small opening angle (*a*) limits the useful ventilation area. In the figure, the useful area is that formed by the rectangle EFCD plus the triangles AED and BFC, as long as the area of the frame of the vent is not exceeded (ABCD) (adapted from Wacquant, 2000).

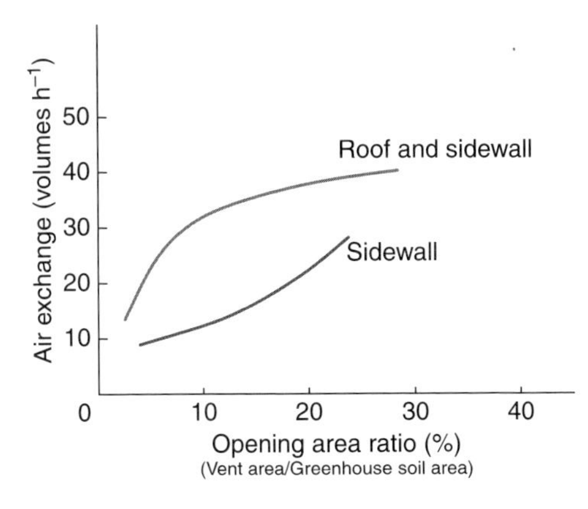

Fig. 8.8. In a tunnel greenhouse, with weak wind, when sidewall and roof vents are used, the increase in the vent opening area (beyond a certain threshold value) does not increase the air exchange. The sidewall ventilation alone is less efficient than the combination of roof and sidewall vents. In a multi-tunnel greenhouse the opening area ratio must be higher than in a simple tunnel with roof and side ventilation, to achieve the same air exchange effects.

span facilitate alternate use depending on the wind direction (Wacquant, 2000):

1. With zero or weak wind – open vents of both sides of each span.
2. With moderate wind – preferably open the vent protected from the wind (leeward), in a first phase to profit from the suction effect. The opening of the vent facing the wind (windward) is delayed until the ventilation requirements are higher.
3. With strong wind – the vent facing the wind opens even less, or even closes if the wind is very strong. In the case of extreme winds, both sides close, to avoid breaks.

Computational Fluid Dynamics (CFD; see section 8.4.5) analysis indicates that windward ventilation is more efficient than leeward ventilation, especially in greenhouses with limited total width. Therefore new greenhouse constructions should have larger openings oriented towards the prevailing winds. In existing designs, outside air may enter and leave the greenhouse without mixing with the internal air. To avoid this problem, the use of deflectors to conduct the entering air through the crop area is strongly recommended (Baeza, 2007).

CFD simulations have also demonstrated that the greenhouse roof slope has a significant effect on ventilation rate; therefore, in the south of Spain, traditional horizontal roof greenhouses are being replaced with symmetrical or asymmetrical greenhouses with a near 30° roof angle (Castilla and Montero, 2008). No further increase in ventilation has been identified for roofs with roof angles greater than 30° (Baeza, 2007).

The combination of side and roof vents is more efficient than the use of a single type of vents, of equal opening areas (Montero and Antón, 2000a). The roof vents located by the ridge are more efficient than those located by the gutter (Muñoz *et al.*, 1999; Montero and Antón, 2000a, b). The hinged vents (with a flap) oriented windward are more efficient than the leeward vents (on the side protected from the wind) (Plate 15), improving the air exchange rate from 35 to 60%, when the wind ranges from 2 to 7 m s^{-1} (Pérez-Parra, 2002), whereas in rolling vents the air exchange rates do not depend on the wind direction, at least in low slope multi-span greenhouses.

In each case, specific locations of the vents are used. For instance, if there are regular winds (land or sea breezes, in coastal areas) the greenhouse may be oriented with its openings facing windward and leeward, such is the case of the Mediterranean coastal area greenhouses (Wacquant, 2000).

In tropical regions, the tunnels are ventilated best by orienting them in the direction of the trade winds with the front walls open, due to their small length.

In the case when sidewall vents are available, wind flows that are too cold or too dry which impinge directly on the plants must be avoided, by avoiding too large an opening.

For protection from insect damage, and mostly, from the virus diseases transmitted by them, the air can be filtered at the openings by means of low porosity screens to avoid their entrance. These screens reduce the air flowing through the vent, notably decreasing ventilation (Plate 15), so the opening area ratio must be increased or a device with a higher screen surface than that of the vent itself must be adopted (see section 8.4.6).

The discharge coefficient of the vents, which measures the reduction of the air flow when passing through a vent, has been studied in Mediterranean greenhouses (Muñoz, 1998; Pérez-Parra, 2002).

The variability in the wind conditions, the greenhouse types, the vents (shape, location, presence or not of screens) induced a simplification of the joint incidence of these effects under a global wind effect coefficient (see Appendix 1).

8.4.4 The crop and air movements

The presence of plants affects the air exchange rate and the convective movements, depending on the density of the vegetation and the arrangement of the crop rows.

Plants, when transpiring, cool the air and modify its density. High ventilation with a LAI above 2 limits the temperature gradient and modifies the chimney effect. The vegetation forms a screen that limits the air movement. Therefore, the location of the crop rows in the same direction as the dominant winds facilitates the air circulation and the ventilation.

In tall greenhouses, in which there is a large air chamber between the crop and the cover, the crop has little influence on the air movements if it does not block the sidewall vents.

When the greenhouse is closed, the wind also affects the internal air movements. At the top, the circulation is parallel and in the same direction with that of the outside wind, returning through the lower part where it is heated and its humidity increased (Wacquant, 2000). The hottest spot is located on the side exposed to the wind (Fig. 8.9).

8.4.5 Measuring the ventilation of greenhouses

The most commonly used method to measure the natural ventilation in greenhouses is the use of a tracer gas, which uses an inert and non-reactive gas (Goedhart *et al.*, 1984). This gas (which is usually nitrous oxide) is homogeneously distributed inside the greenhouse and the evolution of its concentration in the air over a period of time, in the existing ventilation conditions, quantified. The ventilation rate is proportional to the decay rate of the tracer gas content in the air.

Recently, the use of models has facilitated the study of ventilation. The flow visualization techniques using scale models (Montero and Antón, 2000a) have been used with satisfactory results, at a low cost (Plate 16). Recent work with scale models, in a wind tunnel, has permitted the study of ventilation in low-cost parral-type greenhouses (Pérez-Parra, 2002).

Another method of studying ventilation is the use of fluid dynamics simulation programs known as CFD (Computational Fluid Dynamics), illustrated in Plate 17. In recent years increasing attention is being paid to this CFD tool in greenhouse technology studies (De Pascale *et al.*, 2008; Dorais, 2011).

Several models have tried to relate the ventilation rate with the area of open

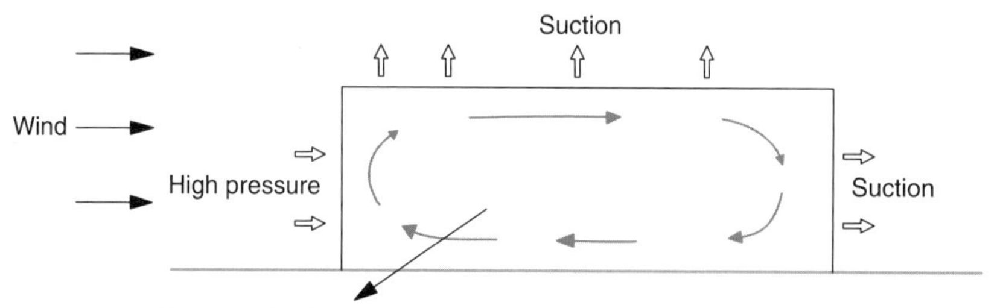

Fig. 8.9. The effect of the external wind on the air movement inside a closed greenhouse (adapted from Wacquant, 2000).

vents. The most simple models establish relationships between the opening angle and the air exchange rate, depending on the wind conditions and the temperature differences. Normally, their use involves previous adaptation to the local conditions (greenhouse characteristics, orientation, etc.).

8.4.6 Anti-insect screens

Anti-insect screens are made of uniform threads which form the screen. The type of screen to be used will depend in each case on the size of the insect to be excluded (Table 8.1). It may be that some biotypes of insects have different sizes: for example the white fly biotype found in Almeria, is smaller than the American one, as it only measures 240 μm (Cabrera, F.J., 2008, personal communication). The nomenclature of the screens in the 'mesh' scale designates the number of threads per inch in each direction (Aldrich and Bartok, 1994). In this way a 64-mesh screen has 64 threads per inch (2.54 cm) in each perpendicular direction. To evaluate the size of the hole the diameter of the threads must be known.

The common nomenclature used in Spain to designate the screens is imprecise, as it does not express their exact characteristics. For instance, a 20 × 10 screen indicates that the screen has 20 threads per centimetre in one direction and 10 threads per centimetre in the perpendicular direction, but this does not specify the thickness of the thread, which is usually around 0.27 mm, although the most commonly used types range between 0.23 and 0.29 mm diameter (Cabrera, F.J., 2008, personal communication).

Table 8.1. Selection criteria for anti-insect screens as a function of insect size and hole size in the screen (Aldrich and Bartok, 1994).

Insect to be excluded	Insect size (μm)	Hole size (μm)
Leaf miner	640	266 × 818
Melon bug	340	266 × 818
White fly	462	266 × 818
Thrips	192	150 × 150

The porosity of the screen is the relationship (per unit or percentage) between the holes area and the total area. The porosity depends on the diameter of the thread and the number of threads per unit area and determines the decrease in ventilation rate when the screen is placed in the vents (see Appendix 1 section A.6.4).

When covering a vent with a screen, the useful ventilation area of the vent is restricted to the area free of threads (net hole area of the screen), which must be taken into account when calculating the useful vent area, correcting it as a function of the screen's porosity. In some cases, screens are used that are larger in total area than the vent they cover, so the area free of screen (net hole area of the screen) equals or exceeds the ventilator area.

When forced ventilation is used, and screens are also present, then it may be necessary to increase the performance of the fans.

The holes of the screens tend to get dirty and may be blocked very easily due to their small size, so a periodical cleaning must be done to avoid ventilation being limited even more.

Although in Mediterranean greenhouses a ventilation area (roof plus sidewall vents) of 15–20% of the greenhouse area has been described as sufficient for a well-developed and irrigated crop (Montero and Antón, 2000b; Pérez-Parra *et al.*, 2003b), the use of anti-insect screens in the vents makes this value insufficient.

Anti-insect screens placed in the roof vents decrease the ventilation rate by 20 and 33%, for flap or rolling vents, respectively, for a 39% porosity screen in a low-cost greenhouse (Pérez-Parra, 2002) (Plate 15).

Screens decrease the ventilation rate by around 40% in the case of anti-bug screens, and 70–80% in the case of anti-thrips screens (Muñoz, 1998; Montero and Antón, 2000a), although this decrease can be higher if the wind velocity is very low (see Appendix 1 section A.6.4).

8.4.7 Screenhouses

Recently, greenhouses covered with a screen as cladding material are being used (Plate 18). They are known as screenhouses. Inside them, the greenhouse effect does not occur and the windbreak and shading effects prevail, besides their restriction to the entrance of insects (depending on the size of holes in the screen and the installation conditions). In fact, they are a variant of the shade houses (see Chapter 4 section 4.6.4).

In screenhouses, ventilation is permanently ensured. Rainfall water penetrates inside them, through the screen, which restricts their use in areas where the average rainfall is high.

More sophisticated variants of screenhouses are greenhouses that have interchangeable roofs, in which the film cover can be substituted by a screen cover (Photo 8.2), and the retractable roof greenhouses (Photo 8.3).

8.5 Mechanical or Forced Ventilation

In order to inject or extract air from the greenhouse helical fans are used, that provide large flows at low pressures (Wacquant, 2000). These fans are built to work at low rotation velocities, because if their velocity is high they are quite noisy and use a lot of electrical power.

Air velocities greater than 1 m s^{-1} (which may affect the plants) must be avoided, so large-diameter fans must be used. They can work blowing air in or sucking air out of the greenhouse, with a pressure differential less than 30 Pa. The air circulation is usually horizontal.

The distance between two fans on the same wall must be less than 8–10 m and the distance between fans and vents must be less than 30–40 m, avoiding obstacles in the direction of air movement at a distance shorter than 1.5 times the diameter of the fan (ASAE, 2002). The vents must close automatically, when the fan stops.

The total flow capacity of the fans must be calculated to ensure 20–30 air renewals h^{-1}, in autumn and spring, and 40–80 air renewals h^{-1} in summer (Wacquant, 2000).

The system generates certain heterogeneity in the temperatures inside the greenhouse, and the energy consumption during the summer is high.

Photo 8.2. Rolling roof greenhouse (plastic film and screen) that allows the cover material to be selected depending on the climate conditions. (Source: J.I. Montero.)

Photo 8.3. The retractable roof greenhouse can provide maximum ventilation.

Greenhouses that are impermeable to insects require forced ventilation. In this case ventilation occurs preferably by drawing air in from outside, to increase the air pressure inside the building relative to outside. The air inlets must avoid allowing access of insects (Photo 8.4).

It is preferable to use several small fans instead of a large one, for the sake of better uniformity. For plastic greenhouses, the practical rule is to use a maximum fan flow of 2.1–3.0 m^3 m^{-2} of greenhouse ground area (Boodley, 1996). With the aim of optimizing the management under different climate conditions, it is very useful if the fans have different rotation speeds that can generate different rates of air renewal, and which in turn may also limit the energy use.

The efficiency of a mechanical ventilation system is of the order of 80%. In the calculation of ventilation the height above sea level must be considered, especially at very high altitude locations, because the air density decreases with height, reducing the efficiency of ventilation. As a correction factor the barometric pressures quotient obtained for the specific place and at sea level can be used (Langhams, 1990).

8.6 Cooling by Water Evaporation

8.6.1 Pad and fan

In a greenhouse with a 'pad and fan' cooling system, one of the sidewalls is equipped with air extraction fans, and the other has porous pads, which are kept wet. The external air is drawn through these pads, evaporating the water in the pads and being cooled, and penetrates inside the greenhouse cooler and more humid than before (Fig. 8.10). This technique allows for temperature decreases of 3–6°C under Mediterranean conditions (Urban, 1997a). Its efficiency depends on the dryness of the external air. With very low humidity (RH < 20%) the temperature can be decreased by up to 10°C (Hanan, 1998).

The distance to be travelled by the air (between fans and pads) is a limiting factor, 40 m maximum, to avoid excessive temperature and humidity differences between different greenhouse zones (Urban, 1997a). A good water quality is also advisable, because otherwise the salts in the water will soon block the pads, which then will have to be replaced, at significant cost. The water, if recycled, must be filtered and treated with biocide for control of algae. A recommended air renewal rate is 60 volumes h^{-1} (ASAE, 1988).

Photo 8.4. The injection ventilation fans must avoid the introduction of insects, by covering the air inlets, where external air is drawn inside, with proper screens.

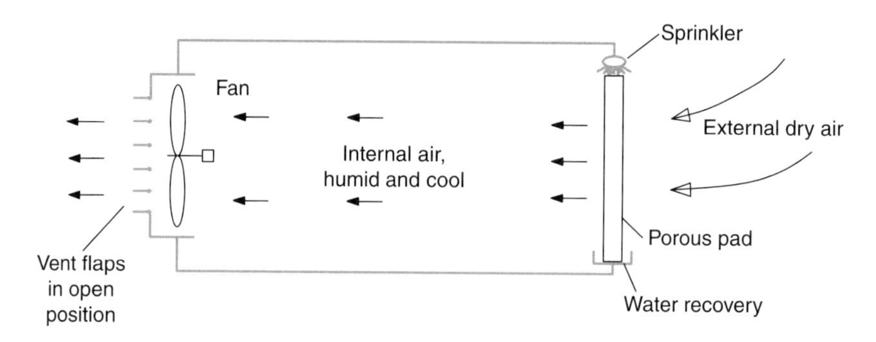

Fig. 8.10. Scheme of a mechanical ventilation system with water evaporation in pads (pad and fan).

8.6.2 Fogging and misting

The aim of fog-mist systems is to generate a fog or mist to cool the interior of the greenhouse (Photo 8.5).

Water droplets must be small enough so they do not wet the plants, to avoid the development of diseases and the deposit of salts contained in the water, when the water evaporates from the surface of the leaves between two fogging episodes. In fogging systems, the optimum range of water droplet sizes is between 0.5 and 50 μm, for maximum efficiency (ASAE, 2002).

The droplets must also be produced at a certain height above the canopy, so that as they fall very slowly they will evaporate before reaching the plants, absorbing energy and decreasing the air temperature. This system may allow for a decrease of up to 6°C under Mediterranean conditions (Urban, 1997a), while under very dry conditions the temperature reduction can reach up to 7–10°C (Conellan, 2002). Under high radiation conditions, both fog systems and shading can be complemented (Plate 19).

In the misting systems the droplets are higher and they fall rapidly and wet the

Photo 8.5. Greenhouse water fogging.

canopy surface, requiring that this excess water be carefully managed to prevent diseases and crop damage (Conellan, 2002).

The first effect of water fogging or misting is the cooling of the air by evaporation, as 2.45×10^3 J g^{-1} of heat energy are extracted. The cooled air (more dense) falls down and induces a convective movement. If the fogging system is properly regulated, the water will not reach the plants. Furthermore, fogging provides additional relief to plants from too high temperatures because it creates some degree of shading.

A negative effect of fine fogging is that it decreases PAR radiation (Urban and Langelez, 1992), but the decrease is to a far less extent than that caused by whitewash or other types of shading.

There are three main fog-mist systems: (i) water at high pressure (fogging); (ii) water at low pressure (misting); and (iii) air/water systems.

High pressure systems (fogging)

Their working pressure is higher than 7 MPa. The pipes must be of copper or steel. There are two types of diffusers (nozzles): (i) the type with a turbulence chamber which provides a droplet size in the order of 1 μm; and (ii) the needle type which produces droplets smaller than 10 μm. The latter is the most commonly used in sophisticated greenhouses.

The flow of these diffusers is around 7 l h^{-1}, with a density of 0.06–0.1 diffusers in a square metre, and the water use can be as high as 2.5–4.2 l m^{-2} day^{-1}, for an average of 6 h of daily operation (Urban, 1997a). Fog system consumption may be as high as half of the water used for irrigation during the summer.

The water must be of very good quality with a pre-filtering from 50 to 100 μm, followed by a filtering from 0.5 to 5 μm. In waters with bicarbonates, acid must be injected to decrease the pH to 6.6–6.8. If the salt content in the water is high (more than 0.7 dS m^{-1}) it will have to be desalinated, for which reverse osmosis is the usual system (Urban, 1997a). In practice, using rainfall water is the most functional, to avoid blockages. Desalinated water contains bicarbonates and its buffer capacity is drastically reduced and when the droplets come in contact the air, they absorb CO_2, decreasing the pH below 5, and could potentially become corrosive (Urban, 1997a).

Any obstructions in the water flow induce an increase in the size of the droplets, so it is necessary to keep the nozzles clean (e.g. by submerging them in an acid solution).

Low pressure systems (misting)

Their working pressure is lower than 0.5 MPa. The nozzles generate droplets whose size ranges from 20 to 100 μm, with flows from 10 to 120 l h^{-1}. With a density of 0.025–0.01 nozzles per square metre, and 6 h of daily operation, they involve a water use of 0.6–18 l m^{-2} day^{-1} (Urban, 1997a).

This system is cheaper and has fewer blockage problems although it may wet the plants, leaving deposits over the leaves, and dripping at the beginning and at the end of each fogging episode.

Air/water systems

In these systems, there are two circuits: (i) a low pressure circuit for the water (with an operation pressure between 0.2 and 0.6 MPa) and another one for compressed air (between 0.2 and 0.35 MPa).

The water and the air are canalized to the interior of an atomizer which spreads the flow into small droplets. The size of the droplets and the flow of the nozzles is a result of the pressure differences between the air and water flows. The air pressure must necessarily be, at least, equal to that of the water to achieve water drops smaller than 10 μm (fogging). If the air pressure is lower, the water droplet size is greater than 50 μm (diameter), with flows of up to 50 l h^{-1}.

The best results are obtained with pressures of 0.2–0.25 MPa for the water and 0.3–0.35 MPa for the air, the consumption being similar to those of the high pressure systems (6–7 l h^{-1}), with densities of 0.06–0.1 nozzles per square metre (Urban, 1997a).

This system has fewer blockage problems and it is easier to install but it is more expensive to install than the high pressure system because it needs a compressor. It must be well regulated, drops fall at the beginning and at the end of each fogging episode and it uses quite a lot of energy.

8.6.3 Cooling by evapotranspiration

The simplest method to evaporate water is to do it through the plants by their transpiration, which involves a non-restrictive water supply and a good air exchange to evacuate the exceeding heat. Under Mediterranean conditions at least 20 volumes h^{-1} must be achieved (R, the air exchange rate) for external radiation values of 700 W m^{-2} (Fuchs, 1990).

8.7 Shading

The limitation of solar radiation, as a means to avoid high temperatures in the greenhouse, is that it involves a concomitant decrease in photosynthesis, which in turn involves a yield decrease.

The use of ‘cooling’ films, which limit the input of IR radiation from the Sun inside the greenhouse without affecting the PAR range, could be a solution when their efficiency in limiting the IR radiation is improved and their use is economical (see Chapter 4).

The shading devices can be outside or inside the greenhouse. External shading screens are preferable from the energy point of view, as they avoid the heat input in the greenhouse (Photo 8.6), but they must resist atmospheric agents (wind, degradation, loads such as dust or dirt, hail or snow carried in the air).

The placement of wood sheets, reeds or similar materials over the cover, which could be rolled back and forth as required, was used decades ago and this technique is still popular in some countries like China (see Fig. 7.5). However, due to their sensitivity to the wind and difficulty of management, in most countries nowadays they have usually been replaced with shading screens that have good mechanical resistance and that are fixed using a large range of fixing systems (Fig. 8.11).

The whitening of the cover with different products that reflect the radiation is a usual practice during high radiation periods. The duration depends on the characteristics

Photo 8.6. External shading device located over the greenhouse. The vents have anti-insect screens.

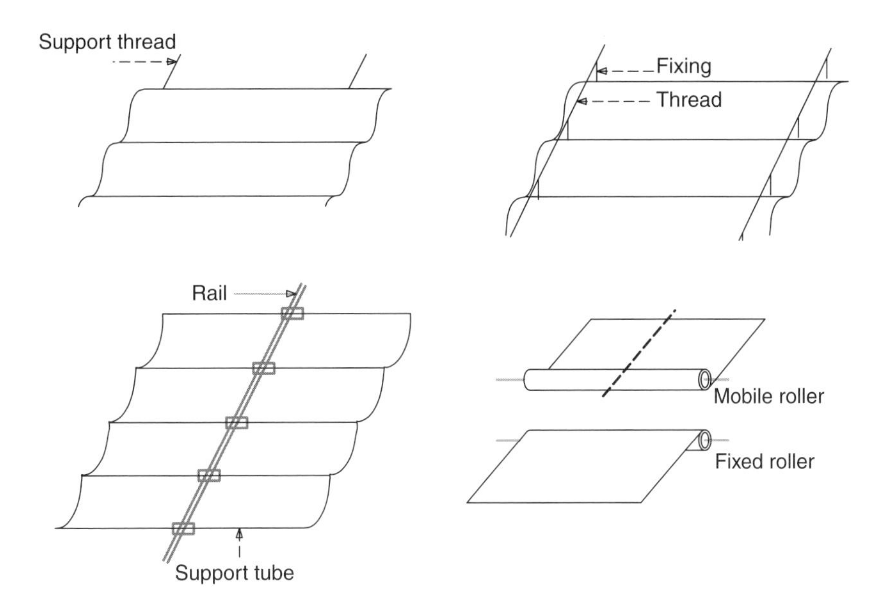

Fig. 8.11. Different systems for fixing and moving the screens (adapted from Urban, 1997a).

of the solution used (additives) and of the rain, which may wash them off.

Internal screens do not have to be as resistant as the external ones and there is a wide range of them on the market (see Chapter 4). The internal shading screens must be permeable to the air to facilitate ventilation. The use of aluminium sheets woven into the screen or use of aluminized plastic film, allows for the reflection of the solar radiation. So, in the case of aluminium-based shading screens, they must be placed with the aluminium side on the upper face, whereas in the case of energy-saving screens the aluminium must be in the inner (downwards) face.

The distinction between a thermal and a shading screen is not well determined, because all screens act against all the radiative losses and inputs, and growers prefer to use a single polyvalent screen, so it is often necessary to achieve a compromise in the performance of the screen to make it useful for both purposes (see Chapter 4).

8.8 Other Cooling Methods

Soil and substrate cooling consists of circulating cold water through a pipe or over a carpet in contact with the soil or the substrate. The water is cooled with a water/water heat pump. It is an expensive system and it is only used in very sophisticated greenhouses.

During the summer months, circulating a water film over the greenhouse cover cools it and limits solar radiation, which is partially absorbed, decreasing the greenhouse temperature by up to 3°C (Breuer and Knies, 1995). The water use can be important in warm climate areas. This technique, which is usual in sophisticated greenhouses, has not been used in Mediterranean greenhouses, where it could be of interest (Photo 8.7).

8.9 Ventilation and Climate Management

8.9.1 Temperature management

The efficiency of ventilation to decrease the temperature depends on the amount of heat to be removed (which in turn depends on the solar radiation input), on the air exchange rate and the state of the vegetation.

To achieve inside greenhouse temperatures that are close to the external values, air exchange rates of 20 volumes h^{-1} in winter, 40 during the spring and 80 or more during the summer are needed, provided there is a crop that is transpiring normally (Wacquant, 2000). The threshold temperature to begin operating natural ventilation is usually set between 23 and 26°C, depending on the climate conditions and the crops. In winter, the threshold temperature to begin ventilation must be higher, 4–6°C higher than the heating threshold temperature, with the aim of avoiding simultaneous heating and ventilation (ASAE, 2002), although this might also be acceptable in certain situations. When the external air is very cold (<5°C) the air must be introduced in such a way that, before reaching the

Photo 8.7. The 'irrigation' of the cover to cool the greenhouse has not been used in the Mediterranean.

plants, it must mix well with the internal air (warmer) to avoid a sudden change in the plant temperature.

In summer, under Mediterranean conditions, when the maximum global external radiation intensity varies between 900 and 1000 W m^{-2} during the hours in the middle of the day, the heat flux to be evacuated from inside the greenhouse is close to 700 W m^{-2}. The crop uses 60–70% of the solar energy for transpiration, leaving the remaining 210–280 W m^{-2} to be removed (Wacquent, 2000), which involves very high air exchange rates.

8.9.2 Humidity management

In the morning, when the Sun comes out, the plants start transpiring, increasing the water vapour content of the air inside the greenhouse. The interior temperature can be even lower than the ventilation set point temperature. The greenhouse air is close to humidity saturation and as the walls are colder, water condenses on them first. Later, condensation will occur in other parts of the greenhouse and even on the coldest parts of the plants such as the stems and fruits. A small opening of the vents will evacuate a large amount of the air saturated with humidity, decreasing this condensation.

Sometimes in heated greenhouses, for energy-saving reasons, low temperature set points are used, which may notably increase the water condensation before dawn. To alleviate this situation it is recommended that the set points are increased just before dawn.

The efficiency of ventilation to decrease the air humidity depends on the state of the entering air. In winter and beginning of the spring, when the external air is cold, and with low humidity, a moderate air exchange rate is sufficient to dehumidify, especially if it is associated with a heat supply.

In summer, when the external air is hot and dry, high ventilation may cause a large drop in humidity inside the greenhouse. To avoid an excessive drop, ventilation can be limited, tolerating a small thermal increase.

The recommended management of ventilation under Mediterranean conditions (Sánchez-Guerrero *et al.*, 1998) establishes a set point of 25°C (or higher, if there is CO_2 enrichment) and RH set points of 75% (daytime) and 85% (at night).

At the beginning of the crop cycle, when there is less transpiration due to the limited development of the plants, water fogging is very advisable to decrease temperatures.

8.10 Dehumidification

8.10.1 Associated heating

When there is no specific dehumidification equipment (such as in the case of most greenhouses), dehumidification is achieved by heating and ventilating. This method is efficient, but it requires very high energy consumption and a very powerful heating system. The energy consumption for dehumidification through this procedure may involve 15% of the total energy used for heating well-insulated greenhouses in the south of France (Baille, 1999). When the air is heated, the RH decreases as the saturation vapour pressure increases. Later ventilation avoids a temperature rise and excess water vapour accumulated is evacuated, introducing fresh air of low water content.

8.10.2 Dehumidification systems

Conventional systems, that use heat pumps or a refrigerator circuit, are not economically feasible (Urban, 1997a). Another option is to circulate the humid air through a hygroscopic fluid (calcium chloride, triethylene glycol) that absorbs part of the water vapour in the air. Periodically, the fluid must be heated to regenerate it. Its profitability is not clear.

8.11 Summary

- In order to avoid excessive temperatures inside the greenhouse the energy inputs must be decreased and the heat losses maximized. The reduction of natural inputs is achieved by limiting solar radiation and by means of shading. The increase in energy losses, as a first step, is achieved with ventilation, natural or forced. If the interior temperature must be further decreased, active cooling methods, most commonly by evaporating water, will have to be used.
- Greenhouse natural ventilation (exchanging interior air for exterior air) allows for the evacuation of the excess heat and decrease of the temperature, modifying the humidity and the gas content of the greenhouse atmosphere.
- The air exchange rate (*R*) of a greenhouse is the ratio between the volume of air exchanged per hour and the total volume of the greenhouse.
- A greenhouse is not an airtight structure, having leaks of air whose value depends on the construction quality of the greenhouse and on the exterior wind speed.
- Natural ventilation of the greenhouse takes place through permanent or temporary openings (normally vents) in the roof, the sidewalls or front walls. The efficiency of ventilation depends, in the first place, on the exterior wind intensity and direction (the wind effect) and the temperature differences between the interior and exterior air (the thermal or buoyancy effect). These two effects generate pressure differences which move the air.
- The characteristics of the openings (vents) and the vegetation (height and plant arrangement) also influence ventilation.
- The efficiency of the roof ventilation depends on the greenhouse height. Due to the chimney effect, tall greenhouses ventilate better.
- In greenhouses with a good ventilator area when the wind velocity is low (less than 2 m s^{-1}) natural ventilation depends, primarily, on the buoyancy effect. With higher wind velocities, the wind effect is more important.
- The opening area ratio is the proportion of the total ventilator area of a greenhouse with respect to the area covered by soil, expressed as a percentage. The recommended values for the opening area ratio in high radiation conditions range from 15 to 25%.
- The use of anti-insect screens in the greenhouse vents to avoid or limit the entrance of insects, notably reduces ventilation, thus the opening area ratio must be increased where screens are used.
- In small greenhouses the combination of sidewall and roof vents is more efficient for natural ventilation than the use of only roof vents.
- Mechanical or forced ventilation, blowing air into or extracting air from the greenhouse, allow for high ventilation rates to be achieved.
- Cooling by water evaporation may be done by means of 'pad and fan' systems (injection of external air cooled as it passes through wet pads) or by means of water fogging or misting inside the greenhouse, renewing the interior humid air in both systems.
- The fog-mist systems can be of different types: low or high water pressure, and air/water systems.
- Water transpiration through the plants is the simplest method to evaporate water in a crop, but it involves a non-restrictive water supply and efficient ventilation.
- Shading allows for a permanent high temperature decrease but it involves limiting the PAR, which decreases photosynthesis and yield. Therefore, it is advisable to use mobile screens, to deploy them only when they are needed.
- Shading devices can be internal or external to the greenhouse, and permanent (fixed) or mobile (movable).

External shading is preferable, but its installation is much more expensive.

- Whitewashing the greenhouse cover is a type of low-cost permanent shading, which is widely used during the high radiation seasons.
- Management of ventilation to maintain a suitable temperature involves achieving high air exchange rates which may be as high as 80 or more greenhouse volumes h^{-1} in full summer, in the Mediterranean area, in a totally developed crop.
- The set point temperature to ventilate (threshold temperature) usually ranges between 23 and 26°C, although if carbon enrichment is used it may be a little bit higher.
- The RH set point to ventilate ranges between 75% (daytime) and 85% (night).
- The decrease of the environmental humidity, when heating is available, can be achieved by heating and ventilating. It is an effective method but it uses a lot of energy.

9

Air Movement in the Greenhouse: Carbon Dioxide Enrichment – Light Management

9.1 Air Movement Inside the Greenhouse

9.1.1 Introduction

In a greenhouse with a hot-air heating system there is a great heterogeneity on temperatures, with important vertical gradients, temperature stratification, so the top air layers of the greenhouse are warmer than the lower ones, because warm air weighs less than cold air. Water heating systems with horizontal pipes provide more uniformity. The same applies to heated soils. In unheated greenhouses temperature stratification is of less importance. For temperature uniformity, air needs to be moved inside the greenhouse.

9.1.2 Air movement: objectives

Besides its contribution to the uniformity of greenhouse air temperature by avoiding stratification, air movement also has a large impact on the morphology, physiology and reproduction of the plants as it affects the temperature of the leaf, gas exchanges and the resistance of the boundary layer and, therefore, photosynthesis, transpiration and water use (Langhams and Tibbitts, 1997). The boundary layer is a thin air film which surrounds the leaf surfaces, where the plant exchanges energy, water vapour and CO_2 with the environment. The thickness of the boundary layer may range from less than 1 mm to a maximum of 10 mm (Hanan, 1998).

Inside a greenhouse, the wind velocity is of the order of 10% of the outside wind velocity, as an average (Day and Bailey, 1999). This limited movement of the air inside the greenhouse induces a thick boundary layer which notably hinders, in relation to open air conditions, the CO_2 and water vapour diffusion through the stomata (Gijzen, 1995a) limiting photosynthesis.

9.1.3 Plant responses

The air movement affects the plant growth, altering the energy transfers, the transpiration and the CO_2 absorption, so that the leaf size is affected, as well as the stem growth and the yield (Langhams and Tibbitts, 1997). The most notable effects manifest themselves as an agent that decreases the resistance of the boundary layer. In the boundary layer the air velocity, temperature and CO_2 and water vapour properties differ from those of the surrounding air (Day and Bailey, 1999).

The air movement through the vegetation triggers 'thigmomorphogenesis', which results in thicker stems and shorter internodes, probably induced by an increase in the ethylene level in the internodes (Biro and Jaffe, 1984). This contributes to the strengthening of the plants (hardening).

Mechanical stress

The mechanical effect of the wind on the plants affects their morphology. Any mechanical stimulation, that is intense enough, has negative effects in any of the growth stages (Jaffe, 1976). Wind velocities above 4.5 m s^{-1} produce mechanical damage (Breuer and Knies, 1995), but normally these values are never reached inside a greenhouse.

Temperature and gas exchange

The main effect of air movement on heat exchange, CO_2 absorption and transpiration and evaporation is explained by its influence on the leaf surface boundary layer (Nobel, 1974a, b).

Air movement decreases the thickness of the boundary layer, affecting all the processes that depend on temperature and gas exchange.

9.1.4 Air movement regulation

Air velocities from 0.5 to 0.7 m s^{-1} are recommended as optimal for plant growth, to facilitate gas exchange (CO_2 and water vapour) at the leaves (ASHRAE, 1989), whereas velocities greater than 1 ms^{-1} around the leaf restrict growth (ASAE, 1984).

Some authors recommend that all the greenhouse air above the canopy should be moved at velocities of 0.2 m s^{-1}, avoiding stratification of the air layers without generating turbulence (Hanan, 1998). For this, fans are commonly installed above the crop, but never more than 0.9–1.0 m from the top of the crop, so that some of the air is also moved inside the canopy (Photo 9.1). The axis of the fan must be parallel to the ground surface and in the direction of the ridge. The required flow is 0.01 m^3 s^{-1} per greenhouse square metre, installing fans in the direction of air movement at distances less than 30 times the diameter of the fan (ASAE, 2002).

When a good air distribution network is available, such as the one used for carbon enrichment, air may be circulated inside the greenhouse injecting air through this network.

Photo 9.1. Destratification fans are increasingly used to mix the air in the greenhouse.

9.2 Carbon Enrichment (CO_2)

9.2.1 Introduction

In a greenhouse, the limited air movement hinders the supply of CO_2 to the stomata of the leaves for photosynthesis. This gas exchange is dominated by the boundary layer resistance, as the air velocity, even with mechanical ventilation, is hardly above 0.3 m s^{-1} (Hanan, 1998). Therefore, it is necessary to achieve a minimum horizontal air movement for CO_2 supply to the leaf stomata. Increases in CO_2 levels generate an increase in photosynthesis and a subsequent increase in yield. Besides, CO_2 enrichment induces an improvement in the water use efficiency (Sánchez-Guerrero *et al.*, 1998).

The nomenclature used for the presence of CO_2 may take different forms: ppm (parts per million), vpm (volumes per million) or partial pressure. In this text the most common form (i.e. ppm) will be used. For more details see Appendix 1 section A.7.

9.2.2 Recommended CO_2 concentrations

The recommended CO_2 concentration depends on the species and the variety, the climate conditions (especially PAR and leaf temperature), as well as economic reasons such as the price of CO_2 and benefits of its use. For vegetables, it has been recommended not to exceed 1500 ppm for cucumber, or 1000 ppm for tomato and pepper (Urban, 1997a). Nowadays, 1000 ppm is considered a suitable maximum limit for all the species, except for cucumber, aubergine and gerbera (Hanan, 1998). In aubergine, 700 ppm must not be exceeded (Nederhoff, 1984). The excess of CO_2 in tomato plants may cause abnormally short leaves or the rolling of the leaves, whereas in other crops it may cause leaf chlorosis (Langhams and Tibbitts, 1997).

An adaptation of the plants to high CO_2 is possible, so that the increases in CO_2 levels do not generate yield increases. This, in turn, may be caused by an accumulation of starch, which would limit photosynthesis, or by an increase in the thickness of the leaf (Hanan, 1998). When conditions occur which induce high photosynthesis rates (high radiation together with high CO_2 levels, for instance; see Chapter 6), sometimes an induced photosynthetic inhibition may occur (feedback effect), due to an accumulation of photosynthates (complex sugars such as starch), as they are produced at a higher rate than that with which they are transported or exported to other organs of the plant. The phenomenon is complex and induces the Calvin cycle to stop (Lambers *et al.*, 1998).

In a greenhouse, under normal stagnant conditions it is very difficult to have a difference of CO_2 content between the external air and the greenhouse air greater than 600 ppm (Seeman, 1974). In practice, CO_2 levels that reach close to 1000 ppm is only feasible (at an affordable cost) with closed greenhouses.

Care must be taken with problems derived from the pollution caused by the supply of toxic gases, depending on the source of CO_2. From the point of view of workers security the maximum limit is 5000 ppm (i.e. 0.5% of CO_2 in the air) (Hicklenton, 1988).

Ventilation is the most economic method to limit CO_2 depletion in the greenhouse air, but it only allows the maximum to reach levels close to those of the external air (350 ppm). Besides, in many cases ventilating is not desirable, for other reasons. Therefore, artificial enrichment is a usual practice.

Maintaining high CO_2 levels involves closing the vents to avoid leakage, which may induce excess temperatures, in some cases. Thus, under Mediterranean conditions, a usual strategy is to maintain levels of 350 ppm by injection and stopping the injection when the vents must be kept open to limit the thermal excesses; when vents are closed, the CO_2 level is increased to 600–700 ppm (Lorenzo *et al.*, 1997c; Sánchez-Guerrero *et al.*, 1998; Segura *et al.*, 2001).

9.2.3 CO_2 enrichment techniques

There are two main CO_2 supply sources: (i) supply in the form of pure gas; and (ii) supply of CO_2 generated by burning organic substances. The generation of CO_2 by decomposition of organic matter, which was important in the past and is not used nowadays, may be considered as a form of that produced by combustion. Carbonic ice (solid CO_2) has only been used in the laboratory, due to its high cost and because it notably decreases the temperature.

Soil mulching was used in the past to increase CO_2 levels, activating the decomposition of the organic matter of the soil (Levanon *et al.*, 1986). In a sand-mulched soil, just after the supply of organic matter, the decomposition of the organic matter during the first months may involve a relevant supply of CO_2 to the crop.

Pure CO_2

This is the ideal method, as it can be applied at any time and in any desired amount, only being limited by the capacity of the equipment. Unfortunately, the cost of pure CO_2 is much higher than that from other sources. In the gas phase CO_2 is colourless, odourless and incombustible. It is heavier than the air (density 1.52 kg m^{-3}, at normal temperature and pressure).

It is supplied in small bottles or from a central tank with a distribution network. The CO_2 is stored as a liquid at low temperature and under pressure.

When exact control of the input as a function of the CO_2 levels in the air is not possible, some authors recommend to add 5.6 g m^{-2} h^{-1} (Hicklenton, 1988).

Combustion gases

The gases must be devoid of harmful components, so the fuels, such as natural gas, paraffin or propane, must have a low sulfur content.

Natural gas is the most used. It produces 1.8 kg CO_2 m^{-3} of gas, at 20°C and standard atmospheric pressure and its combustion requires 1.77 m^3 of oxygen m^{-3} of gas (equivalent to 8.5 m^3 of air at equal temperature and pressure) (ASAE, 1988). In practice, 60% surplus of air is supplied (air factor of 1.6).

In natural gas boilers the heat of the combustion gases can be recovered as well as the CO_2. Natural gas, such as propane and butane, almost don't have problems of harmful gases (NO_x and SO_2). A complete combustion must be achieved, to avoid the formation of toxic CO (carbon monoxide) and other gases such as ethylene and propylene. It is very important (and convenient) to install a CO analyser/monitor. When the time of heating and CO_2 supply do not coincide, the hot water is stored.

Other devices used are CO_2 generators that produce CO_2 by combustion (these are in fact open-flame natural gas burners). Their cost is low and they only produce a little heat but they increase the RH of the air (Photo 9.2). As a guide, propane supplies three volumes of CO_2 and four volumes of water vapour per volume of burned gas; butane supplies four volumes of CO_2 and five volumes of water vapour per volume of gas burned; whereas natural gas supplies one volume of CO_2 and two volumes of water per volume of natural gas burned (ASAE, 2002).

Biogas coming from fermentation generates CO_2 which can be used for carbon enrichment. If the biogas comes from anaerobic fermentation, combustion would be used first to remove the methane (Urban, 1997a).

Enrichment with small burners

The main use of burners that send the combustion gases directly inside the greenhouse may be the supply of CO_2 or the supply of CO_2 and heating simultaneously. In the first case, the equipment has lower capacity.

In relation to the air supply, there are two types of burners: (i) without a fan; and (ii) with a fan. In all of them, the air supply is critical for good combustion. Besides, if the burner is used often and there is no renewal of the air (ventilation) the lower oxygen level may result in incomplete combustion. Therefore, it is usual to increase

Photo 9.2. CO_2 generator that produces CO_2 by combustion.

the air supply (by a factor of 1.6) to avoid incomplete combustion. The supply of external air to the burner (by means of a fan) avoids these problems.

The use of small burners is usually imprecise. If they are used just to heat the air, they may produce very high CO_2 levels, which are undesirable sometimes; whereas if there is no need to heat, their use involves an undesired thermal supply.

CO_2 enrichment from a central boiler

In a centralized heating system, the combustion gases originating in the boiler can be used for CO_2 supply, if the gases are pure enough. The advantage is that the supply of CO_2 and the heating can be done separately in the greenhouse. As the boiler is larger, a better control of the combustion is possible. Besides, the water vapour contained in the combustion gases can be extracted, avoiding their entrance in the greenhouse.

When CO_2 is produced by combustion in a central boiler, the destination of the heat can be: (i) to use it directly for heating; (ii) to dissipate it inside the greenhouse to get rid of it (maintaining a low pipe temperature); and (iii) to store it during the daytime and use it for heating at night. Dissipating it to the exterior (air, underground water) is not advisable due to the environmental impact.

9.2.4 Distribution of CO_2

It is important to achieve a homogeneous distribution of CO_2 in the whole greenhouse, without differences between the beginning and end of the supply lines (horizontal gradient), and also avoiding vertical gradients (low concentration inside the canopy). The distribution ducts may be located inside the crop, laying over the soil, so the enriched air crosses the canopy before it reaches the roof vents.

There are two main methods for the distribution of pure CO_2. In the first instance, the liquid CO_2 is evaporated by means of specialized equipment and then forced by its own pressure through a distribution network of pipes and delivered through perforated PE tubes along or under (when raised gutters are used) the rows of plants; this method is popular in northern countries (Nederhoff, 1995). The second method is to inject CO_2 in the airflow of a fan, which is connected with a pipeline of large-diameter perforated air-circulation

PE tubes; this method is popular in southern warm countries, most notably along the Mediterranean.

A special form of distribution is to dissolve CO_2 in the irrigation water (0.6–0.8 g CO_2 l^{-1}) by the system called 'carborain'. It is admitted that this method does not significantly improve photosynthesis, but has other positive effects on root growth and nutrient absorption (Nederhoff, 1995).

The CO_2 that comes from combustion gases requires a proper transport pipeline with aluminium pipes if the temperature is high or PVC if it is low. The distribution takes place through PE pipes. A network of PE film ducts (of 50 mm diameter) drilled with holes (of 1 mm diameter) every 20–120 cm is a usual solution, avoiding ducts longer than 40 m and using a recommended pressure at the beginning of the duct of 750 Pa, by means of a fan (Hicklenton, 1988). The holes in the distribution ducts must be more frequent near the ends than near the beginning.

The CO_2 must be injected near the plants. In a greenhouse with crops in paired rows, one small-diameter perforated PE tube is placed for each double row, either on the soil, or under the raised gutter (when available, in the case of soilless culture).

9.2.5 CO_2 balance

The CO_2 balance depends on: (i) the CO_2 supplied; (ii) the CO_2 exchanged with the external air; (iii) the CO_2 assimilated in photosynthesis; and (iv) the CO_2 originating in organic matter. The last item is usually neglected.

The photosynthesis rate ranges from 1 g m^{-2} h^{-1} of CO_2, or less under cloudy weather, to 4–5 g m^{-2} h^{-1} under good light and CO_2 conditions, sometimes even reaching 7 g m^{-2} h^{-1} (Nederhoff, 1984).

As a rule of thumb, a minimum supply of 4.5 g m^{-2} h^{-1} CO_2 is recommended, or its equivalent as natural gas combustion gases (Van Berkel and Verveer, 1984), to maintain high levels (up to 1000 vpm CO_2) in closed greenhouses and to avoid important CO_2 depletions in ventilated greenhouses. For economic reasons, the supply must not be greater than 4.5 g m^{-2} h^{-1} of CO_2.

9.2.6 CO_2 control

Under good photosynthesis conditions, the consumption of CO_2 ranges from 3 to 4 g CO_2 m^{-2} h^{-1} (Bordes, 1992), but an important amount is lost by leakage or by ventilation, which may be as high as 75%; the average consumption in sophisticated greenhouses, being estimated at 8–13 kg CO_2 m^{-2} $year^{-1}$ (Baille, 1999).

CO_2 may be supplied from dawn until dusk, but in many cases it is usually limited for economic reasons to the hours around noon. In winter, with low PAR levels (up to 100 W m^{-2}) it has been recommended not to exceed 2 g CO_2 m^{-2} h^{-1}, whereas during the spring and summer, with PAR levels from 100 to 400 W m^{-2}, supplies of 2–8 g CO_2 m^{-2} h^{-1} are recommended (Chaux and Foury, 1994b), although in each case, the economic reasons to fix the supply must prevail.

If the CO_2 is free of pollutants, the CO_2 levels do not cause problems between 1000 and 2000 ppm (0.1–0.2 kPa) (Langhams and Tibbitts, 1997).

Under normal crop conditions in a low-tech greenhouse, measurements of CO_2 depletion in the air have been recorded of up to 37% (Sánchez-Guerrero *et al.*, 1998, 2005, 2008), reaching values of 55% under extreme conditions (Lorenzo *et al.*, 1997c). The increases in yield due to CO_2 supply (700 ppm with vents closed, 350 with vents open) have been of the order of 19–25% in cucumber and from 10 to 15% in green bean (Lorenzo *et al.*, 1997c; Sánchez-Guerrero *et al.*, 1998, 2005, 2008), supplying the CO_2 from the beginning of the morning until 1 or 2 h after noon.

The combined effect of heating and CO_2 enrichment has allowed for yield increases in cucumber in an autumn–winter cycle of the order of 50% (Sánchez-Guerrero *et al.*, 2001), duplicating the increases obtained by the use of heating alone. In green bean, the increases are smaller (Lorenzo *et al.*, 1997c). Nevertheless, the most economic use of both techniques may involve different yield increases, depending on the different heating and CO_2 enrichment set points used, and the associated costs of the inputs in relation to the corresponding values of the produced yields.

Nowadays, in sophisticated facilities, it is possible to carry out dynamic control of the CO_2 supply. The CO_2 enrichment threshold varies depending on the heat demand conditions, radiation, wind speed and opening of the vents and this is represented in Plate 20 (Nederhoff, 1995). A high CO_2 set point (line A in Plate 20) is used if the heating is independent of ventilation and radiation. Line B is taken as the set point when heating is not required and radiation exceeds a certain threshold. If the radiation is lower than a pre-fixed value and heating is not used the CO_2 set point follows line C. A last option for a CO_2 set point is chosen when the greenhouse is highly ventilated (from a certain percentage, the 'Min.' line). How far the vents are open (the percentage), which is determined by the wind speed among other parameters, influences the CO_2 thresholds (Plate 20); an increase in the wind speed moves lines B and C towards the left.

The grower can maintain a higher CO_2 level when using carbon enrichment by avoiding ventilation but it does not seem advisable.

9.3 Light

9.3.1 Introduction

Light regulation is practised in a greenhouse for the following reasons: (i) to alter the length of daylight hours (increasing or reducing them); (ii) to interrupt the darkness at night (briefly, for regulation of photoperiod); (iii) to extend or reduce the dark period of the night using artificial light or darkening screens; (iv) to increase photosynthesis (complementing the naturally available light and/or extending the length of the day with artificial light, Photo 9.3); and (v) to decrease the light intensity (e.g. with shading screens when the air temperature gets too high).

The objective is to maximize photosynthesis by maximizing the light interception (PAR) by the greenhouse, which involves optimizing its design and orientation. In order to make the radiation useful for photosynthesis it must be intercepted by the crop, which will require the crop rows to be appropriately orientated (north–south) and a proper arrangement and density of the plants (lower in winter than in high radiation seasons), depending on the species, cultivar and crop conditions.

Photo 9.3. Artificial light is used in areas where there is a deficit of solar radiation.

Under normal conditions, the LAI (leaf area index) is an indicator of the light interception (see Chapter 6). During the first stages of the crop a high plant density allows for better light interception, so early production will increase (in relation to a normal density). Once the crop covers all the available space the plant density is less relevant. A high planting density involves a decrease in the quality of the product in most species, and beyond a certain threshold a decrease in yield, when expressed on a per unit area basis.

When the solar radiation is insufficient, it may be complemented with artificial light, to increase the PAR level above the radiation compensation point and maintain an active growth. The positive effect of an increase of the PAR on the growth is more relevant at low PAR levels (Hanan, 1998). Artificial light may also be used to extend the period of photosynthetic activity in the winter season. In vegetable cultivation in the Mediterranean area, artificial light is not economically feasible in most cases. The vegetables usually grown in greenhouses are insensible to photoperiod, under normal conditions, but tomato, as an example, will become chlorotic when the day length exceeds 18 h.

Whitening the greenhouse cover is efficient at decreasing excess temperatures in the high radiation season but it notably limits radiation, which involves a decrease in the potential yield (Morales *et al.*, 2000). It is preferable to improve the ventilation system to limit thermal excesses. The use of permanent shading screens has similar effects to whitening (Pérez Parra *et al.*, 2003c). The use of mobile screens, which are deployed only during the hours around midday when the radiation is too high (and so is the temperature), is another option of possible interest.

9.3.2 Light increase

Inside the greenhouse, various techniques have been used to improve the availability of light to the crop, such as: (i) painting the greenhouse structural components white; (ii) applying a white plastic film as soil mulch (Hernandez *et al.*, 2001); and (iii) in general making extensive use of other light-reflecting materials.

A usual practice is to use reflecting walls, such as in the lean-to greenhouses in China (Fig. 7.5). Several reflection devices have been proposed to increase radiation, but they are usually uneconomic. However, the reflectors perform well with direct light and not with diffuse light, and unfortunately the highest interest for increasing the light availability is in the winter months when diffuse radiation prevails (Hanan, 1998).

Artificial light is the most reliable and effective method to increase the light availability.

9.3.3 Artificial light to increase the illumination

The use of supplementary artificial light is common in sophisticated greenhouses for high added value crops, in latitudes above 40°N in America and 50°N in Europe (Nelson, 1985). The main goals of supplementary light are to increase photosynthesis (daytime illumination) and to extend the length of the day (photoperiod) which allows for an increase of the accumulated daytime radiation (Huijs, 1995). Traditionally, its most popular use has been in cut flower crops (rose and chrysanthemum) and during the first growth stage of young plants, but in the last decade the use of artificial light has also spread to other flower species (Moe *et al.*, 1992) and, to a limited extent, to high-tech greenhouse vegetable production areas of northern countries.

The types of greenhouse lamps are: (i) incandescent; (ii) fluorescent; and (iii) high-intensity discharge lamps. The recent innovation on the use of LEDs (light emitting diodes) for lighting appears very promising, but needs further research on its use. The incandescent lamps have a very low energy efficiency in converting electricity into PAR (around 6%) emitting most of the energy in the IR range (Baille, 1993); they are used to control the photoperiod, or to complement other lamps, as they can induce a morphogenic response.

Fluorescent lamps are more efficient than the incandescent, with around 20% efficiency of conversion into PAR (Baille, 1993). They usually produce white light, although there are different types. Fluorescent lamps are used effectively for germination and during the initial stages of growth in growth chambers (or rooms) because they can be placed close to the plants, but are rarely used in greenhouses, as they are not compact and cause large shadows, limiting the daytime radiation (Hanan, 1998).

The high-intensity discharge lamps are used when high-intensity radiation levels are required. Among these types of lamps we can find the mercury, the halogen, low pressure sodium, high pressure sodium and xenon (Baille, 1993). The ones with better energy efficiency (of PAR conversion) are the halogen lamps and the sodium (low and high pressure) lamps, which reach efficiencies of 26–27% (Baille, 1993).

A primary aspect to consider is the radiation emission spectrum of the lamp, in the PAR interval, so that the emitted radiation is as similar as possible to the PAR.

When selecting lamps for complementary illumination, the characteristics of radiation emission of morphological significance to plants (red light and far red light) must be considered, besides their energy efficiency (PAR conversion) and the proximity of the emitted light spectrum with that of the PAR. The most commonly used are the high-pressure sodium lamps, with 400–450 W of power. They are usually installed so that there is one lamp for up to every 10 m^2, and this provides an installed power of 50 W m^{-2}, and a useful PAR level of 10 W m^{-2} (Urban, 1997a). These lamps are usually positioned at between 1.5 and 2 m high, in frames of 2.2 × 2.2 m and 3.2 × 3.2 m, to cover between 5 and 10 m^2 per lamp (Hanan, 1998; Photo 9.4).

The uniformity in the light distribution at plant level must be assessed and the shadows caused by the lamps limited as much as possible, with the aim of minimizing any decrease in solar radiation caused. Therefore, it is preferable to use rectangular reflectors which provide a rectangular light distribution, rather than circular reflectors which distribute the light in circular shapes hindering light

Photo 9.4. Lamps for complementary artificial illumination.

uniformity (Hanan, 1998). Specialized computer software is available to assist in the optimization of the positioning of the electric lamps across the greenhouse area and above the plants.

The illumination system is usually switched on when the natural PAR radiation level is below 10 W m^{-2} (Baille, 1999) or 15 W m^{-2} (Kamp and Timmerman, 1996), extending the day length up to a total of 12–16 h, but not exceeding 18 h as this would be harmful for some crops, although some species such as lettuce will benefit from constant illumination (Nelson, 1985). However, constant illumination may cause leaf abscission in some species, and is counterproductive in tomato (Hanan, 1998). The recommended illumination levels in vegetables range between 12 and 24 W m^{-2} PAR in cucumber, pepper and tomato, and between 12 and 48 W m^{-2} PAR for aubergine and lettuce (Hanan, 1998). When the PAR is very low, the artificial light may double the photosynthesis rate; as for example when increasing PAR from 15 W m^{-2} (without complementary light) to 30 W m^{-2} (Kamp and Timmerman, 1996).

Pioneering work on the effects/benefits of artificial light on greenhouse crops (first on flowers and then on vegetable crops) that started in Canada and the Scandinavian countries in the last couple of decades of the 20th century (Tsujita, 1977; Blain *et al.*, 1987; Gislerød *et al.*, 1989) has now spread to many countries. Many new findings have been reported in proceedings of frequent international scientific symposia in the first decade of the 21st century (Dorais, 2002, 2011; Moe, 2006; De Pascale *et al.*, 2008).

The supply of power for illumination is usually expensive, so it is frequent to resort to cogeneration (simultaneous production of heat, used for heating, and electricity), which is more efficient and cheaper, although it requires a higher initial investment. Using inter-lighting, instead of lights only on top of the crop, and LEDs can substantially increase light and energy efficiency (Heuvelink and González-Real, 2008; Montero *et al.*, 2010).

The use of artificial illumination has not spread in the south of Europe. Artificial light is more used in the north of Europe, where, in the proximity of the cities, its use is forbidden in the early morning and at night for environmental reasons (including a potential alteration of the biological rhythms of humans). If its use is restricted it becomes difficult to make it profitable.

In order to lengthen the useful life of the lamps, they must not be frequently switched on and off. It is recommended to keep them on for at least 20 min, in the case of photosynthesis lamps, and switching them off for 10–15 min, before switching them on again (Van Meurs, 1995).

The lamps return a great deal of the energy consumed as heat (in the order of 75%), which decreases the heating requirement, an aspect to consider in the management of heating systems.

The lamps must be protected from the fog systems. The profitability of the supplementary illumination depends highly on the ability of the grower to optimize the growing conditions, with the aim of avoiding other factors limiting productivity.

9.3.4 Partial light reduction

There are many methods to decrease the solar radiation in greenhouses. The most simple is to whitewash (i.e. spray the outside with white paint) the greenhouse, but this does not allow for easy reversal of the shading effect (Photo 9.5).

The usual objective is to reduce the radiation to limit high temperatures in the high radiation season. But if this radiation reduction involves a PAR reduction, as normally happens with conventional shading screens or with whitewash, it generates a decrease in growth and yield. Therefore, whenever possible, it is preferable not to shade but to increase the capacity of the cooling system, which in most cases means more and better ventilation.

It may be that, due to fruit quality problems (for instance, sun scald in tomato or

Photo 9.5. Whitening of the greenhouse cover.

pepper), we wish to shade, in which case a PAR decrease must be avoided using mobile shading screens (Hanan, 1998).

The main problem of the screens is that they are a nuisance when they are folded and limit the light transmission. There are several types of rails, folding (by hanging wire or by rails) and rolling equipment. They can be placed inside or above the greenhouse, and must be mobile so they can be deployed to move the screens when the light levels are low.

If they are used only for shading, textiles with white or aluminium bands and with holes for air exchange are commonly used. Their light reduction usually ranges from 20 to 80%. The light distribution may be unequal, but it is better in the screens without aluminium bands. If they are also used for energy-saving purposes, lower porosity screens must be used that still have openings to allow the air to pass to avoid high humidity levels.

The use of coloured screens alters the quality of the light (Mortensen and Roe, 1992). The use of light diffusing additives in the plastic films decreases direct light while increasing diffuse light.

For more details on screens, see Chapter 4.

9.3.5 Control of the duration of day/night

In greenhouses the duration of the day can be manipulated, altering the photoperiod, to control the flowering of some ornamental species, by means of their action on phytochrome.

In relation to photoperiod, there are two categories regarding the response of plants to the alternation of day (light) and night (dark): (i) short-day plants, which flower or accelerate their flowering when the duration of the day is shorter than its critical photoperiod, normally less than 12 h; and (ii) long-day plants, in which flower induction occurs only when the day length is longer than its critical photoperiod, usually more than 12.5 h. In some species of long-day plants the certain threshold in the 24 h cycle is related to the accumulated radiation. Those plants whose flowering is not dependent on the photoperiod are the day-neutral plants. The response mechanism to photoperiod is based on the existence in the plants of a sort of 24 h clock, marking the so-called circadian rhythms (Hanan, 1998).

In assessing the duration of the day and the night, the plant's biological clock may distinguish a 5-min difference in a 24-h cycle (Nelson, 1985). The signal of passing from day to night is the decrease in light intensity. For the majority of the plants, the night corresponds to a radiation intensity lower than 0.05–0.1 W m^{-2} (Hart, 1988).

Darkening screens

To achieve long nights darkening screens are usually placed over the crops, to obtain short daylight conditions (Photo 9.6). The solar radiation transmission in these screens must be less than 0.1%, which can be achieved with black PE films or with black textiles (Bakker *et al.*, 1995).

The shading screens must be dark enough as to avoid light intensities greater than 0.1 W m^{-2} PAR (see Table 2.2). In order to limit excess warming of these screens, materials that are reflective (aluminized or white in the upper part and black in the lower part) and also slightly permeable to air must be used. They must be deployed after noon.

The typical long-night treatment consists of deploying the darkening screen for 12 or 14 h, to limit the light below 0.1 W m^{-2} PAR (Nelson, 1985).

Photoperiodic artificial light

With the aim of altering the photoperiod of sensitive plants the length of the day can be extended or the duration of the night interrupted with low intensity photoperiodic illumination, typically of 0.4 W m^{-2}, obtained normally with incandescent lamps or sometimes with fluorescent lamps (Urban, 1997a). Nelson (1985) estimated a minimum light threshold to achieve this was 0.5 W m^{-2} PAR.

The photoperiodic illumination is used cyclically: for instance, in chrysanthemum, switching on from 6 to 10 min every half an hour, between 22 p.m. and 3 a.m. (Urban, 1997a), which allows the greenhouse to be divided into sectors (between three and five sectors of sequential and consecutive lighting) decreasing installation costs. The radiative supply of the incandescent lamps in the PAR range is low (6–12% of consumed energy), and the rest is transformed into heat. The light produced by incandescent lamps is approximately of the same wavelengths as solar radiation, although its distribution is different (Seeman, 1974), but a high proportion is in the red light range, which is

Photo 9.6. Low-cost greenhouse equipped with a darkening screen and source of illumination for photoperiodic control of a chrysanthemum crop.

required by phytochrome, making it suitable for its use in photoperiodic control (Nelson, 1985).

9.4 Summary

- The air movement in the greenhouse, besides avoiding temperature stratification, is of great importance to the crop, affecting photosynthesis, transpiration and water use, and therefore, growth and yield. An absence of air movement has a negative effect on crop production.
- The optimum values of air velocity in greenhouses are of the order of 0.5–0.7 m s^{-1}. To achieve such values, fans are used that move the interior air, with flows of 0.01 m^3 s^{-1} per square metre of greenhouse.
- Increases in the CO_2 air content generate an increase in photosynthesis, with a subsequent increase in yield, whose value depends of the CO_2 level and the climate conditions. Anticipated yield increases are higher with the joint use of CO_2 enrichment and heating. Nevertheless, the profitability in their use must be determined by the specific management of the CO_2 enrichment, in each case.
- In low-tech greenhouses, notable depletions of CO_2 are observed, in high radiation conditions, frequently exceeding 20–30%.
- Ventilation is the cheapest method to limit CO_2 depletion (due to the plants' use of CO_2 for photosynthesis) in the greenhouse air, below the normal levels (350 ppm).
- Nowadays, the maximum appropriate level of CO_2 in the air, in practice, is 1000 ppm for the majority of crops.
- In Mediterranean greenhouses, the most efficient carbon enrichment strategy is to maintain levels of 350 ppm of CO_2, by injection, when the vents are open and, when closing the vents, to raise the level to 600–700 ppm.
- Greenhouse CO_2 enrichment is usually done by means of pure CO_2 injection or CO_2 produced by combustion. The generation of CO_2 by decomposition of organic matter, used in the past, is now very rare. The supply of pure CO_2 is the most expensive method.
- CO_2 produced by combustion must be free of harmful gases, so the fuels used must be 'clean', especially with a low content of sulfur. The combustion must be complete, to avoid the formation of other harmful gases (carbon monoxide, ethylene).
- The gases most commonly used to generate CO_2 are natural gas and the LPGs (propane and butane). In heating boilers that use natural gas, the combustion gases of the boiler are normally used as a source of CO_2.
- CO_2 supplies to the greenhouse, for economic reasons, do not normally exceed 4.5 g m^{-2} h^{-1} of CO_2 during working hours.
- Greenhouse light regulation allows for altering the length of the day or to interrupt the duration of the night (by use of darkening screens or artificial light) as well as to achieve higher light levels to increase photosynthesis, complementing natural light and lengthening the duration of the day with artificial light. The use of shading systems allows for a reduction of the intensity of solar radiation.
- The increases in the light available for the crop can be obtained by means of cheap techniques, such as the use of reflecting materials (white mulch and others) or painting the surface of the structural elements white. Normally artificial supplementary light is not used under Mediterranean conditions.
- The partial reduction of solar radiation is practised, mainly, as a means to limit high temperatures in poorly ventilated greenhouses. It is normally done by means of whitening or using shading screens. The use of permanent shading (whitening or permanent screens) generates a continuous decrease of

radiation, which involves a yield reduction. Therefore, it is preferable to use mobile screens.

- The control of the duration of day/night, to regulate photoperiod, is done by means of darkening screens, which allow for shortening the length of the day, and with photoperiodic artificial light (of low intensity), which interrupts the length of the night. Most edible vegetables are insensitive to photoperiod (see Chapter 6).

10

The Root Medium: Soil and Substrates

10.1 Introduction

Normally the location of the greenhouses is based, primarily, on the local climatic conditions, with little relevance to the soil qualities of the chosen location. If the soil was not appropriate, an extreme remedy would be to create an artificial agricultural soil, a frequent situation in some areas of south-east of Spain and some parts of the Canary Islands, where the crop profitability allowed for such an intervention.

The problem of soil exhaustion, in many cases due to monoculture, decades ago induced the system of greenhouse 'rotation', displacing them by means of special devices (Photo 10.1) or simply changing their location (usual in tunnel-type greenhouses, in the north of Africa).

A later innovation was the use of inert substrates and the development of soilless crops, which insulate the root media from the soil, notably limiting soil-borne disease problems.

The water and nutrient reserves of a good horticultural soil are higher than those of a substrate, and so the margin of error is much smaller when growing in substrate.

10.2 Desirable Characteristics of Horticultural Soils

10.2.1 Physical and hydraulic characteristics

Although modern high-frequency irrigation techniques notably limit the basic function of soil being a water and nutrient store, a loamy or loamy-sandy textured soil would be most appropriate for horticultural crops, ideally composed of around 50–60% sand and well supplied with organic matter. A texture of this type with no gravel, stones and boulders and that is well balanced chemically, normally has good hydraulic and chemical characteristics (see Chapter 11). A proper structure provides for good porosity (basic for the root aeration) contributing to a balanced permeability which is so necessary in protected crops.

An aspect that is often neglected is the drainage conditions of the greenhouse soil. If necessary, the soil should be provided with an artificial drainage network.

10.2.2 Chemical characteristics

A proper cation exchange capacity (CEC), a balanced pH (from 6.0 to 7.5, if possible)

Photo 10.1. Decades ago, in certain cases in England, to avoid soil-borne problems greenhouses were moved from one location to another, in a peculiar form of crop-rotation system.

and the absence of salinity or alkalinity problems (electric conductivity of the saturated extract (EC_e) lower than 4 dS m^{-1} with an exchangeable sodium percentage (ESP) lower than 10) would be desirable generic qualities, but these are not always achievable in a horticultural soil used for protected cultivation.

The soil depth in horticultural crops is usually not as limiting as in other crops, due to their smaller root development, provided that good drainage is available. The soil depth must be at least 30–40 cm.

The higher the CEC, the higher will be the buffering capacity of the soil, limiting the risks of fertilization errors. The organic matter content is very closely linked to the climatic conditions, which determine the speed of mineralization, and which is very relevant to the important nitrogen supplies it produces. In Mediterranean climates, organic matter decomposes very quickly and it becomes a source of nitrogen losses.

As the soil is the source of the essential nutrients for the plants (N, P, K, Ca, Mg, S and micronutrients) it is necessary to know their concentration and availability to plan a proper fertilization programme.

The salinity of the soils affects plant growth because it increases the osmotic potential of the soil solution, and for horticultural cultivation it is advisable to discard soils whose EC_e exceeds the threshold of 3–4 dS m^{-1} (Dasberg, 1999a), except in extreme situations and for very tolerant crops.

10.2.3 Considerations on the management of greenhouse soils

In horticultural protected cultivation, besides the usual soil cultural practices, the biological status of the soil must be especially considered (presence of parasites and soil-borne diseases). The incorporation of organic matter in the soil (primarily to improve its physical and hydraulic characteristics) must include efforts to monitor the salinity and alkalinity conditions along with the availability of nutrients.

The use of transparent plastic mulch on a bare soil during the high radiation season (solarization) has proved to be effective in reducing the phytosanitary problems originating in soils, and, in avoiding or decreasing the use of pesticides.

High soil salinity is very common in arid regions, especially in coastal areas, where it is common that the irrigation water is also of high salinity. The decrease in the evaporation of the soil water with mulching techniques limits the capillary rise of water and salts, decreasing the problems derived from salinity. The absence of rain in greenhouses (except in shelters with perforated covers) decreases the possibilities of salt leaching. In addition, the high doses of fertilizers used can generate saline conditions.

The use of a sand layer over the soil, characteristic of the 'enarenado' technique (used in the south-east of Spain), plays the role of mulching, decreasing evaporation from the soil and allowing the use of slightly saline irrigation waters, without yield reduction.

In this 'enarenado' (sand mulching) technique manure is also applied, in quantities of up 100 t ha^{-1}, incorporating part of it into the soil and leaving the rest in a uniform layer over the soil. The addition of a sand layer, of 7–10 cm thickness, over the manure finalizes the preparation of the 'enarenado' (Castilla *et al.*, 1986).

The root development in the sand–organic matter–soil interfaces reaches high values of root density (much higher than those measured in the soil). These high densities can be of great interest for the crop, providing good aeration of the roots in soils of low permeability (Castilla *et al.*, 1986).

10.3 Soilless Cultivation

10.3.1 Introduction: systems

Soilless cultivation is the system in which the plant develops its root system in media (normally solid or liquid) confined in a limited and isolated space, away from the soil (Abad and Noguera, 1998). Nowadays, the term hydroponics which properly refers to water culture (hydroculture), is confused with all the methods and techniques to cultivate plants out of soil in artificial substrates or in well-aerated nutrient solutions (hydroculture) (Pardossi, 2003).

Hydroponic crops may be classified as: (i) a function of the type of substrate or container; (ii) a function of the method used to supply the nutrient solution (drip irrigation, sub-irrigation, circulating water, trays or floating tables or aeroponics); or (iii) a function of the use of drainage (open or free drainage systems and closed or recirculation systems) (Winsor *et al.*, 1990).

The proper hydroponic systems most commonly used are: (i) the nutrient film technique (NFT); (ii) the deep flow technique (DFT); (iii) the floating raft technique (FRT) in which cultivation takes place on floating polystyrene boards; and (iv) aeroponics. All of these are closed systems. In the NFT a thin layer of nutrient solution flows through gutters that contain the roots (Cooper, 1979). The method known as NGS (new growing system) can be considered as a variant of NFT. In the floating systems, trays made of light-weight synthetic materials float over a nutrient solution. In the aeroponics systems the plants are cultivated in perforated plastic pads, which separate the aerial part from the roots, which remain in the dark in an enclosure where nutrient solution is fogged with a very high frequency. The aeroponic systems are seldom used commercially.

However, the substrate-based systems are the ones that have received widespread use in horticulture (Plate 21); a brief description follows.

10.3.2 Advantages and disadvantages of substrate-grown crops

The main reasons for the expansion of crops grown on substrates have been to avoid soil-borne diseases (in contaminated greenhouses, mainly, by monoculture) and for the good agronomic performances of the crops with these systems (Urban, 1997b). Furthermore, with the development of substrate technologies and improvement in their management they may improve the efficiency in the use of water and nutrients (closed systems), and also contribute to a reduction in several crop cultural practices in the management of the soil (Penningsfeld and Kurzmann, 1983; Savvas and Passam, 2002).

The higher production and earliness of soilless crops is derived from improvements in the water and nutrients supply, and the good root oxygenation which results in good quality products, if properly managed (Morard, 1995).

The main disadvantages of substrate-grown crops are: (i) their higher initial cost (relative to conventional soil cultivation); (ii) the requirement for highly technical crop management by the grower; and (iii) the low buffering capacity of the systems. Because of the limited volume of the substrate involved, the limited availability of water and nutrients requires continuous monitoring, to avoid failures in the continuous water and nutrient supply, which can have disastrous results. The cost of the substrate cultivation system is, nowadays, lower than the implementation of an 'enarenado' crop.

Another negative aspect of substrate cultivation is the generation of a high volume of solution that drains from the substrate, which has to be removed, as well as the substrate waste, the recycling of which depends on the type of substrate.

10.3.3 Substrate cultivation systems

In substrate cultivation systems the nutrient solution is supplied in excess from above, by means of drippers or micro-sprinklers, or from below the substrate, so that it ascends by capillarity action. The excess supply of nutrient solution must be removed by drainage.

Another method to irrigate is by sub-irrigation, which is used in pot plants.

In relation to the positioning of the substrate in the greenhouse, it may be done in several ways among which we can highlight (see Fig. 10.1): (i) in a ditch, isolated from the soil by a plastic film, used with heavy substrates such as sand; (ii) in a gutter, laid over the soil (or sand, perlite, etc.); (iii) in slabs of rockwool or coconut fibre pre-packed in plastic film; (iv) in a plastic bag filled with substrate and laid over the soil (or perlite); and (v) in pots or containers.

The substrate must be laid out over a well-levelled surface, with a certain slope (not higher than 1%, if possible) to facilitate drainage; otherwise the drainage system may be simple or complex depending on the characteristics of the installation. The separation of the drained solution from the root zone is important to avoid possible disease infections (Fig. 10.1).

10.3.4 Characteristics of the substrates

Physical properties

Among the relevant physical characteristics of the substrate are: (i) porosity; (ii) water retention and availability; and (iii) air content.

The total porosity (or total porous space) is the total volume of the substrate that is not occupied by organic nor by mineral particles; its optimum level is greater than 85% (Abad and Noguera, 1998). It is important to distinguish between the capillary pores, which retain water, and the macropores, which allow for aeration (Bunt, 1988).

The available or useful water (see Chapter 11) in substrates, given their physical and hydraulic characteristics, lies within very narrow ranges of matrix tension. Figure 10.2 shows the water retention curve of a substrate considered as ideal. The matrix tension values of 1, 5 and 10 cb define: (i) the readily available water (RAW) contents (between 1 and 5 cb); (ii) the reserve water (RW) (between 5 and 10 cb); and (iii) the not readily available water (NRAW) (above 10 cb) of the ideal substrate (Caldevilla and Lozano, 1993).

The aeration capacity is the proportion of the substrate volume that is occupied by air, once saturated and drained, which usually corresponds to 20–30% in volume (Abad and Noguera, 1998). The height of the container or of the substrate slab has a great influence on the air content of the substrate, because the capillarity of the water dictates that with low height the air content is lower, and vice versa.

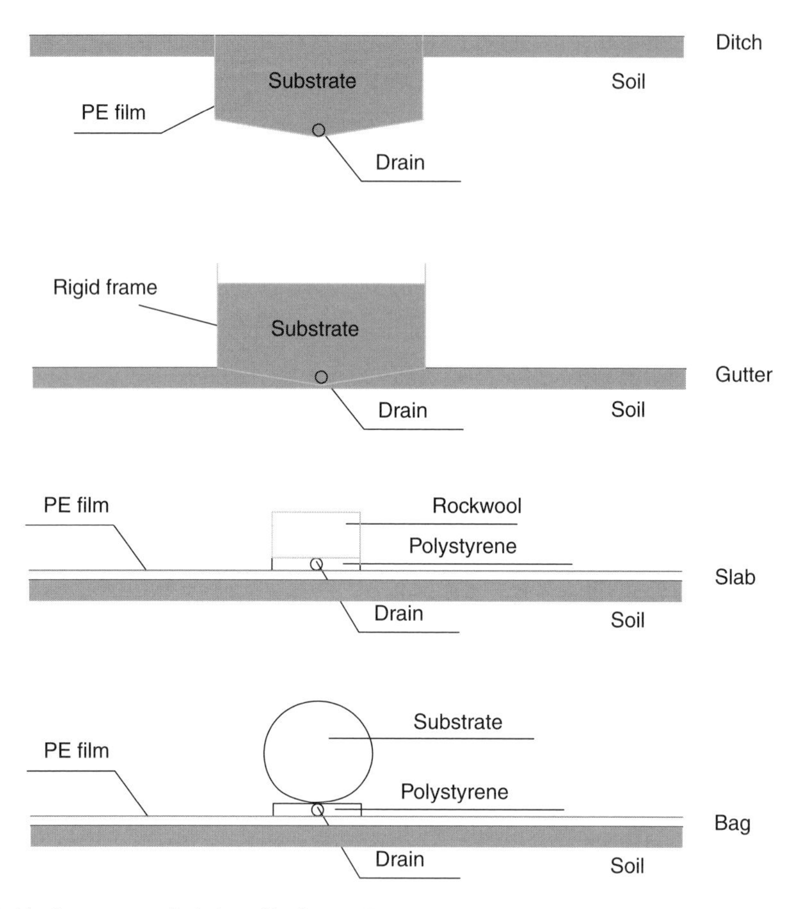

Fig. 10.1. Most common substrate cultivation systems.

The volumetric relations between different geometrical dispositions of perlite (P-2 type, grain size between 0 and 5 mm) and of rockwool are shown in Plates 22 and 23, respectively. The value of limiting the height of the substrate to maximize water content can be observed.

Other important characteristic of those substrates that are a mixture of particles, which influences the porosity, is the size distribution of the particles. A very important physical property is the hydraulic conductivity of the substrate, as it has a crucial influence on the availability of water to the crop. Also of interest are: (i) the capacity of the substrate to re-wet; (ii) its apparent density; and (iii) how much it contracts in volume (Raviv and Lieth, 2008).

Chemical properties

The ideal substrate must not only be devoid of harmful substances, especially of heavy metals, but also it must be chemically inert, which is not the case in some organic substrates.

The CEC (cation exchange capacity) defines the quantity of cations that can be fixed per unit volume or weight of the substrate. Substrates with no or very low CEC will be the most convenient. The CEC is important in organic substrates and it is advisable to saturate the substrate before its use with calcium supplies, with the aim of minimizing the CEC, so it does not affect the availability of nutrients scheduled for use in fertilization.

Some materials can be acidifying (e.g. peat), or cause a basic reaction in

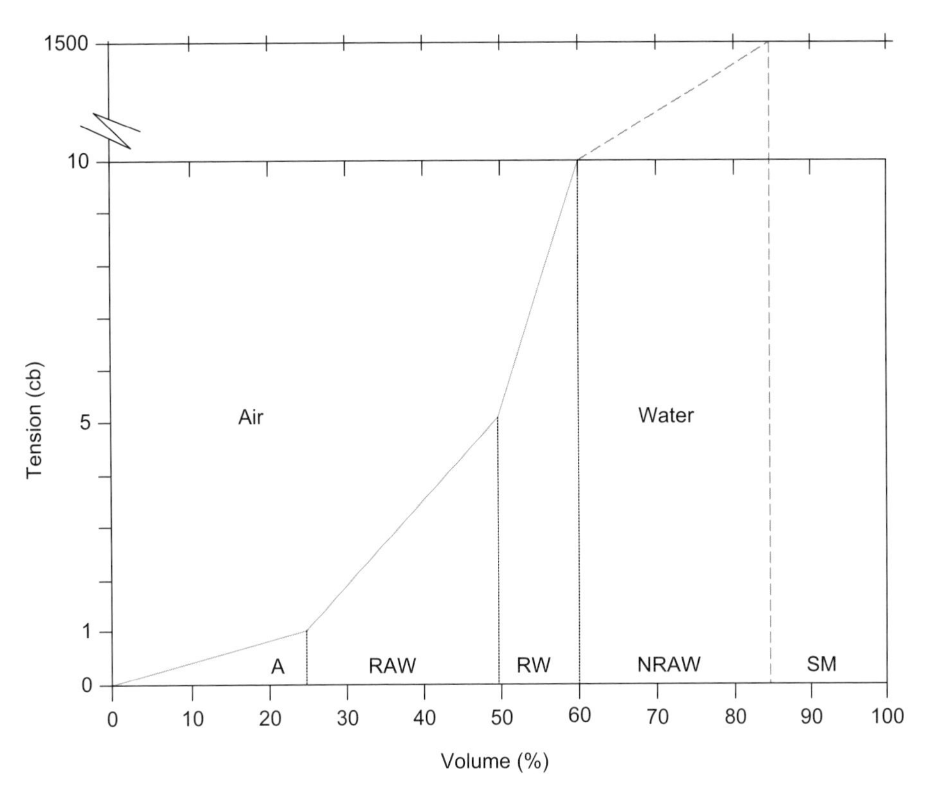

Fig. 10.2. Water retention curves in an ideal substrate (adapted from Caldevilla and Lozano, 1993). A, percentage of air (25%); RAW, percentage of readily available water; RW, percentage of reserve water; NRAW, percentage of not readily available water; SM, percentage occupied by solid matter (15%).

the solution (e.g. rockwool) at the beginning of the cultivation, which can be corrected by accordingly adjusting the pH of the supplied nutrient solution. The optimum pH level of the nutrient solution, for horticultural crops, ranges between 5.5 and 6.5 (Pardossi, 2003).

In general, a good substrate must be chemically stable, which avoids any release of elements that can generate problems of salinity or phytotoxicity, or induce undesired precipitations in the solution.

The salinity of the substrate can be altered by an imbalance between absorption (plus leaching) and supply, or by a high CEC of the substrate, so monitoring the EC of the solution is of paramount importance.

In soilless cultivation, the availability of mineral elements is essential. Therefore, the substrate that does not interfere with this availability will always be preferable.

Biological properties

Substrates of mineral origin are biologically inert (at least at the beginning), which is not the case with organic substrates, which are biodegradable and able to release ammonia, or phytotoxic substances or growth regulating substances. The use of organic substrates with high biodegradability must be avoided as they contain a lot of plant roots which eventually degrade and become a problem.

The carbon/nitrogen (C/N) ratio has been proposed as a biological stability index for organic-based substrates. A C/N ratio between 20 and 40 is considered appropriate for substrate cultivation (Abad and Noguera, 1998).

10.3.5 Types of substrate

Classification of substrates

From the point of view of their horticultural use, substrates can be classified as organic and inorganic (or mineral). The organic substrates can be of natural origin (peats) or synthetic (polyurethane foams), and also include several by-products of natural origin (sawdust, coconut fibre, cork residues). The mineral substrates can be of natural origin (sand, gravel) or artificially transformed (rockwool, perlite), including in this group several industrial by-products (blast furnace slag).

Criteria for substrate selection

Although the best substrate for cultivation will vary in each case depending on the specific use conditions, a good substrate must have good physical characteristics (with high capacity of readily available water retention, enough aeration, low apparent density, high total porosity and stability of characteristics and of the structure) as well good biological and chemical properties (scarce or no CEC, reduced salinity, slightly acid pH and biological stability) (Abad and Noguera, 1998).

In addition, obviously, it must have a cost (in which transportation is very important) that is in line with its performance. It is also important to consider its availability in each local market. Finally, it must not be forgotten that the choice of substrate must be in agreement with the technological level of the greenhouse, especially with the capabilities of the fertigation system.

Most common substrates

The most common substrates are rockwool, peat, coconut fibre (coir) and perlite.

Rockwool has excellent characteristics of water retention (Plate 23), being used in slabs of 7.5 and, sometimes, 10 cm height (usual dimensions 100 × 20 × 7.5 cm or 100 × 15 × 10 cm). For ease of raising transplants seeds are sown in rockwool cubes. Rockwool can be considered an inert substrate, with no CEC and a slightly alkaline pH (that is easily neutralized and controlled if the slabs are wetted with an acidic solution before the beginning of the crop cycle). It has a homogeneous structure and low density (which eases its transport) and good porosity (Smith, 1987).

Nowadays, it is one of the most widely used substrates in Europe, and elsewhere. A variant of rockwool is fibreglass, which has a laminar structure (milfoil type) to ease the lateral diffusion of the water.

Expanded *perlite* is a very light, very porous and well-aerated material. Several sizes of perlite are available commercially, the most popular comprising particles between 1.5 and 2.5 mm (Morard, 1995). Its main problem is its mechanical fragility which over time degrades its good porosity and aeration characteristics, as the grains break up increasing the proportion of fine elements. Although initially it has a basic reaction, such as rockwool, after some time it becomes chemically neutral.

The water retention capacity of perlite depends on its size. Plate 22 shows the volumetric relations of the water content for different geometrical dispositions of P-2 perlite (grain size of 0–5 mm). In practice, P-2 perlite bags usually have a diameter of 15–20 cm in cross-section, are 1.2 m in length and contain 40 l (Caldevilla and Lozano, 1993). For economic reasons smaller volumes are also used (33 l).

The *peats* are organic materials originating from the decomposition of swamp plants. They are usually free of pathogens, despite their organic origin, but have the inconvenience of a high CEC, the ability to greatly contract when they dry out, and being very difficult to re-wet.

The blonde peats have good physical properties and are easy to re-wet, so they are very convenient for soilless cultivation in bags. The black peats are more decomposed than the blonde types and their physical properties are inferior to those of the blonde ones. Both peats are commonly used mixed with very porous substrates, such as pozzolana or pumice (Penningsfeld and Kurzmann, 1983; Urban, 1997b).

The *barks* are light and well-aerated substrates, but retain little water. Their C/N ratio is high and they absorb a lot of N as they decompose. *Other agricultural and forest by-products* (e.g. wood waste, pressed grape rests) are used in mixes for pots.

Coconut fibre (coir) has properties closer to rockwool rather than to forest by-products, being widely used in the form of plasticized slabs and in polystyrene containers for substrate-grown horticultural crops, although its CEC complicates its management, in relation to inert materials (García and Daverede, 1994). There are important qualitative differences in the coconut fibre depending on its origin.

The use of *pozzolana (pumice) and volcanic gravels* is limited to those areas in which they are naturally found, such as in the Canary Islands, where they are known as 'picon', or in Mexico ('tesontle'). Their properties vary depending on their origin and texture.

The *sands and gravels* of siliceous origin are preferable to those of calcareous origin. As they are abundant everywhere, they were used when substrate cultivation first began to expand. Their main disadvantage is their heavy weight. The most applicable sand sizes range between 0.2 and 2.0 mm, and the best gravel varies from 2 to 5 mm (Urban, 1997b). The gross materials (gravels) require a high irrigation frequency due to their low water retention capacity. Sands and gravels are commonly used in mixtures for open field pots, because, due to their weight, they provide the pots with stability against the wind.

Vermiculite is an industrial transformation of mica, and is light, porous, well aerated and with good water retention capacity (Zuang and Musard, 1986). It is usually used in mixtures.

Expanded clay has good physical characteristics, but its low water retention capacity, which forces a higher irrigation frequency, and its high price have restricted its use to pot crops (Zuang and Musard, 1986).

Polyurethane foam is very durable, is light, inert and recyclable, but it has a low water retention capacity.

The joint selection of the substrate and the soilless growing system

The choice of the substrate determines the system to be used and vice versa, so they must be considered together.

The simplest cultivation system is the use of *ditches* (Fig. 10.1), whose large volume provides greater inertia than other systems. Scheduling of irrigation (see Chapter 11) and fertilization can be done by a timer. The best adapted substrates are sands, gravels, perlite, pozzolana and volcanic gravels and barks, alone or mixed with peat.

Gutters (Fig. 10.1) that are located over the soil or suspended have a smaller volume of substrate than the ditches, so irrigation scheduling must be more precise than with ditches, using a demand tray (see Chapter 11) or calculated as a function of the intercepted radiation. The most suitable substrate materials are perlite, bark, expanded clay and pozzolana.

In the *light bags or slabs* (Fig. 10.1) irrigation is scheduled by means of a demand tray or as a function of the accumulated radiation and the drainage, usually requiring the use of a computer to manage the fertigation. The most usual substrates are rockwool and coir (coconut fibre) for the slabs and perlite, blonde peat, expanded clay, bark and volcanic materials for the bags. Sometimes, bags and slabs are isolated placing an expanded polystyrene board below to ease the runoff of the leachate (Fig. 10.1) and, possibly, the incorporation of a root heating pipe.

Closed systems (with recirculation)

With the aim of avoiding contaminating the aquifers with the drainage solution from soilless growing systems, systems have been developed whereby the drainage water is recirculated. These are also known as closed systems, and they require good quality water. In these systems, where irrigation is coupled with drainage and recirculation of the excess solution (usually in the order of 20–30%), the environmental impact is avoided or minimized but the danger of propagating diseases with the recirculating solution increases, necessitating disinfection of the return solution (Dasberg, 1999a; Marfá, 2000). The drainage percentage is limited by the capacity of the disinfection system and by the risk of root asphyxia. The more drainage the better will be the control of salination of the recirculating solution.

When the water is of good quality it is possible to use closed systems, but if the water is of medium quality it will be advisable to use semi-closed systems (flushing out or periodically discarding the concentrated recirculating solution) and when the water is of poor quality it will be necessary to use open systems, or to clean the source water by reverse osmosis. In open systems it is essential to have good irrigation scheduling (doses and frequency), the absorption of nutrients being less relevant, whereas in closed or semi-closed systems it is the opposite (Sigrimis *et al.*, 2003).

The simplest closed system (Fig. 10.3) uses a mixing tank where the nutrient solution is added to the drainage water that is being recirculated. Then, irrigation is performed from the mixing tank. The system has, at least, one output (water and nutrients absorbed by the plants) and one input (water and nutrients injected in the mixing tank), if all of the drainage water is used.

Recirculation systems require good levelling of the land, with uniform slopes of 0.5%, and may adopt different arrangements (Fig. 10.4).

In order to prevent the propagation of diseases it is necessary to disinfect the drainage water. For this, the most usual procedures are (Kempes, 2003): (i) to heat the nutrient solution, at least, to 95°C for 30 sec; (ii) to apply UV radiation, within the range of 200–315 nm in darkness, to inactivate the pathogens; (iii) to filter the drainage water, by means of a membrane or sand beds; the majority of pathogens are eliminated with pore sizes between 0.01 and 10 μm; and (iv) chemical treatment with chloride, ozone, hydrogen peroxide or bleach.

When salts accumulate excessively in the mixing tank (Fig. 10.3), it will be necessary to discard the drainage water (i.e. to flush it out), in order to eliminate the salts.

Salinity in soilless growing systems

The total salt content in the irrigation water, expressed by the electric conductivity (EC),

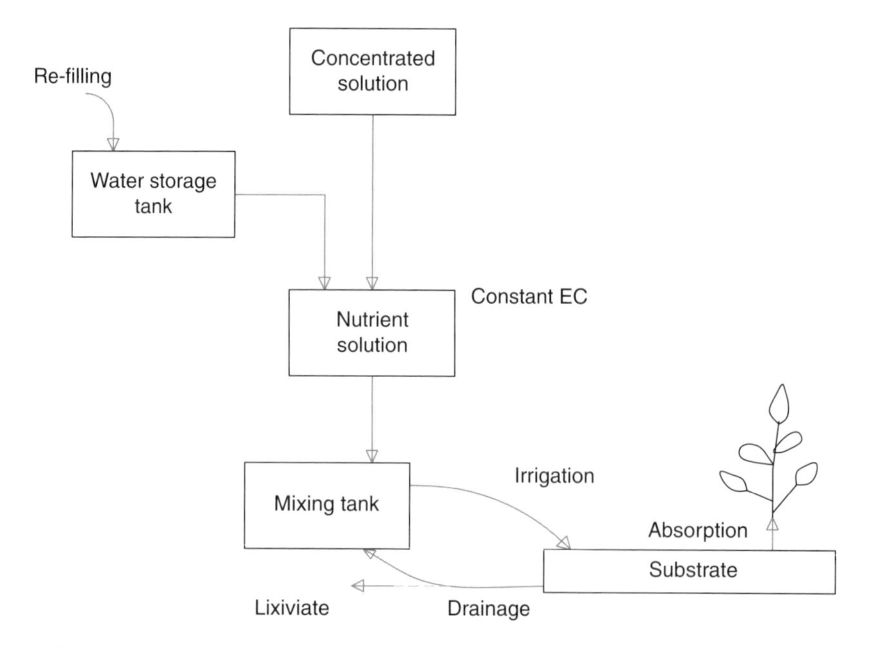

Fig. 10.3. Scheme of a simple soilless closed growing system (with recirculation).

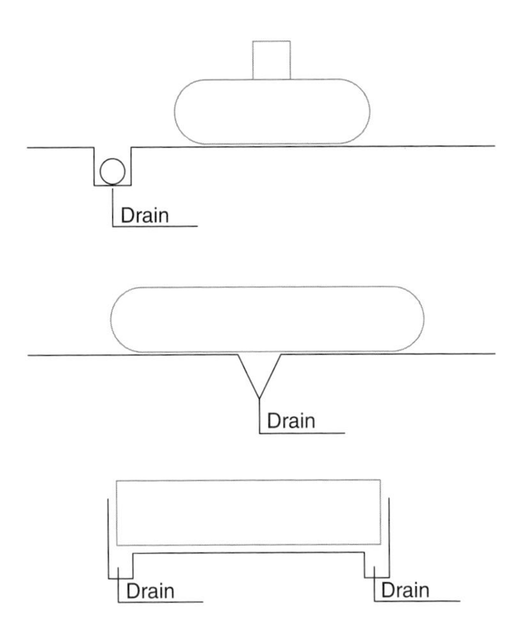

Fig. 10.4. Different arrangements for drainage collection in substrate-grown crops.

is an index of the quality of the irrigation water. The main effect of salinity in the water or the nutrient solution is osmotic, because the higher the salt content the higher is the osmotic pressure of the solution.

$$\text{Osmotic pressure (MPa)} = 0.33\ \text{EC (dS m}^{-1}\text{)} \qquad (10.1)$$

As the osmotic pressure increases, the water stress increases as the water is less available for the plants. The effects of salinity in the production, both quantitatively and qualitatively, are complex as they are influenced also by different growing conditions (Sonneveld, 1988). Other salinity effects are those influencing the nutrition as the absorption of specific ions is altered, such as the antagonism of sodium with calcium and magnesium, or generating toxicity, in the case of ions without osmotic relevance (Dorais *et al.*, 2001b; Sonneveld, 2003).

The salinity decreases the size of the fruits and the fresh weight production, because the high osmotic pressures hinder the water supply to the fruits (Ehret and Ho, 1986), although in crops like tomato it contributes to improving its organoleptic quality (Magán, 2003; Dorais *et al.*, 2008).

Salinity also limits the leaf expansion (Hsiao, 1973) and increases the incidence of blossom end rot (BER) in the fruit (Dorais *et al.*, 2001a).

In the majority of crops the minimum average concentration of the nutrient solution required is in the order of 1.5 dS m^{-1} (Sonneveld, 2003), although the production of high quality vegetable fruits may require EC levels as high as 2.5–3.0 dS m^{-1} in the nutrient solution (Pardossi, 2003), or even higher (Dorais, 2001a, b, 2008) under certain conditions. The postharvest useful life of some fruits, such as tomato or cucumber, improves if they are cultivated at high salinity (Welles *et al.*, 1992), which is not the case for pepper.

The negative effect of salinity on crops can be mitigated by limiting the transpiration rate, by means of VPD and radiation reduction (Li *et al.*, 2001), for which under Mediterranean conditions positive effects have been observed with shading (Lorenzo *et al.*, 2003) and fogging (Montero *et al.*, 2003), although this may involve an increase of BER in fruits like tomato.

10.4 Changes in the Management of the Root Medium

Traditionally, the root medium fulfilled three main functions: (i) water storage; (ii) nutrient supply; and (iii) plant support (Dasberg, 1999a). These three functions have lost relevance in modern greenhouse production systems.

The plants are not supported by the root system, but with a complex framework of threads and training nets. The nutrient and water supply is regulated more and more by the grower (and his or her computers) in substrate growing systems in accordance with the changing needs of the plants. These conditions made possible the reduction of the variability in the root medium, relative to conventional soil cultivation, but need continuous monitoring and control.

There is a clear trend towards the use of water and fertilizers using closed systems (recirculating) aimed at limiting the environmental impact of the drainage solution/water, although using such a system requires the availability of good quality water and/or high investment.

10.5 Summary

- The choice of the location of greenhouses has been based, mainly, on climatic conditions and if the soil was not suitable in the chosen location, it was improved, even creating an artificial soil.
- Although modern high frequency irrigation techniques limit the function of the soil as a water and nutrient reserve, a soil with good physical, chemical and hydraulic characteristics is always preferable.
- High salinity is a common feature of arid region soils, where many of greenhouses are located, and this is aggravated by the use of more-or-less saline water for irrigation. The absence of rain inside greenhouses decreases the possibilities of combating salinity. Some techniques like mulching or 'enarenado' (sand mulching) allow cultivation in soils with slightly saline water without compromising production.
- Soilless cultivation allows for the development of the plant roots in a medium (solid or liquid) isolated from the soil, and thus, avoiding soil-borne diseases.
- The reasons for the expansion of soilless crops, besides the prevention of the soil-borne diseases, has been the good agronomic performances derived from better control of the water supply and mineral nutrition, and good root oxygenation.
- Soilless crops, in addition to their higher installation costs, require precise monitoring to avoid failures in the water and nutrient supply, which could be disastrous given the low inertia of these systems.
- Crops grown in solid substrates are the most prevalent in greenhouses, with perlite, rockwool, sand and gravel as the most popular substrates. Other substrates used are coir (coconut fibre), peat, bark and expanded clay.
- The main physical characteristics of a substrate are porosity, water retention and availability, and air content.
- The readily available water in substrates is retained in very limited and narrow margins of matrix tension.
- A good substrate, besides having good physical characteristics and stable chemical properties (or ideally being inert), should be biologically stable and inexpensive.
- The most common substrate growing systems in Mediterranean greenhouses are ditches, gutters (supported or elevated), slabs and bags. The choice of the substrate determines the system to be used and vice versa.
- Closed systems, which recirculate the drainage water, require good quality water. These systems avoid the environmental impact of eliminating leachate, but the recirculating water must be disinfected, to prevent the propagation of diseases.
- The negative influence of salinity in the production of soilless crops may be minimized with proper climate control.
- The main traditional functions of the root media (water storage, nutrient supply and plant support) have lost relevance in modern (soilless) greenhouse production systems.

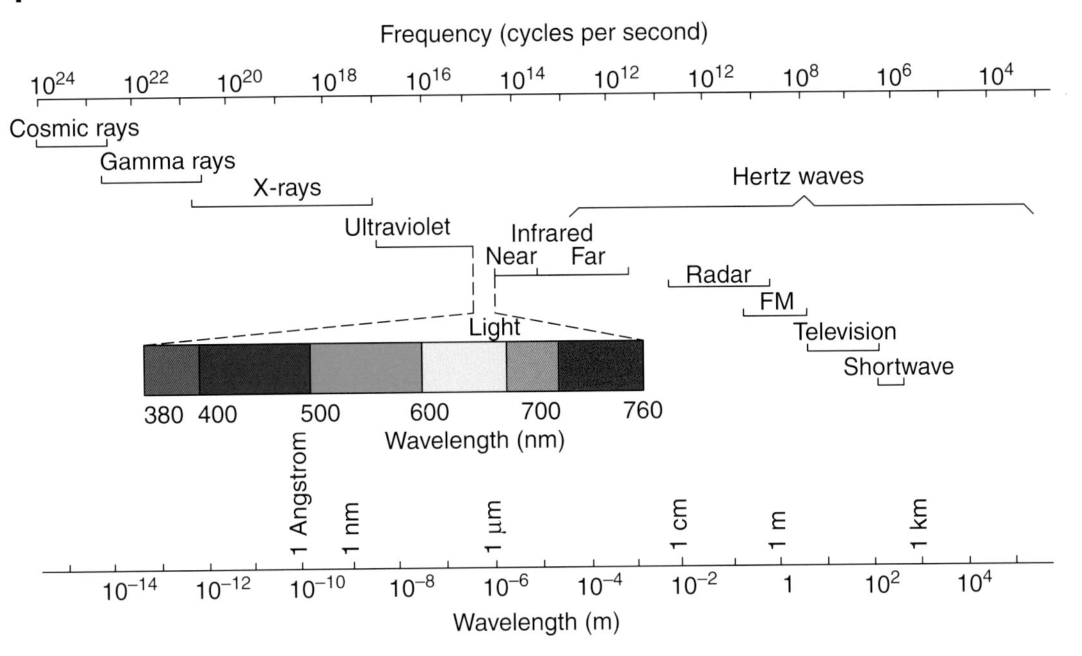

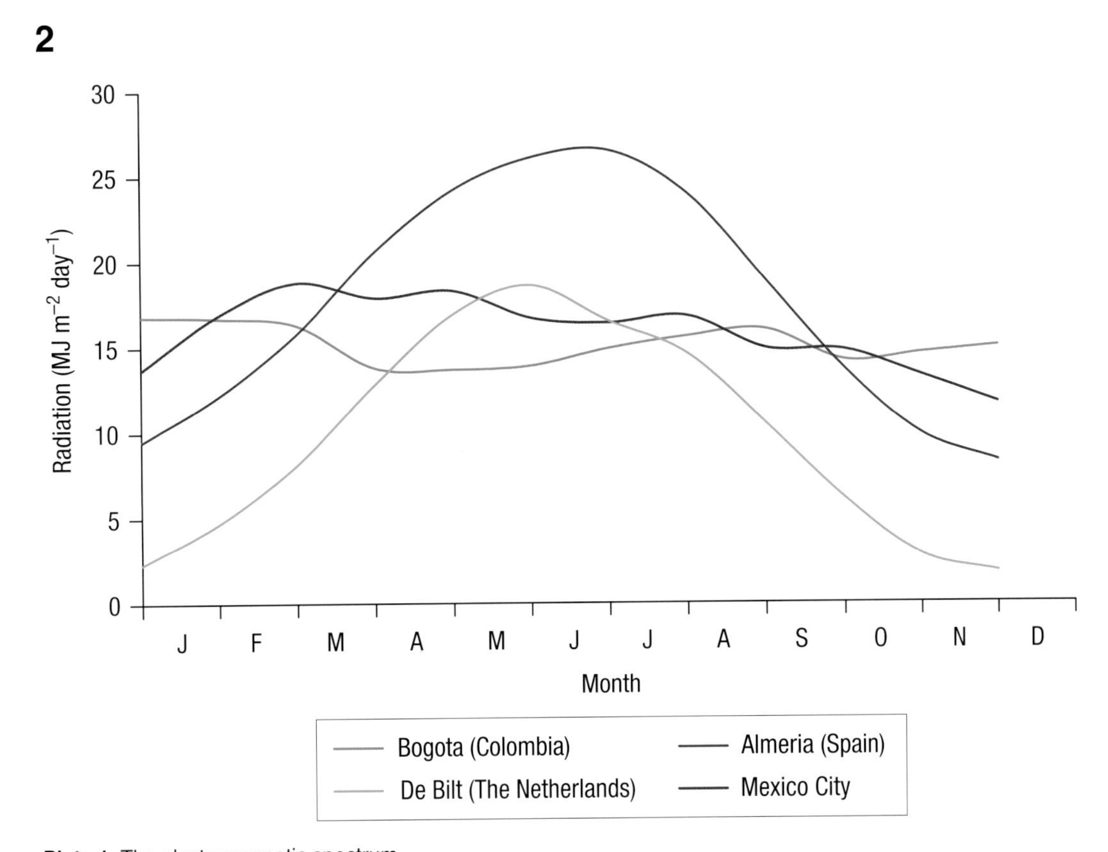

Plate 1. The electromagnetic spectrum.

Plate 2. Evolution of the total daily solar radiation in several locations through the months of the year: Almeria (Spain), Bogota (Colombia), Mexico and De Bilt (The Netherlands).

3

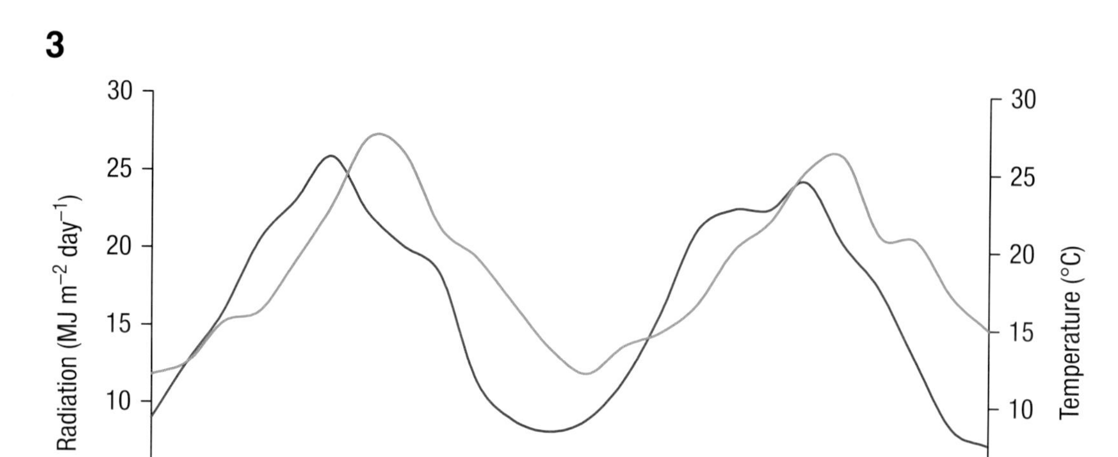

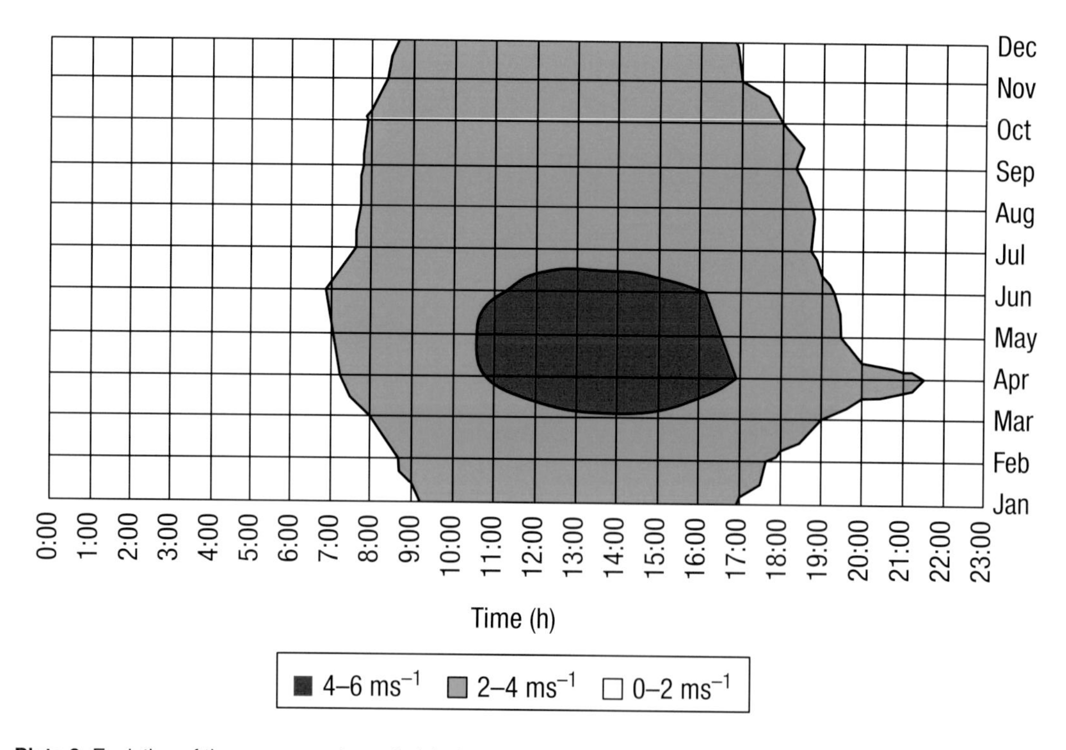

Plate 3. Evolution of the average values of global solar radiation and outside air temperature throughout the year (Almeria).

Plate 4. Daily distribution of the average wind velocity in Almeria, Spain (Experimental Station of Cajamar Foundation-Cajamar; from Pérez-Parra, 2002).

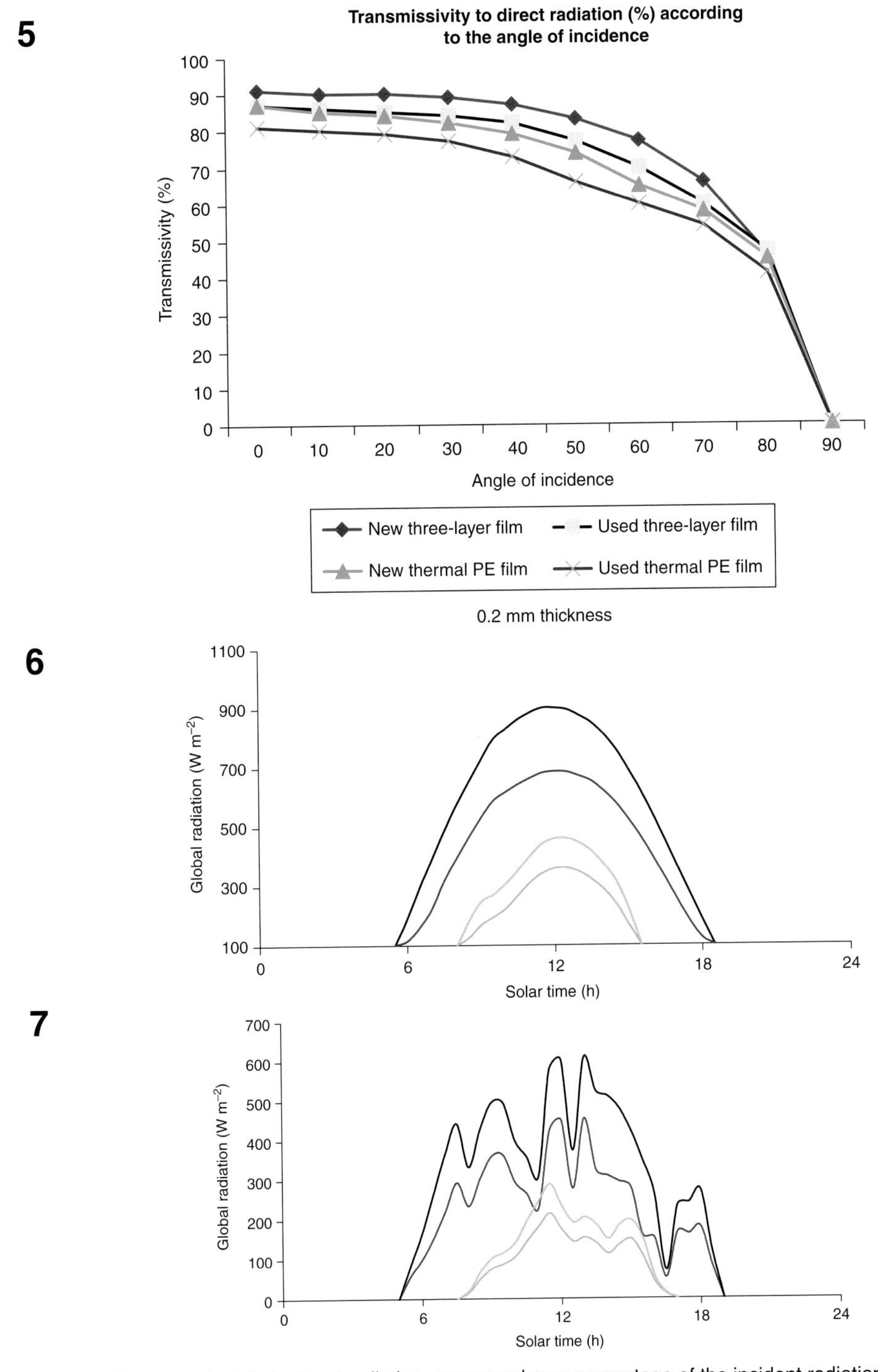

Plate 5. The transmissivity to direct radiation, expressed as a percentage of the incident radiation, decreases as the angle of incidence increases (see Fig. 3.3). The transmissivity varies depending on the characteristics of the material. The ageing of plastic films, influenced by their use, decreases transmissivity (Montero *et al.*, 2001).

Plate 6. Evolution of solar radiation intensities in the open air and in the greenhouse around the winter (21 December) and summer (21 June) solstices, on sunny days (coast of Granada, Spain). The daily total radiation in each case is the area of the surface delimited by each curve and the abscissa axis. Black line, open air, summer; pink line, greenhouse, summer; green line, open air, winter; blue line, greenhouse, winter.

Plate 7. Evolution of solar radiation intensities in the open air and in the greenhouse on a not completely cloudy day in spring and another one in winter (coast of Granada). Black line: open air, spring; pink line: greenhouse, spring; green line: open air, winter; blue line: greenhouse, winter.

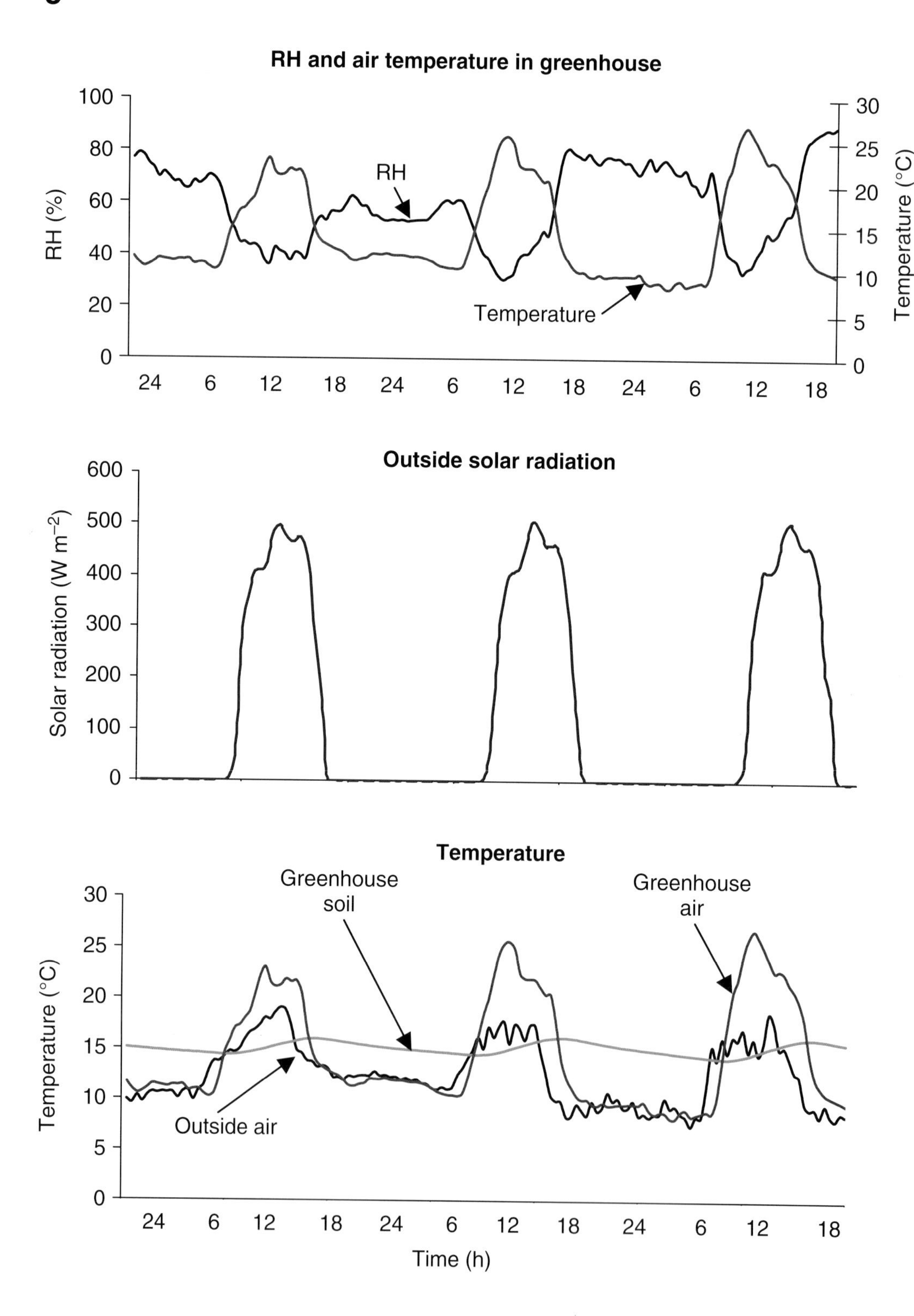

Plate 8. Evolution of a set of greenhouse and outdoor climate parameters throughout several sunny days at the end of the winter (Motril-Granada, Spain). Unheated greenhouse.

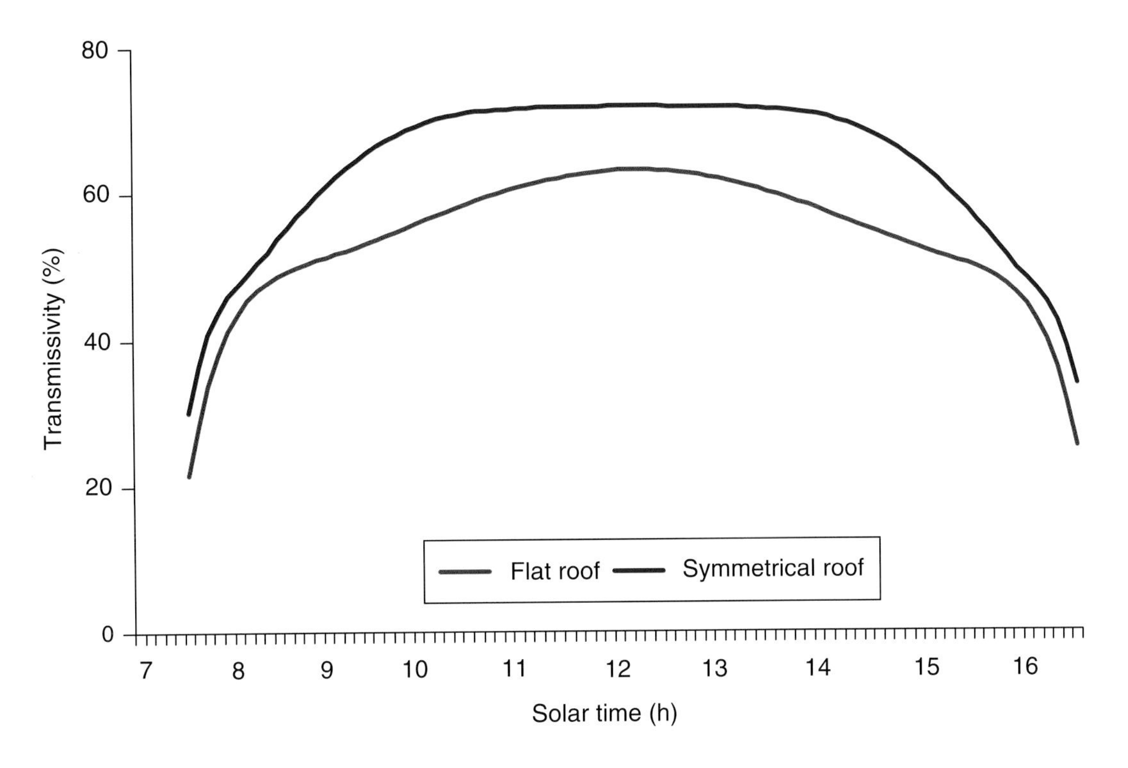

10

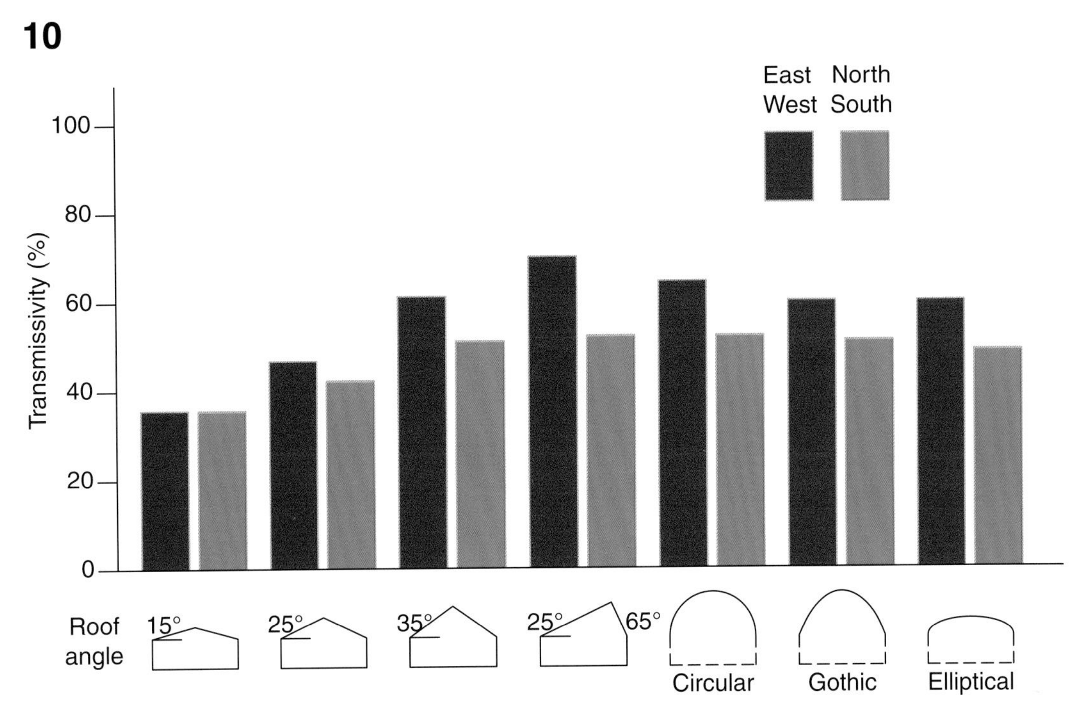

Plate 9. Hourly evolution of transmissivity on 21 December in a low-cost, flat-roof greenhouse and in a greenhouse with a symmetrical roof with a 220 μm cover of thermal PE and a 15° roof angle, oriented east–west.
Plate 10. Transmissivity of different greenhouse roof geometries of single-span greenhouses at the winter solstice (21 December), at latitude 51°N (Belgium), depending on the orientation (east–west or north–south) (adapted from Nisen and Deltour, 1986).

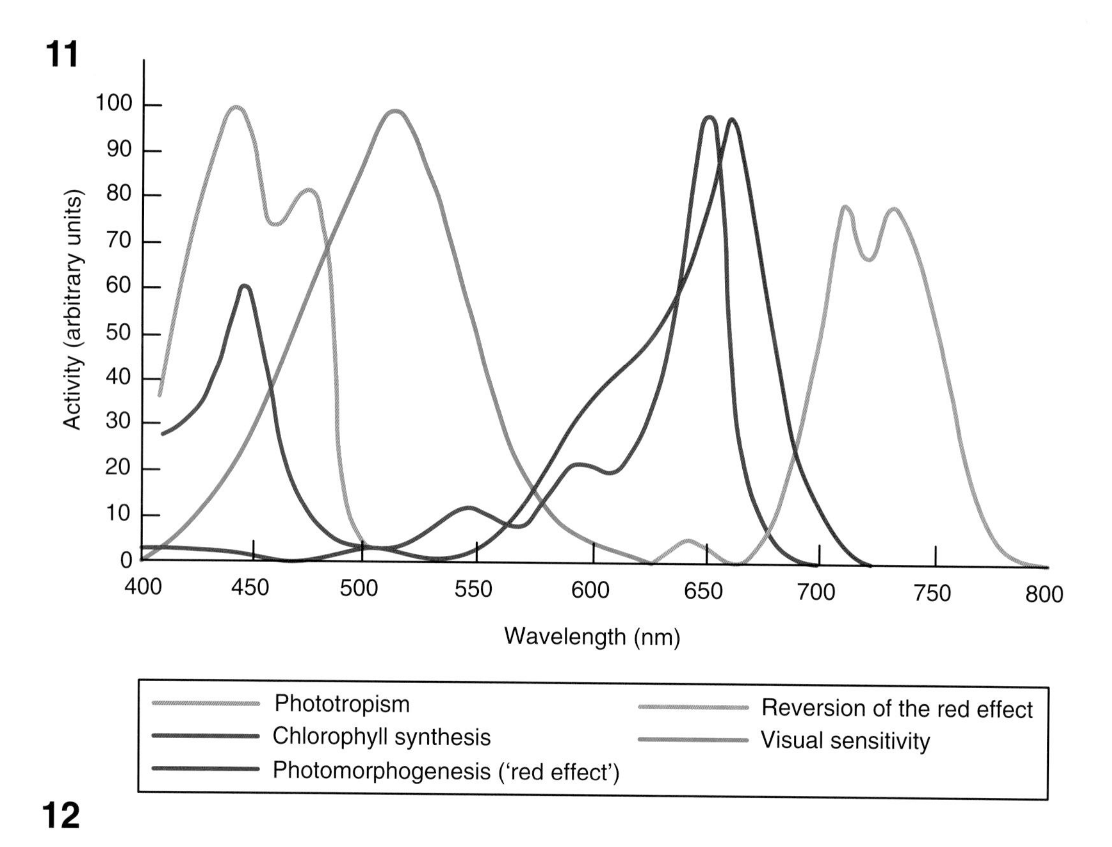

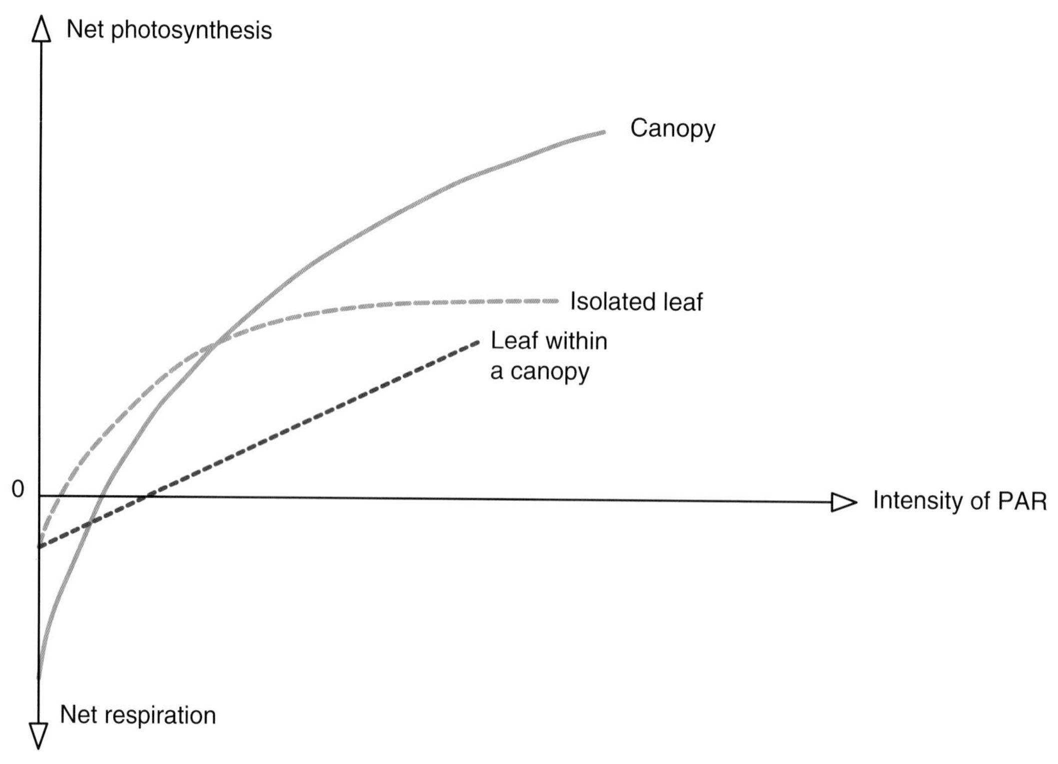

Plate 11. Spectral activities of different photo-biological processes (adapted from Whatley and Whatley, 1984).
Plate 12. Net photosynthesis of an isolated leaf, of a leaf within a canopy and of a whole canopy (adapted from Urban, 1997a).

13

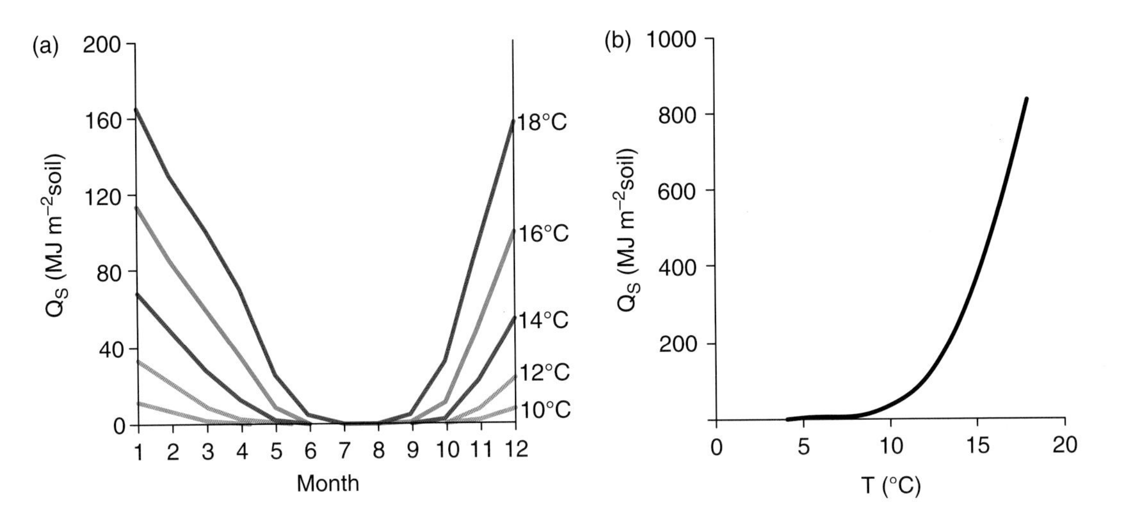

14

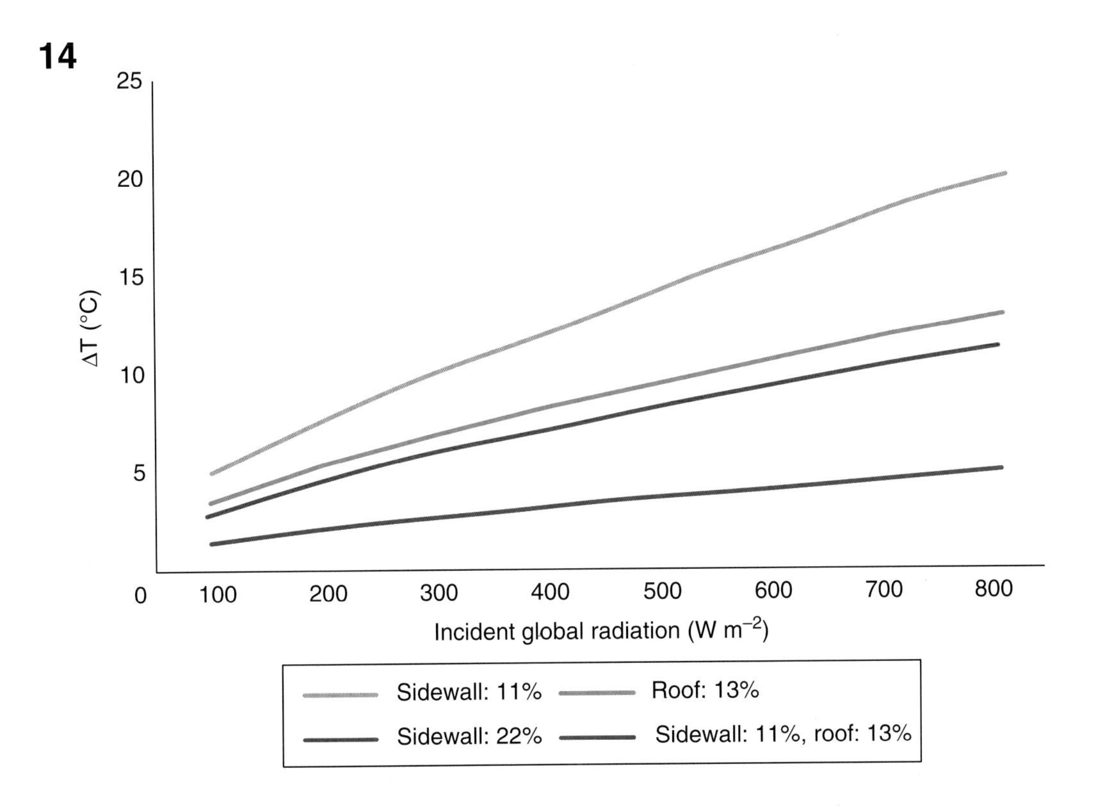

Plate 13. (a) Monthly energy consumption (Q, in MJ m^{-2}) for different temperature set points (T_C) in a low-cost greenhouse in Almeria. (b) Yearly accumulated energy consumption (Q_S, in MJ m^{-2}) for different temperature set points (T_C). Data refer to an average year using hot-air heating (adapted from López, 2003).

Plate 14. Increases in the air temperature of a greenhouse with respect to the outside air, depending on the incident global radiation for different ventilation conditions. Key shows location and percentage of ventilation area (source: J.I. Montero).

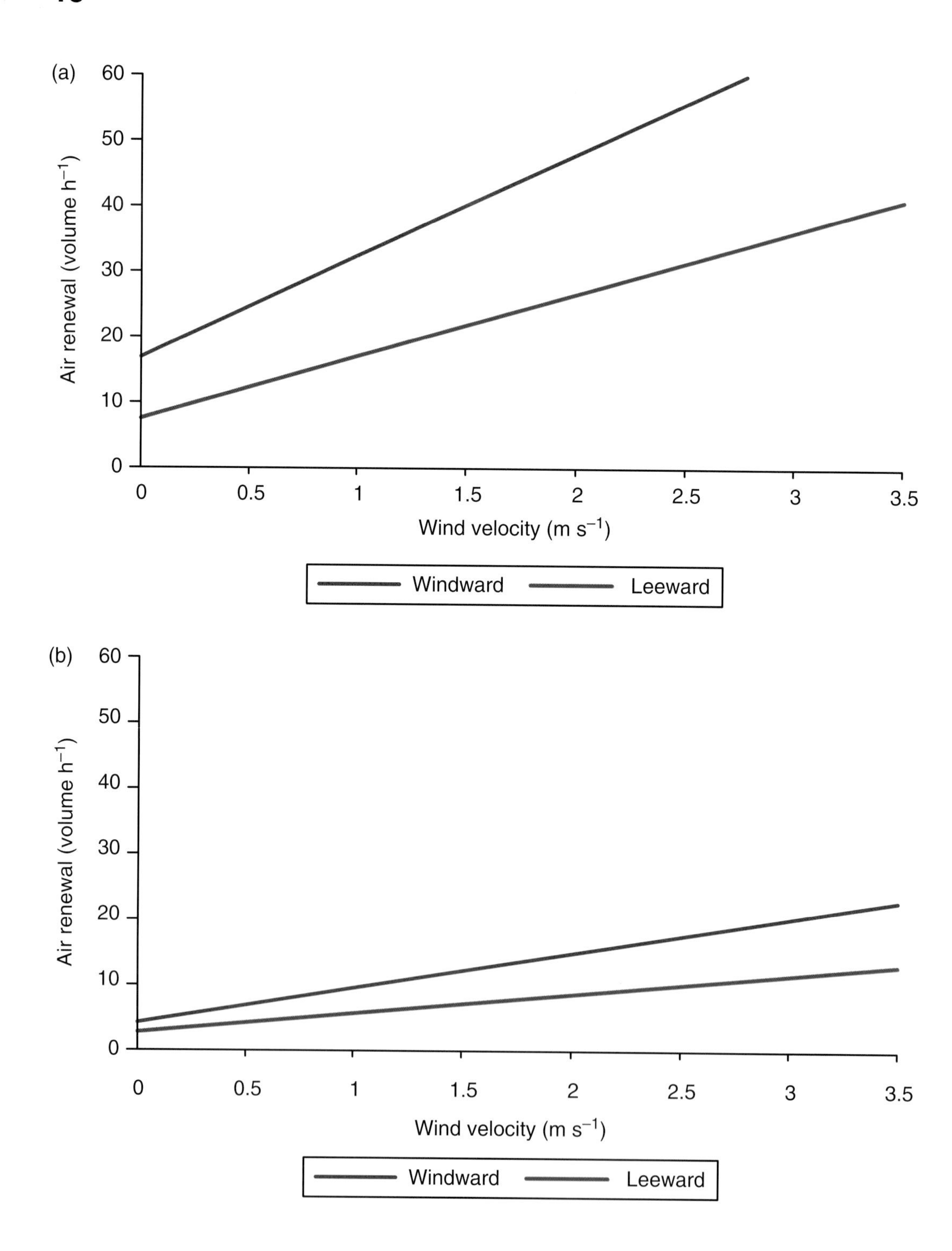

Plate15. Air renewal rate as a function of velocity and wind direction in a multi-tunnel greenhouse with hinged-type vents (with flap) without obstacles (a) and vents implemented with an anti-thrips screen (b) (from Muñoz, 1998).

16

17

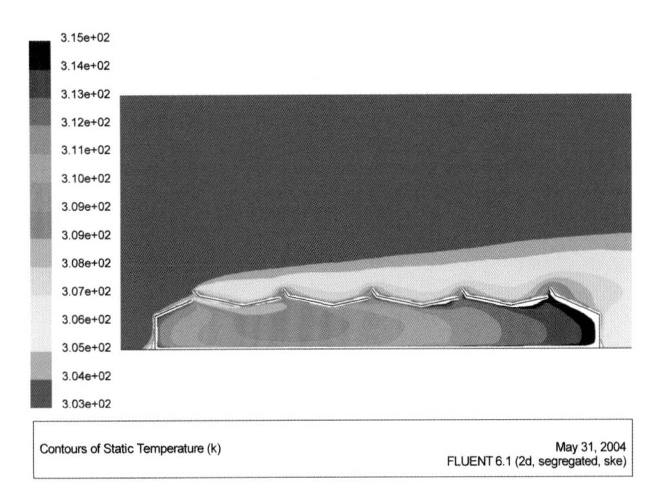

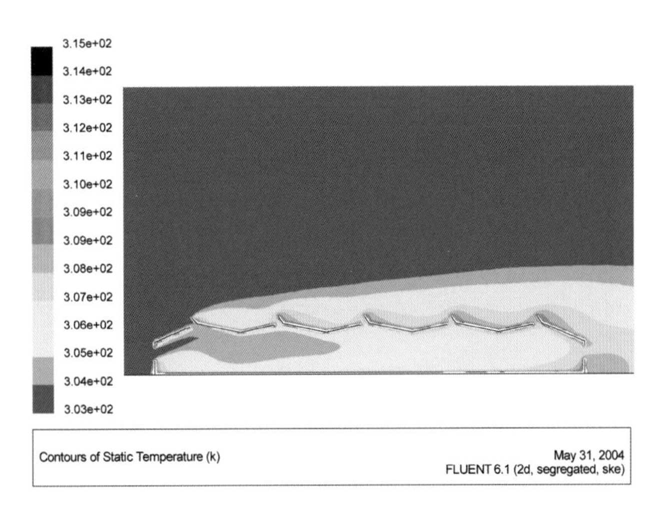

Plate 16. Ventilation studies of scale models of greenhouses, by means of a flow visualization technique, using liquids of different colours and densities. The scale model is mounted downwards (source: J.I. Montero).
Plate 17. Scheme of a scalar temperature field of a low-cost greenhouse of five spans, with only roof vent (top) or combined roof and side ventilation (bottom), which allows for the visualization of the best ventilated zones (blue colours) and the worst ventilated zones (red colours). Windward wind, 3 ms^{-1}. CFD technique, simulation of fluid dynamics (data provided by 'Las Palmerillas' Experimental Station, Cajamar Foundation, Almeria, and by J.I. Montero).

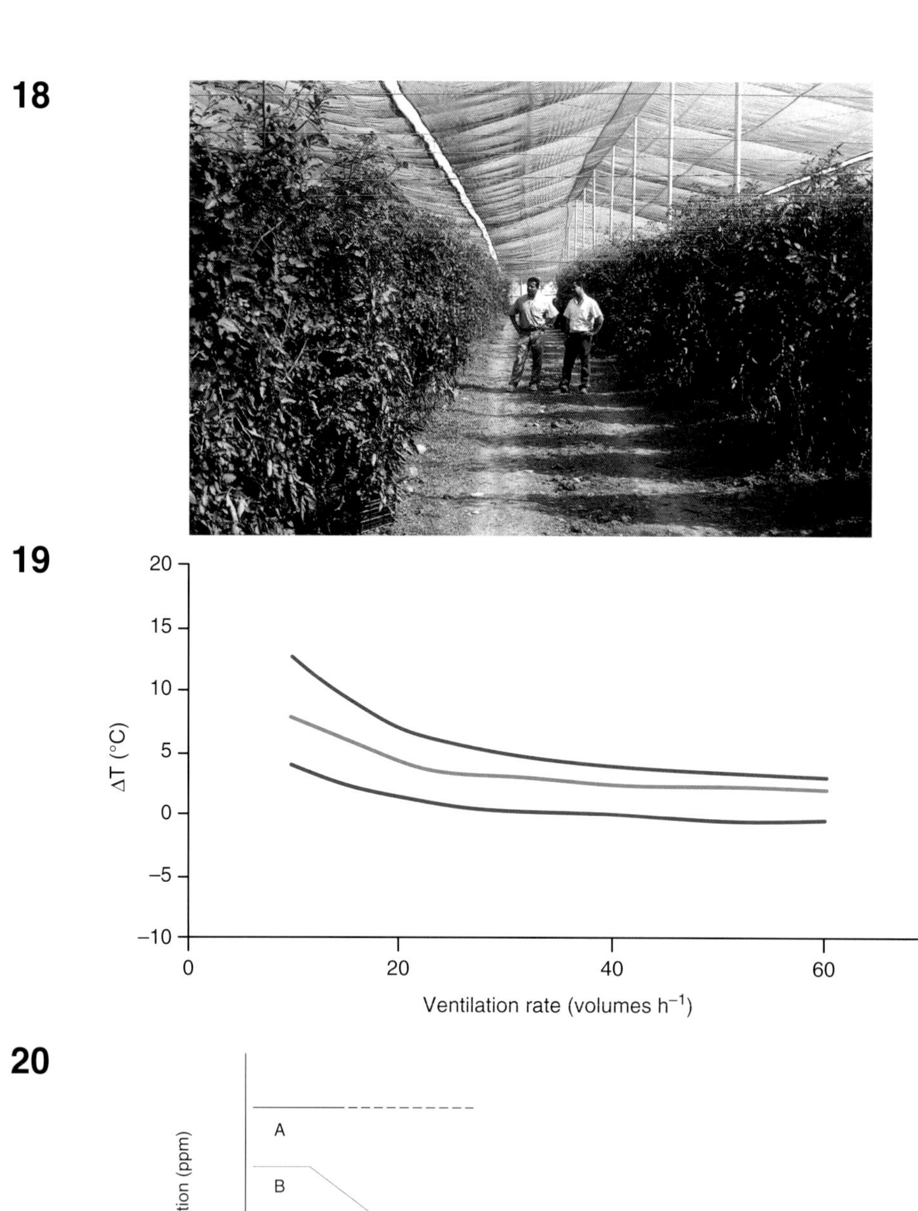

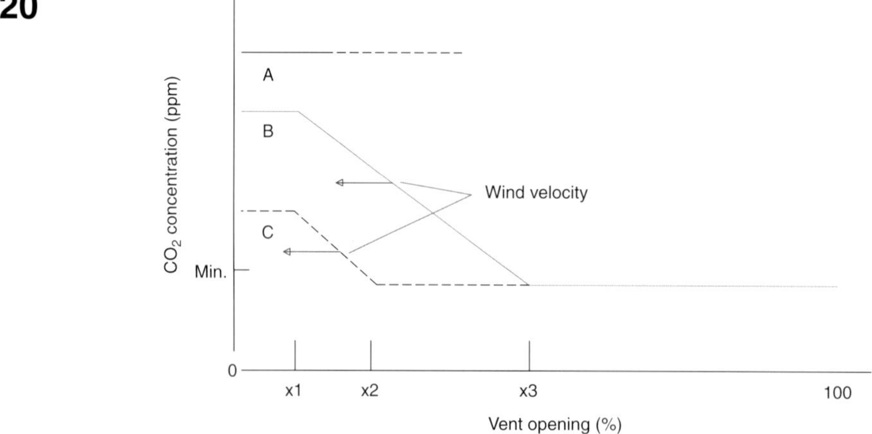

Plate 18. The greenhouse that uses a screen as cladding material is known as a screenhouse.
Plate 19. High greenhouse air temperatures (with respect to the exterior temperature) as a function of the ventilation rate with a well-developed crop, under different conditions. Green, with shading; blue, with shading and air humidification; red, without shading or air humidification (source: J.I. Montero).
Plate 20. Set point values for carbon enrichment (CO_2) depending on the demand of heat (A) and the solar radiation intensity level (A and B) depending on the vent opening conditions (adapted from Nederhoff, 1995). x1 represents the lower degree of vent opening that determines changes in the CO_2 enrichment set point. x2 and x3 represent the vent openings that determine the minimum CO_2 enrichment set point when radiation is below a preset level (x2) or when radiation exceeds a preset level (x3). See explanation in the text (Chapter 9).

21

22

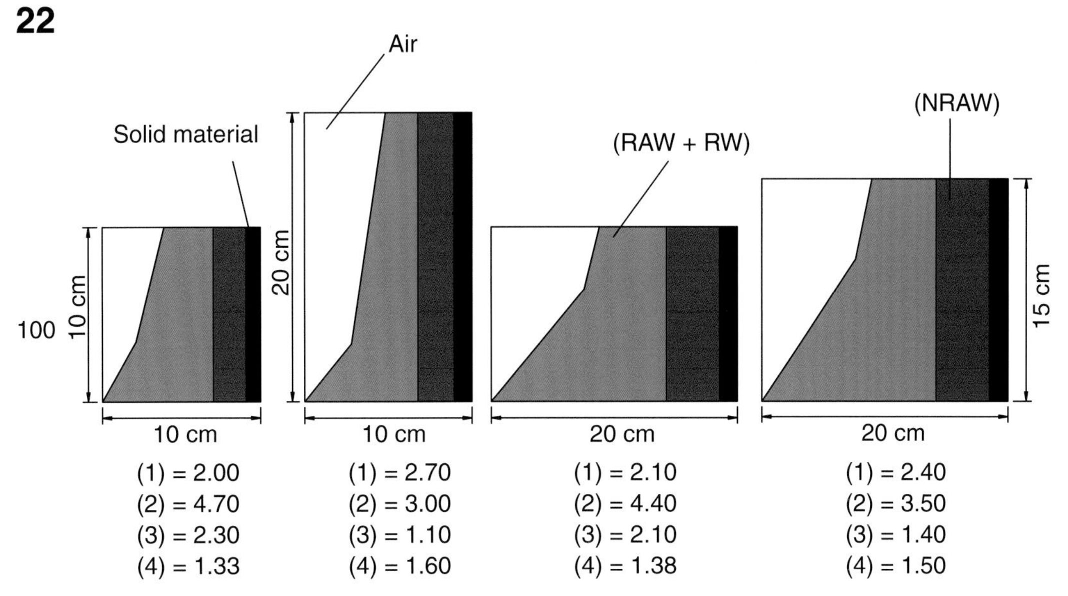

Considered relations:

(1) Total volume of substrate / (Volume of RAW + RW)
(2) Total volume of substrate / Air
(3) (Volume of RAW + RW) / Air
(4) Total volume of substrate / Total water volume

Plate 21. Substrate cultivation has spread widely.
Plate 22. The volumetric relations of the water and air content of a P-2 type perlite (grain size between 0 and 5mm), depending on the height and width of the container holding it, represented in cross-section (adapted from Caldevilla and Lozano, 1993). RAW, readily available water; NRAW, not readily available water; RW, reserve water.

23

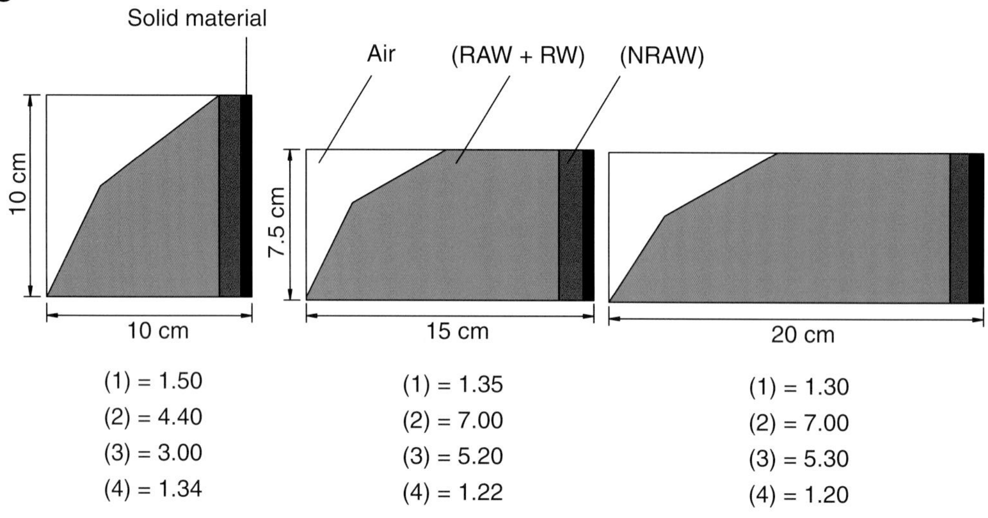

Considered relations:

(1) Total volume of substrate / (Volume of RAW + RW)
(2) Total volume of substrate / Air
(3) (Volume of RAW + RW) / Air
(4) Total volume of substrate / Total water volume

24

Plate 23. The volumetric relations of the water and air content of rockwool, depending on the height and width of the slab, represented in cross-section (adapted from Caldevilla and Lozano, 1993). RAW, readily available water; NRAW, not readily available water; RW, reserve water.
Plate 24. Mechanized irrigation in the nursery.

25

26

Plate 25. Bumblebee (*Bombus* spp.) pollinating a tomato flower.
Plate 26. Tomato flowers with brown trace (left behind) typical of having been pollinated by bumblebees.

27

28

Plate 27. The perishable nature of vegetables is a critical factor in their marketing process.
Plate 28. Distribution is the last step in the marketing process, getting the products to the consumers.

29

30

Plate 29. The improvement in the technological level of Mediterranean-type greenhouses to improve the quality of the products is an increasingly popular trend among horticulturists.
Plate 30. Cut flowers are another greenhouse product option.

31

32

Plate 31. When considering crop options, selection of species with lower thermal requirements (lettuce is shown in the image) prevail in colder areas or during the cold season.

Plate 32. The location of greenhouses is a key issue, both for its influence on production costs (depending on the climatic conditions) and for its effect on the transport costs to the markets. Image shows the Poniente area (Almeria, Spain).

11

Irrigation and Fertilization

11.1 The Plants and Water

Water serves a number of basic functions in a plant's life, constituting up to 95% of fresh weight (Sutcliffe, 1977). Water dissolves several substances and is the transport vehicle for the nutrients in plants. By means of cell turgor it provides rigidity and gives shape to several plant organs. It is necessary for photosynthesis and participates in a large number of chemical reactions of plant metabolism. In addition, it allows plants to be cooled through its evaporation, by means of transpiration, absorbing heat and cooling the leaf surfaces.

11.2 Transpiration

Transpiration may reach up to a maximum of 98% of the total amount of water absorbed by cultivated plants in their life cycle (Sutcliffe, 1977), but normally it represents 95% (Kramer, 1983), the rest being used in plant metabolism.

Transpiration takes place, mainly, through the stomata of the leaves, which must open to capture the required CO_2 for photosynthesis, and transfers water vapour from the plant to the atmosphere (see Chapter 6). One way of looking at it is that water losses by transpiration are the unavoidable cost for the plant in order to be able to fix CO_2 from the air, essential for photosynthesis and plant growth.

Transpiration requires energy (normally solar) for the water evaporation process. If the energy decreases, transpiration decreases. The energy supplies from the greenhouse heating systems also contribute to the evaporation process.

The majority of the water is absorbed passively by the roots, as a result of transpiration. When transpiration stops, there is no passive absorption of water and nutrients, which can have negative effects for the plant. Therefore, in areas of low solar radiation some energy is applied to the greenhouse by maintaining a certain minimum pipe temperature to promote transpiration.

The roots can obtain energy, burning sugars, and absorb water actively, when there is no transpiration or during the night. This active absorption can be enhanced by increasing the soil temperature, and thus, the roots.

Normally, on a sunny day, transpiration increases quickly immediately after sunrise and, as the roots cannot absorb the transpired water at the same pace, the plant transiently uses the water stored in its tissues. When transpiration decreases, the root absorption restores the water deficit of the tissues, completing rehydration during the night.

The maintenance of the turgor of the cell and tissues, by means of water, is fundamental for the elongation of the tissues and growth. Therefore, proper levels of water content must be maintained in the plant, to avoid inducing growth limitations that negatively affect the yield.

Transpiration depends on the intercepted solar radiation and on the environmental humidity at the level of the leaf boundary layer (see Chapter 6). Transpiration shows a hysteresis in relation to solar radiation, from sunrise until noon. At night, the rehydration may involve, on cloudy days, a relevant percentage of the daily water use (Medrano, 1999).

When air humidity decreases, the VPD (water vapour pressure deficit) increases and transpiration increases. If this transpiration increase is large, and the water supply by the roots is low, the stomata will progressively close to avoid tissue dehydration. Therefore, in climate controlled greenhouses, an excessive increase of the VPD must be avoided. The water movement through the soil–plant–atmosphere continuum is governed by its overall potential, so the alteration of the water potential conditions in the soil or in the plant affect the whole set.

With high ambient humidity and low levels of radiation the transpiration rate is very low. As nutrient absorption is linked to the transpiration rate, these conditions (low radiation and high ambient humidity) may induce nutrient deficits, especially for elements like calcium whose mobility in the plant is very strongly linked to transpiration.

Under Mediterranean conditions, the maximum transpiration value for a greenhouse crop such as tomato is 6 mm day^{-1} (Jolliet, 1999), although some authors place it in general between 6 and 9 mm day^{-1}, or even 1–1.5 mm h^{-1} (Kempes, 2003).

11.3 Evapotranspiration

Evapotranspiration (ET) is the sum of the water evaporation from the soil surface (E) and the transpiration (T) or water evaporation through the plants.

The evaporation (E) from the soil is high when the soil is wet (e.g. after irrigation) but decreases quickly as the soil surface dries out and it is not rehydrated. When crop development is slow, and the plants only slightly shade the soil (intercepting little radiation), the evaporation can be very important. When the crop is well developed and shades the soil completely, the evaporation is very limited, because the plants intercept most of the solar radiation preventing it from being used in the evaporation of water from the soil.

Soil mulching, such as sand mulch ('enarenado') or with plastic materials, totally or partially avoid the evaporation of water from the soil.

In greenhouses, with crops grown in the soil, drip irrigation minimizes the evaporation (E), wetting much less of the soil surface than for example sprinkling irrigation systems.

In soilless crops, the substrate is usually covered by a plastic film, so the E component of the ET is virtually nil.

11.4 The Water in the Soil

11.4.1 Introduction

The water status of the soil is characterized by its capacity to retain water and by its water potential (energy status of the water in the soil).

11.4.2 Characterization of the soil water stress

Water potential

The most precise way of quantifying the water available in the soil for absorption by the plant's roots is by means of the water potential in the soil (ψ_{soil}). The water potential is normally measured in pressure units, using MPa and kPa (1 kPa = 0.001 MPa).

The water potential is a measure of the free energy of the water and has four components (matrix (matric) potential, gravitational potential, osmotic potential and pressure potential). The matrix potential (ψ_m) is caused by the forces that retain the water in the soil. The osmotic or solutes potential (ψ_s) is caused by the salts dissolved in the soil solution. An approximate relation is ψ_s = 0.036 EC, expressing ψ_s in MPa, and EC in dS m^{-1}. The gravitational potential (ψ_g) depends on the elevation of the particular point. The pressure potential is caused by the external pressure exerted by the soil's atmosphere and is usually neglected (except in waterlogged soils).

Normally, under saline conditions and around field capacity, the main component of the water potential is the matrix potential (ψ_m). Its value is negative. In practice, the absolute value of ψ_m is used, which is called matrix tension.

The characteristic moisture curve of a soil, or water retention curve, represents the relationship between matrix potential (or tension) and water content (Fig. 11.1).

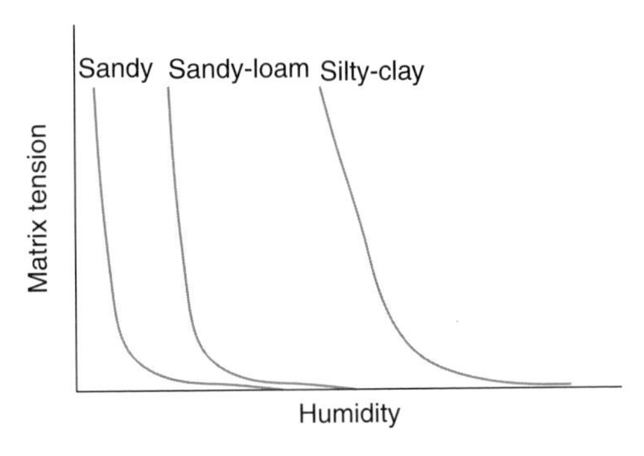

Fig. 11.1. The retention curve, or characteristic moisture curve, of a soil represents the relationship between the moisture content and the matrix tension (absolute value of the matrix potential).

Field capacity (FC)

The field capacity of a soil is the amount of soil moisture or water content held in the soil after excess water has drained away freely for 1 or several days. As most soils do not drain until they have retained a certain amount of water and then retain it indefinitely, the definition of field capacity is idealized and the concept applies more to soils with a gross texture (i.e. with a large particle size) (Table 11.1).

Given the existence of soil layers that interfere with the movement of the water in the soil as well as possible water tables, it is convenient to measure the field capacity *in situ* (Castilla and Montalvo, 1998).

Permanent wilting point (PWP)

The permanent wilting point of a soil is the water content below which the plant wilts and water absorption from the soil ceases.

Like the field capacity, PWP it is not a constant of the soil and does not depend only on it. There is no unique value of water content for which the plants stop extracting water. For instance, a plant under a low evaporative demand can extract more water from a soil than if the demand is higher, because it has more time for absorbing water. On the contrary, if the evaporative demand is high the plant can temporarily wilt with water content in the soil that is higher than the permanent wilting point.

Table 11.1. Field capacity (FC), permanent wilting point (PWP) and readily available water (RAW) for soils with different textures.

	Moisture content (% dry weight)			Available water holding capacity
Soil texture	FC	PWP	RAW	(mm m^{-1})[b]
Sandy	6–12 (9)[a]	2–6 (4)	5	85
Sandy-loam	10–18 (14)	4–8 (6)	8	120
Loam	18–26 (22)	8–12 (10)	12	170
Clay-loam	25–31 (27)	11–15 (13)	14	190
Silty-clay	27–35 (31)	13–17 (15)	16	210
Clay	31–39 (35)	15–19 (17)	18	230

[a]Average value in parentheses.
[b]Available water holding capacity expressed in millimetres of water per metre of soil depth.

The field capacity and permanent wilting point concepts are idealized. The best way to define them, although not perfect, is by the water tension in the soil (e.g. 0.33 atmospheres for the FC and 15 atmospheres for the PWP).

Readily available water (RAW)

The readily available water is the difference between the water content at field capacity and at permanent wilting point.

$$RAW = FC - PWP \quad (11.1)$$

Tables 11.1 and 11.2 show the FC, PWP and RAW values for different soils.

When considering substrates other than soil, due to their special characteristics FC and PWP are not used. The RAW is found under very limited ranges of matrix tension (see Chapter 10).

Allowable soil water depletion

Although plants can extract available water retained in the soil or substrate, as the available water decreases the plant has difficulties in extracting all the required water and starts suffering water stress. This threshold moisture value is the allowable soil water depletion and varies depending on the soil conditions, evaporative demand and crop development stage. The allowable soil water depletion is usually represented as a percentage of readily available water (RAW). In tomato, for instance, for an open field well-irrigated crop the allowable soil water depletion has been established between 30 and 50% of the readily available water (Castilla, 1995).

Table 11.2. Readily available water (RAW, in mm m^{-1}), as a function of the soil matrix tension for different soils.

	Soil matrix tension (atmospheres)			
Soil texture	0.2	0.5	2.5	16
Clay	180	150	80	0
Silty-clay	190	170	100	0
Loam	200	150	70	0
Silt-loam	250	190	50	0
Silty-clay-loam	160	120	70	0
Sandy-clay-loam	140	110	60	0
Sandy-loam	130	80	30	0
Fine sand-silt	140	110	50	0
Fine sand-medium	60	30	20	0

11.4.3 Measurement of the soil water content

Introduction

The soil water content can be measured by direct procedures, taking soil samples and drying them in a stove to calculate their gravimetric moisture (in weight), but this is a time-consuming process that does not allow for continuous data monitoring.

If the gravimetric moisture has to be transformed into volumetric moisture, the apparent density of the soil is required.

Several indirect procedures allow for the evaluation of the volumetric moisture in the soil. Among them, we can highlight the neutron probe method (nowadays in disuse) and time domain reflectometry (TDR) and frequency domain reflectometry (FDR), which, to some extent, will be discussed later on. A full description of all available methods can be found in Raviv and Lieth (2008).

The continuous monitoring of the matrix potential is carried out, normally, by tensiometers, although electric resistance sensors can also be used. Monitoring the osmotic potential is commonly practised by measuring the EC of the soil or substrate solution. A practical review of their use in greenhouses can be found in Thompson and Gallardo (2003).

Description of soil moisture sensors

Soil moisture sensors measure the volumetric water content of the soil or matrix potential. The matrix potential is very close to the total water potential, if the soil is not saline.

MATRIX POTENTIAL SENSORS

Tensiometers. Tensiometers are cheap, simple, easy to use, and require minimum maintenance (Photo 11.1). Tensiometers use gauges, but in some models these are being substituted by pressure transducers, which are more precise but more expensive, allowing for continuous monitoring.

Tensiometers perform well within the 0–80 kPa matrix tension range. To achieve exact readings, the correction by gravity in the water column must be taken into account, because 10 cm of water column equals 1 kPa, which in greenhouse irrigation management practice is of little relevance. Nowadays, there are tensiometers in the market that allow for the automation of the readings.

Electric resistance sensors. The principle behind electric resistance sensors is that the electrical resistance between two electrodes is a function of the water content. The most usual types are plaster blocks, which are of no use where high-frequency drip irrigation is used as they do not work efficiently with high levels of soil moisture. In addition, they do not last long.

Other similar sensors (granular matrix sensors) have improved their performance with respect to the plaster blocks, but tensiometers are more suitable for greenhouse vegetable growing (Thompson and Gallardo, 2003).

VOLUMETRIC MOISTURE SENSORS. In the past the neutron probe was used. Nowadays, it has been substituted by dielectric sensors, which measure the dielectric constant of the soil matrix, deducing the value of the volumetric moisture content in the soil from it. Two methods are used: (i) TDR; and (ii) FDR.

TDR is based on measuring the transmission time of an electromagnetic signal along a metallic probe introduced into the soil. FDR sensors use the capacitance (the ability of a body to store an electrical charge) to measure the dielectric constant of the soil matrix. FDR sensors have better performance than those of TDR. A review by Thompson and Gallardo (2003) on their use in greenhouses summarizes several aspects about their use.

Management of tensiometers (in the soil)

In greenhouse high-frequency drip irrigation systems, the recommended intervals of water tension in the soil lie between 10 and 20 kPa (for soil with a gross texture, i.e. with

Photo 11.1. Tensiometers are increasingly used in greenhouses.

large-sized particles), between 10 and 30 kPa (for soil with medium texture, i.e. medium-sized particles) and between 20 and 40 kPa (for fine textured soil), measured in the maximum root density zone. These values are for guidance, and must be adjusted depending on each case's specific conditions.

Normally, during the autumn and the winter in the Mediterranean coast the matrix tension must be maintained between 20 and 40 cb, because very low values (10–20 cb) can generate problems of root asphyxia; when the evaporative demand increases, in spring and summer, it is advisable to maintain it between 15 and 30 cb.

Good tensiometer management, besides fixing the irrigation frequency, allows the quantities of applied water that are used to be checked thus avoiding unnecessary leaching and wastage of water.

Use of moisture sensors in substrates

In order to measure the moisture content of substrates very sensitive tensiometers are used, whose measuring limits are, normally, between 1 and 10 kPa of matrix tension and with special ceramic capsules as the 'laptometers' (Terés, 2000), which give a very fast reading. They must be installed so that there is good contact between substrate and sensor, which is difficult in porous substrates, locating them at the right depth, given the variability of water content in the substrates.

11.4.4 Quality of the irrigation water

In greenhouse crops the assessment of the irrigation water quality must not be limited to the conventional parameters (see Appendix 1 section A.8), but must include evaluation of: (i) the solid elements content (if drip irrigation is used); and (ii) its temperature (particularly where significant volumes of water are applied (i.e. surface irrigation) in unheated greenhouses during the cold season, to avoid a thermal shock due to low water temperature).

11.5 The Water in the Plant

11.5.1 Introduction

In herbaceous plants the water constitutes normally more than 80% of their fresh weight.

11.5.2 Characterization of the water in the plant

The water content and availability in the plant can be characterized by direct or indirect methods. Among the direct indicators are the *relative water content*, which quantifies the water content of a plant tissue, in relation to its maximum possible value, and the measurement of the *plant water potential*. Neither of these direct methods is of practical application.

Among the indirect indicators of the plant water status, besides visual symptoms (leaf rolling, colour changes, wilting) that only appear under severe water stress conditions, we can include: (i) the stomatal conductance; (ii) changes in stem diameter; (iii) the sap flux; and (iv) the plant temperature (Gallardo and Thompson, 2003a).

In response to dehydration of the phloem caused by transpiration and later rehydratation, the stems and trunks of plants experience contractions and dilations in 24 h cycles. Quantifying these is a good indicator of the plant's water status (Huguet *et al.*, 1992).

The sap flux that ascends through the stem (due to transpiration) is another indicator of the water status of the plant.

When the plant's water status is good, with a normal transpiration that cools the plant as water evaporates, the plant's temperature is usually lower than the surrounding air. If water stress occurs, transpiration is limited by stomatal closure increasing the temperature (Jackson, 1982); this can be measured using an infrared thermometer. The 'crop water stress index' (CWSI) has been proposed as an indicator of the water status of the plant (Idso *et al.*, 1981) (see Appendix 1 section A.8.1).

11.5.3 Water stress

Water vapour losses from the sub-stomatal cavities to the atmosphere in the transpiration process are compensated for by water absorption from the soil.

A plant is considered to suffer water stress, or water deficit, when the water potential in its tissues decreases to the extent that it negatively affects the performance of the physiological processes.

The causes of water deficit can be:

1. Low ψ in the soil, due to low water content or to salinity.
2. High transpiration rate.
3. High resistance to the water flux in the soil or in the plant.

During the hours around midday, in well-irrigated plants, short-term water deficit due to high evaporative demand may occur.

Long-term water deficits are usually caused by progressive depletion of the water in the soil.

11.5.4 Effects of water stress in the plant

The most sensitive processes to water deficit are: (i) cell elongation; (ii) cell wall synthesis; and (iii) protein synthesis (Hsiao, 1973). Leaf expansion, that determines the useful leaf area for photosynthesis, is very sensitive to water deficit (Hsiao, 1973). The growth of the aerial part is much more sensitive to water deficit than the roots' growth.

It is now known that stomatal closure may occur in response to the water status of the soil independently of that of the plant, by means of sending root signals (production of abscisic acid), which also influences leaf expansion (Gallardo and Thompson, 2003a).

11.5.5 Saline stress

The sensitivity to salinity varies depending on the species, cultivar and the age of the plant. A crop can survive under high salinity conditions, but the production will be seriously affected, quantitatively and qualitatively.

Excessive fertilization, the accumulation of ions in the vicinity of the roots and the salts supplied by the irrigation water, together with the natural salinity conditions of the soil, are the main causes of saline stress (Ehret and Ho, 1986) (see section 11.7.4).

In greenhouse horticultural crops, the primary objective of irrigation is to avoid water stress, but also to avoid undesirable conditions of salinity at the root level.

11.6 Greenhouse Irrigation

11.6.1 Introduction

Surface irrigation systems, mainly by furrows, that were traditionally used in greenhouses are now no longer used. Equally, there has been a decline in the use of microsprinkler systems (at height), and now different high-frequency irrigation systems (drip, exudation) are widely used.

11.6.2 Components of the drip irrigation system

The main components of a drip irrigation system are: (i) the irrigation/fertilization control centre; (ii) the main pipelines and secondary lines; and (iii) the micro-tubes and emitters or diffusers.

The irrigation control centre (Photo 11.2) is basically composed of filtering and fertilization equipment, the pressure and flow control elements, and the automatic control equipment (schematically represented in Fig. 11.2; Montalvo, 1998). It may also incorporate the water pumping system although this will not be the case if the water has enough pressure.

The filtering equipment is important in order to avoid physical blockages, particularly in the emitters. Usually it consists of sand filters, disc and mesh filters. In cases where there is risk of using water with a

Photo 11.2. The control centre of a localized irrigation system.

1. Electric valve
2. Hydrocyclone
3. Manometer
4. Sand filter
5. Gate valve
6. Fertigation equipment
7. Disc or mesh filter
8. Controller (computer)
9. Flow meter

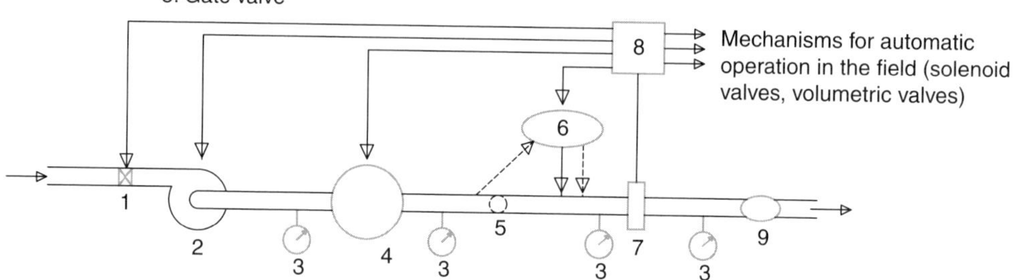

Fig. 11.2. Scheme of a control centre of a localized irrigation system.

high solid content, mainly sand, it is common to use a separating hydrocyclone to pre-filter the water.

The fertilization equipment is an essential component for fertigation; fertilizer tanks, venturi injectors and metering pumps being its most common components. In more sophisticated irrigation control centres, pH and electric conductivity probes usually complete the system.

The main and secondary pipelines, made of plastic materials (PVC or PE), carry the water towards the irrigation pipes (of PE), to which the emitters or diffusers are connected, through which the water is supplied to the soil or substrate.

Pressure meters are required to monitor the performance of the system. These are placed, at least, after the pump and after the filters. Flow meters are compulsory in some automated facilities (pump controllers) and are convenient in all types of facilities in order to provide precise information on the amount of water supplied. The most usual features that are automated are the irrigation controllers (which manage solenoid valves at a distance) and the volumetric valves.

11.6.3 Management of drip irrigation

Characteristic features of drip irrigation systems, although not exclusive to them, are high-frequency irrigation and localized water supply directed to only a part of the potential root zone of the crop (Vermeiren and Jobling, 1980). They are the most common form of high-frequency localized irrigation (HFLI) systems, and require different management to that of surface or microsprinkler systems.

Good management of drip irrigation requires a proper knowledge of the water and salts movements (affected by the slow and localized supply of small water volumes) to avoid salination of the root zone on the long term and to achieve, especially in areas where water is a scarce resource, efficient water use, which will only be attained with a suitable irrigation schedule. This requires a good knowledge of the crop water requirements.

The wet soil volume reduction achieved with drip irrigation may generate a corresponding adaptation of the root system and induce nutrient limitations in the crop; it is therefore highly recommended that fertilizer is incorporated in the water with drip irrigation. The small diameter of the emitters and the slow water flux lead to the accumulation of materials that can potentially cause partial or total blockages. It is therefore necessary to filter the water properly and to avoid blockages by means of preventative maintenance and/or by injecting different chemical products (e.g. nitric, phosphoric or sulfuric acid, for pH adjustment; or special chemicals for microbial control) depending on the type of expected blockages.

11.6.4 Water and salts movements with drip irrigation

The movement of water and salts and distribution patterns in the soil with drip irrigation are of great importance (Photo 11.3), not only for the selection of the type of

Photo 11.3. Knowledge of the characteristics of water movement (a sandy soil is shown) is of great interest for ideal management of high-frequency localized irrigation systems.

emitters and their density, but also to optimize their management (Bressler, 1977; Fereres, 1981; Castilla, 1985).

11.6.5 Greenhouse irrigation scheduling (soil-grown crops)

As water is a limiting resource in many agricultural areas, it must be a basic objective of its management to optimize its productivity by means of adequate (i.e. avoiding water deficits in the root zone) and efficient irrigation (i.e. maximizing the fraction of the applied water that remains stored in the rooted soil profile and is used later by the crop) to obtain maximum yields.

Two questions are essential in irrigation scheduling:

1. When to irrigate? (frequency)
2. How much water to apply?

The amount of water to apply must compensate for the evapotranspired water corrected as a function of the application efficiency (assuming that the water content of the soil is quite stable under drip irrigation given its high frequency). Where saline water is used, the supply must be increased to cover the leaching requirements as described by several authors (Ayers and Westcot, 1976; Doorenbos and Pruitt, 1976; Vermeiren and Jobling, 1980; Veschambre and Vaysse, 1980). Other components of the water balance are irrelevant for greenhouse drip irrigation (Castilla, 1987), except for rain in the case of a perforated greenhouse cover.

Various methods can be used to calculate the irrigation schedule in a greenhouse.

Method based on calculating the water balance in the soil

The most simple expression for soil water balance is:

$$\theta_1 - \theta_2 = \Delta\theta = R_n + P_e + AC - ET_c \qquad (11.2)$$

where:

$\Delta\theta = \theta_2 - \theta_1$ = The difference of moisture content at the beginning (1) and the end (2) of the considered period

R_n = Net water supplied by irrigation (the part of the irrigation water which remains stored in the root volume and is available for the crop)

P_e = Effective rain (part of the rain that remains stored by the root volume and is available for the crop)

AC = Water that enters the root volume by capillary ascension

ET_c = Water evapotranspired by the crop

The soil water balance method is often used to calculate how much water to apply in surface or sprinkler irrigation.

Where HFLI is practised, considering the time interval as the time passed between the end of two consecutive irrigation episodes, the initial water content and the final water content are almost the same ($\Delta\theta = 0$) and the net amount of water that must be added by irrigation, R_n, becomes:

$$R_n = ET_c - (P_e + AC) \qquad (11.3)$$

This is an equation that can be used to carry on the previously mentioned accounting, starting from the moment at which the soil stores all the retainable water.

The effective rain (P_e) is non-existent in a greenhouse and the capillary ascension (AC) is negligible in the Mediterranean area, because the water table is deep. Therefore, $R_n = ET_c$.

In HFLI, when the density of drippers is high, such as in greenhouse horticultural crops, the volume of wet soil in many cases is close to 100%.

In low frequency irrigation systems, the time for irrigation comes when the 'allowable soil water depletion' is achieved in the soil (section 11.4.2). When irrigating, water is replaced up to field capacity, providing the soil with the maximum amount of useful water.

In HFLI the irrigation frequency is much higher and is fixed as a function of other parameters, mainly by matrix tension in practice (as described later).

Determination of the ET_c

The ET depends on: (i) the climate parameters; (ii) the availability of water in the soil; and (iii) the crop. When the ET requirements are not fulfilled, the crop can suffer water stress and yield losses when the deficit is large.

The quantification of the crop evapotranspiration (ET_c) or maximum ET of the crop, which would involve the maximum yield under non-limiting water supply conditions, is:

$$ET_c = K_c \times ET_0 \quad (11.4)$$

where K_c is the crop coefficient, whose value depends on the crop (size and development stage, transplant or sowing date) and ET_0 is the reference evapotranspiration which is taken as standard and depends on the existing climate conditions.

In order to estimate the ET_0 several methods have been proposed by the Food and Agriculture Organization of the United Nations (FAO) (Doorenbos and Pruitt, 1976). In greenhouses, the Class A evaporation pan method (Photo 11.4) is easy to apply. It estimates the value of ET_0 as a function of the evaporation from a water surface (E_0):

$$ET_0 = K_p \times E_0 \quad (11.5)$$

where:
ET_0 = Reference ET
K_p = Pan coefficient
E_0 = Evaporation from a water surface

The evaporation pan acts by integrating the climate conditions (radiation, wind, temperature and humidity). It consists of a cylindrical pan, made of galvanized steel, diameter 121 cm and height 25.5 cm, supported by a wooden platform that stands 15 cm over the soil. It is filled with water that must be clear and its level must always be maintained between 5 and 7.5 cm below the edge of the pan. Measurement of the water level is done by means of a limnimeter provided with a micrometric screw and the difference between two consecutive measurements is the evaporation in the period between the measurements. There are also other models.

Daily readings are recommended, in the early hours of the day. For the previous calculations the average of the daily evaporation readings must be used, over the period of at least 1 week.

The value of the K_p depends on the general climate conditions and the surrounding environment (Doorenbos and Pruitt, 1976).

Therefore, the ET_c in the pan method would be:

$$ET_c = K_c \times K_p \times E_0 \quad (11.6)$$

Photo 11.4. Class A evaporation pan.

The coefficient K_c depends on the crop and its vegetative stage and changes through the crop cycle, increasing from lower values of the initial stage (sowing or transplanting) through the vegetative growth period, reaching the highest values in the maximum development period (when the crop covers the soil, intercepting all the solar radiation) and decreasing during senescence.

In Mediterranean plastic greenhouses, values of the product ($K_p \times K_c$) for several crops have been estimated and are summarized in Table 11.3.

The similarity of the values of the product ($K_p \times K_c$) with the K_c values in several horticultural crops, according to the literature (Doorenbos and Pruitt, 1976; Veschambre and Vaysse, 1980) allowed for the estimation that the K_p value in the greenhouse is normally around 1.0. Later studies (Fernández *et al.*, 2001) stated that K_p was lower (around 0.8), but that the K_c values were higher than those provided for open field crops, so their product is very similar to the figures provided in Table 11.3.

The evolution of the K_c values will depend on the sowing or transplanting date as a function of the climate conditions, but it can also be estimated as a function of the thermal integral of the leaf development when the dates are different to those indicated in Table 11.3 (Fernández *et al.*, 2001; Fernández, 2003).

Other methods for calculating ET_0 that are of interest in greenhouses are the radiation and the adapted Penman–Monteith methods, detailed in Appendix 1 section A.8.3. In order to have proper greenhouse climate data it is necessary to have a meteorological station in the area (Photo 11.5).

Gross irrigation requirements

Of the irrigation water applied only a part is available to the roots, due to: (i) the runoff losses (there may be none, for example in a well-designed and well-managed drip irrigation system); and (ii) the almost unavoidable percolation or deep leaching required where saline waters are used in order to remove the salts from the root zone.

Table 11.3. Evolution of the product of the crop coefficient by the pan coefficient ($K_p \times K_c$) per fortnight, of some horticultural species, in unheated plastic greenhouses in Almeria, Spain, for indicated sowing (S) or transplant (T) dates (Castilla, 1989).

Period (days)	Tomato	Pepper	Cucumber	Melon	Watermelon	Climbing bean	Aubergine
	T: 16 Oct	T: 1 Sept	S: 16 Sept	S: 16 Jan	S: 1 Feb	S: 16 Sept	T: 1 Oct
1–15	0.25	0.20	0.25	0.20	0.20	0.25	0.20
16–30	0.50	0.30	0.60	0.30	0.30	0.50	0.35
31–45	0.65	0.40	0.80	0.40	0.40	0.70	0.55
46–60	0.90	0.55	1.00	0.55	0.50	0.90	0.70
61–75	1.10	0.70	1.10	0.70	0.65	1.00	0.90
76–90	1.20	0.90	1.10	0.90	0.80	1.10	1.10
91–105	1.20	1.10	0.90	1.00	1.00	1.00	1.05
106–120	1.10	1.10	0.85	1.10	1.00	0.90	0.95
121–135	1.00	1.00	–	1.10	0.90	–	0.85
136–150	0.95	0.90	–	1.00	–	–	0.80
151–165	0.85	0.70	–	–	–	–	0.80
166–180	0.80	0.60	–	–	–	–	0.80
181–195	0.80	0.50	–	–	–	–	0.80
196–210	0.80	0.50	–	–	–	–	0.80
211–225	–	0.60	–	–	–	–	0.80
226–240	–	0.70	–	–	–	–	0.60
241–255	–	0.80	–	–	–	–	0.60

Photo 11.5. Meteorological station inside a greenhouse.

The lack of uniformity in applying the water will involve an extra water supply (in total R_b, gross water requirements) to cover the net water requirements (R_n). The water application efficiency coefficient (E_a, lower than 1.0) expresses the ratio between the water stored in the soil profile available for the roots and the applied water:

$$E_a = K_s \times E_u \quad (11.7)$$

where K_s is a coefficient that quantifies the soil's water storing efficiency (which is of the order of 0.9 in sandy soils and 1.0 in loamy or clay soils) and E_u is a coefficient that reflects the uniformity in the emission of water (in a well-designed and well-managed irrigation system, E_u = 0.85–0.95). The calculation of the uniformity coefficient of a certain facility is easy to perform (Castilla and Montalvo, 1998; Castilla, 2000).

In the case of using saline water, it is necessary to add a complementary amount of water to ensure the removal of the salts. This leaching fraction (dependent on the salinity of the water used, represented by *LF*) is the minimum amount of drainage required to maintain the soil salinity between certain limits that do not involve yield loss.

In surface or sprinkler irrigation (Ayers and Westcot, 1987):

$$LF = \frac{EC_w}{5\ EC_e - EC_w} \quad (11.8)$$

where:
EC_w = Electric conductivity of the irrigation water (dS m^{-1})
EC_e = Electric conductivity of the soil's saturated extract, adapted to the degree of tolerance expressed as the expected yield (as a percentage of the maximum yield) in Table 11.4.

In the case of HFLI (Ayers and Westcot, 1987):

$$LF = \frac{EC_w}{2\ \text{Max}\ EC_e} \quad (11.9)$$

where:
Max EC_e = Maximum electric conductivity tolerable of the soil's saturated extract for that specific crop (see Table 11.4).

Once *LF* is known, the gross water requirement (R_b) is:

$$R_b = \frac{R_n}{E_a\,(1 - LF)} \quad (11.10)$$

In systems of low uniformity, scarce supply of water or saline waters, with the aim of reducing the large losses due to

Table 11.4. Tolerance level of some crops to salts (dS m^{-1}), expressed as the expected yield (in percentage of the maximum yield). (Source: Ayers and Westcot, 1976.)

	Percentage of maximum yield								
	100%		90%		80%		50%		
Crop	EC_w	EC_e	EC_w	EC_e	EC_w	EC_e	EC_w	EC_e	Max. EC_e[a]
Climbing bean	0.7	1.0	1.0	1.5	1.5	2.3	2.4	3.6	6.5
Broccoli	1.9	2.8	2.6	3.9	3.7	5.5	5.5	8.2	13.5
Melon	1.5	2.2	2.4	3.6	3.8	5.7	6.1	9.1	16.0
Cucumber	1.7	2.5	2.2	3.3	2.9	4.4	4.2	6.3	10.0
Potato	1.1	1.7	1.7	2.5	2.5	3.8	3.9	5.9	10.0
Lettuce	0.9	1.3	1.4	2.1	2.1	3.2	3.4	5.2	9.0
Onion	0.8	1.2	1.2	1.8	1.8	3.2	2.9	4.3	8.0
Pepper	1.0	1.5	1.5	2.2	2.2	3.3	3.4	5.1	8.5
Spinach	1.3	2.0	2.2	3.3	3.5	4.9	5.7	8.6	15.0
Strawberry	0.7	1.0	0.9	1.3	1.2	1.8	1.7	2.3	4.0
Tomato	1.7	2.5	2.3	3.5	3.4	5.0	5.0	7.6	12.5

[a]Max. EC_e, maximum electric conductivity tolerable of the soil's saturated extract for that specific crop.

leaching that would follow the use of the previous formula (Eqn 11.10), some authors propose that the higher of the terms, E_a or $(1 - LF)$ of the formula (Eqn 11.10) is deleted, although this would result in an incomplete control of the salts in the less irrigated zones of the plot. The drainage of these salts would be entrusted, after the irrigation campaign, to the rain or to possible complementary irrigations (soil disinfection, pre-sowing).

Methods based on soil parameters

Soil moisture sensors can be used as a unique method to schedule irrigation or in combination with the water balance methods, with plant sensors or, even, as a complement to an irrigation strategy based on experience.

The evaluation of the volumetric water content in the soil using reflectometry (TDR) techniques is rarely used, difficult to operate and expensive.

In practice the measurement of the water tension in the soil (which is equal to the absolute value of matrix potential) is the most affordable procedure. The use of plaster blocks is not common, among other reasons due the need for a good calibration depending on the composition of the soil's solution. Tensiometers, on the other hand, have received wide use and nowadays are common in HFLI systems. Their detailed use is described in several papers (e.g. Castilla and Montalvo, 1998).

In order to use the soil moisture measurements for irrigation scheduling, it is necessary to know the desired moisture thresholds, expressed in water volumetric content or in matrix potential at the rooted depth. The upper limit of soil moisture is close to field capacity, and the lower limit is the 'allowable soil water depletion' (see section 11.4.2). The lower limit indicates when to start irrigating and the upper limit when to stop it. The difference between the two limits indicates the maximum amount of water than can be supplied. Once the limits are set, irrigation management can be done manually or automatically.

When installing the moisture sensors the fact that soil moisture is usually very heterogeneous must be taken into account. This heterogeneity necessitates placement of the sensors in representative places, and replicating the number of sensors at least twice or three times, to obtain representative measurements.

For common greenhouse vegetable crops, when a single sensor is being used, it must be located between 10 and 20 cm deep, 10–15 cm from the base of the plant, and 8–10 cm from the dripper. It is advisable to use one deeper sensor, at an equal distance from the plant and dripper, to control the deep water content and the drainage (Fig. 11.3). Furthermore, another sensor can be placed at the border of the wet bulb and at the same depth as the first one, to check the proper distribution of water.

Methods based on plant parameters

The use of plant sensors is a direct method that indicates when to irrigate, knowing the moment at which the plant starts to suffer water stress. After irrigation, the plant sensor can detect if the performed irrigation was insufficient or not, making it possible to adjust the water supplies.

In the past, in order to know the water status of the plants, manual measurements were performed (leaf and stem water potential, stomatal conductance) and plant sensors were only used in research. Nowadays, there are sensors that measure indirectly the water status of the plants, automatically, and it is possible to use them in commercial crops (phytomonitoring). This will be discussed later in this chapter.

WATER POTENTIAL. Water potential is measured with a pressure chamber and characterizes, directly, the water status of the plant. In order to schedule irrigation it is necessary to know the threshold values from which the crop suffers water stress. Measurement of water potential has had no use in horticultural crops, as it is not adapted to automation, among other reasons.

CANOPY TEMPERATURE. The canopy or leaf temperature is another indirect indicator of the plant's water status. The crop water stress index (CWSI) has been proposed as a stress index (Idso *et al.*, 1981) (see Appendix 1 section A.8.1).

The measurement of the leaf temperature is usually done with an infrared thermometer, although thermistors have also been used to plug in to the 'phytomonitoring' equipment (section 11.6.5)

The crop temperature is not an instant indicator of water stress, as the temperature increase due to partial closure of the stomata takes place much later than the incidence of stress in other processes such as leaf expansion, so its use in horticultural crops sensitive to water stress, such as vegetables, is of little interest (Gallardo and Thompson, 2003b).

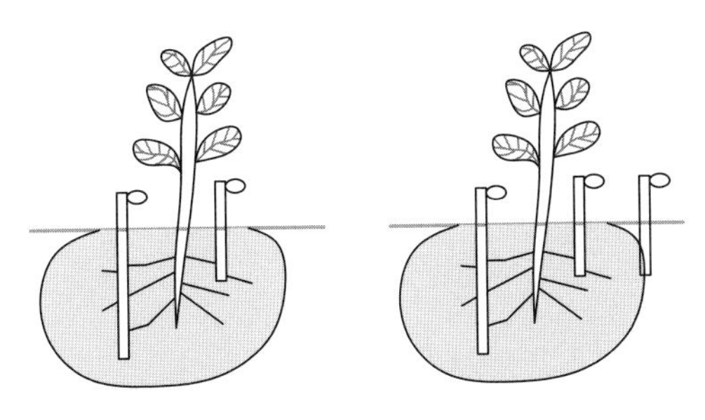

Fig. 11.3. Suggested location of soil moisture sensors in vegetable crops.

STEM DIAMETER. When the plant starts to transpire, at the beginning of the day, a water content decrease in the leaves and stems is caused, as the water absorption by the roots is slower than transpiration, causing a gap that is maintained throughout most of the day. When the evaporative demand decreases, in the afternoon, this gap decreases and the water absorption by the roots continues until all the tissues are rehydrated (Kramer, 1983). The variations in the stem water content caused by this gap between transpiration and root water absorption can be measured with sensors called dendrometers or lineal variable displacement transducers (LVDTs). These operate continuously and automatically.

Over a 24 h period the greatest stem diameter occurs at the end of the night, when hydration is at the maximum, and the minimum value is reached at noon (Gallardo and Thompson, 2003b). In order to use the stem diameter as an indicator of water stress for irrigation scheduling, it is necessary to have proper data interpretation methods.

SAP FLUX. The water flux through the stem is a direct measurement of the plant's transpiration and a very sensitive indicator of its water status. It allows the crop water requirements to be detected, although its use in vegetables is still limited to the research stage.

PHYTOMONITORING. A phytomonitoring device is a data acquisition system supplied with a number of climate sensors (solar radiation, temperature and air humidity), plant sensors (sap flux, stem diameter, fruit diameter and leaf temperature) and soil sensors (water content and temperature), together with analysis and data interpretation software.

The high cost of such systems has limited their use to research labs and high value crops in high-tech greenhouses.

11.6.6 Irrigation scheduling in soilless crops

Introduction

As seen previously, the irrigation strategy when growing in soil is usually based on irrigating when the 'allowable soil water depletion' is reached in the soil or when the water tension in the soil reaches a pre-fixed threshold. In soilless crops, the limited storage capacity of water in the substrates and the difficulty to rehydrate them, after they dry out, necessitates a much higher irrigation frequency than required in soil-grown crops. For rockwool, a number of irrigations between one and four per MJ m^{-2} of global radiation (Jolliet, 1999) has been recommended, although other authors suggest a threshold of 2 MJ m^{-2} per irrigation episode (Urban, 1997b), which is approximately equivalent to applying 0.5 l m^{-2} in each irrigation, under Mediterranean conditions.

A high irrigation frequency and surplus water supplies limit salinity problems, as in soilless cultivation the plants suffer saline stress first and then water stress. This is because there is an increase in salinity in the vicinity of the roots if the nutrient solution film surrounding the roots is not refreshed frequently (Pardossi, 2003). This situation can be avoided if the water circulation is permanent, such as in the NFT system.

Covering the substrate with a plastic film limits water consumption exclusively to transpiration needs, plus any drainage losses.

The management of open irrigation systems (free drainage) depends on the performance of the irrigation system, especially the controller. If the controller is a simple programmer, which does not process received information, the management will be limited to pre-fixed doses and frequency, or to start and stop the irrigation system as a function of the signals received from the different sensors. If the controller is computerized, the management is more dynamic, as it retrieves real-time information from different sensors, processes the data with pre-established models and adopts management decisions.

The irrigation dose (volume applied at each irrigation episode) will be determined by the substrate's water retention capacity (see Chapter 10).

The oxygenation of the roots is important in soilless crops for root respiration, especially as the temperature rises, due to the decrease in available oxygen when temperature increases.

Methods for irrigation scheduling in substrates

In practice, the most usual methods for irrigation scheduling in substrates are outlined below (Medrano *et al.*, 2003).

Measurement of the potential (or tension) of the water in the substrate. The availability of water decreases enormously as the tension rises from 0 to 3 cb (kPa). Therefore, the advisable tension threshold is 1.5 cb (Raviv *et al.*, 1993). The use of tensiometers with pressure transducers has proved effective into these measurement ranges (Terés, 2000), although they are more expensive than conventional tensiometers, which are not effective in substrates due to the measuring range.

The use of *solution level sensors* placed in a tray which supports several substrate units has greatly expanded in commercial greenhouses. Also known as a *demand irrigation tray*, it works by activating the irrigation when the solution level decreases below a pre-fixed threshold.

The *weighing lysimeter*, which registers the weight losses due to transpiration, the *continuous monitoring leaching volume control* and estimation of the *substrate water content* (by means of the dielectric constant) are other methods that are rapidly becoming popular in modern greenhouses using soilless media and hydroponic methods of production.

The control of leachates and their conductivity allows for ensuring the adequacy of the scheduled water supplies while maintaining a fixed leaching percentage. Leaching control is frequently the only reference for scheduling irrigation.

When using these methods, the proper choice of test plants and their location is of primary importance, so that the measurements are representative of the whole greenhouse area, because radiation differences within the greenhouse can be very influential (Soriano *et al.*, 2004a). When collecting the drainage, it is recommended to establish several control points with a minimum of 4 m^2 per sampling point (Urban, 1997b).

The use of models to estimate transpiration, which allows for short-term forecast of the water requirements of the substrate–plant system, is becoming more common. These models are implemented through the irrigation computer. The simplest types estimate the ET as a function of the solar radiation and the most complex models are based on the crop energy balance (see Appendix 1 section A.8.3).

When models are used for estimating the ET, the values are usually increased by around 20% in winter and 30% in summer, under Mediterranean conditions, plus any eventual salinity corrections if any need to be taken into account.

ET estimation methods based on the accumulated solar radiation do not consider the irrigation that takes place at night, which must be included – at least one in winter and two in summer (Urban, 1997b).

The methods based on plant parameters are not used widely in commercial greenhouses. Frequently, two or more of the described methods are used for comparison purposes and for optimizing irrigation management.

When closed or semi-closed substrate-growing systems are used, which recirculate the drainage water, precision in the management of the irrigation doses is not as necessary as in the open systems (free drainage). If enough water is available, it is essential in the closed systems to correct the composition of the nutrient solution, as a function of the nutrient absorption and the salinity increase (Sigrimis *et al.*, 2003).

The small water inertia of substrate crops, derived from the limited volume of water available to the roots, makes it imperative that the required precautions are taken to avoid failures in the system and irrigating in excess. Such precautions include: (i) controlling the drainage; (ii) inspecting the drippers and filters to avoid blockages; and (iii) monitoring any failures in the power supply. Nowadays, alarm systems are available on the market for early detection and correction of such possible failures.

11.6.7 Water use efficiency

The water use efficiency (WUE) has turned into an agronomic expression that is widely used (Howell, 1990; Stewart and Nielsen, 1990), defining the production (photosynthetic, biological or economic) per water unit (transpired, evapotranspirated or applied).

When it refers to transpirated water, the WUE of the transpirated water is the product of the transpiration efficiency (dry matter fixed per water unit transpirated) by the crop index.

The transpiration efficiency is determined, essentially, by genetic factors, being estimated, in open field conditions, at around 67 l of water kg^{-1} of dry matter in CAM type plants, 250 l kg^{-1} in C4 type plants and 500 l kg^{-1} in C3 plants (Fereres and Orgaz, 2000; Hsiao and Xu, 2007; Steduto *et al.*, 2007).

In greenhouses, the WUE is considerably higher than in open-field crop production due to: (i) the lower evapotranspiration (derived from the lower radiation and wind than in open field); (ii) the yield increases (derived from the better climate control);

and (iii) the general use of very efficient irrigation techniques, such as drip irrigation or the recirculation in soilless crops (Jolliet, 1999). Proper climate management can improve the WUE in greenhouses (Sánchez-Guerrero *et al.*, 2008).

The water use to produce 1 kg of tomatoes in the Mediterranean area is of the order of 60 l in intensive open field cultivation (Stanhill, 1980) and between 32 and 44 l in unheated low-tech greenhouses (Castilla *et al.*, 1990b), whereas in the glasshouses with climate control of Northern Europe, with substrate and recirculation of the drainage water, it goes down to 15 l (Stanghellini, 2003). We can, therefore, state that the greenhouse WUE is, at least, double or triple that of open field.

These efficiency differences between open field and greenhouse crops can be much higher in extensively irrigated crops, where the general water distribution (general irrigation networks, channels, etc.) is usually quite inefficient.

Regarding the economic efficiency of the use of water, expressed as total income for the grower (euros) per water unit used (cubic metre), vegetable open field cultivation in Almeria reached a value of €1.60 m^{-3}, whereas in greenhouses it was €6.12 m^{-3} (Colino and Martinez-Paz, 2002).

From the social point of view, we can talk about the social WUE (in terms of hours of labour generated per cubic metre of water used). For vegetable cultivation in the south of Spain the figure for labour generated per cubic metre of water used in greenhouses is 0.24 h m^{-3} compared with 0.09 h m^{-3} in the open field (Colino and Martinez-Paz, 2002).

Several growing techniques contribute to the improvement in the WUE, such as carbon enrichment (Sánchez-Guerrero *et al.*, 2003) or the use of properly managed mobile shading (Lorenzo *et al.*, 2003).

The use of fogging in the greenhouse mitigates the negative effects of the irrigation water salinity on yield, but requires good quality water and, although it decreases the ET, the total water use is higher (Montero *et al.*, 2003).

11.6.8 Quality of the irrigation water

The quality of the irrigation water is determined by the dissolved concentrations of the different ions. The most common indices to classify irrigation waters are the electric conductivity of the water (EC_w), which is a function of the total ionic content and quantifies the water salinity, and the sodium absorption ratio (SAR), which measures its alkalinity. See Appendix 1 section A.8.2.

11.7 Fertilization

11.7.1 Introduction

In protected cultivation, the cost of fertilization is small in relation to the vegetable production costs (see Chapter 14), so fertilization has been usually high, as growers have no incentives to save fertilizers and pretend, mistakenly in many cases, that the crop did not suffer any kind of nutrient deficiency. Nevertheless, nowadays, the trend to minimize the environmental impact has resulted in the adoption of the so-called 'good agricultural practices (GAP) code' (see Chapter 16).

11.7.2 The nutrients cycle (soil cultivation)

In horticulture, and especially in greenhouses, there is more leaching of nitrates than in other agricultural systems, due to the high supplies, the high contents of organic matter in the soil and the surplus irrigation in relation to the ET_c (Dasberg, 1999b). Its environmental impact can be notable, mainly, in the surface and underground aquifers.

The applications of nitrogen fertilizers in the greenhouses normally exceed the crop's requirements, increasing the risks of nitrate leaching to the aquifers (Thompson *et al.*, 2002).

Directive 91-676 of the European Union regulates the protection of waters against pollution caused by nitrates in agriculture and defines the limit at 50 mg of NO_3^- l^{-1} of water. Several areas of the Mediterranean Basin have been declared 'vulnerable to nitrates' in accordance with this directive. The leaching of nitrates comes both from substrate cultivation (open systems) as well as from soil cultivation, in this last case recently documented by Thompson *et al.* (2002).

Phosphorus does not usually cause pollution problems, except in exceptional cases of soils with low phosphorus fixation capacity or when large quantities of animal manure are applied over many years.

The potassium leachate is, normally, limited and does not cause important problems of environmental impact, as it is retained in the soil in high proportions.

Other macronutrients, such as calcium and magnesium, do not cause environmental problems, as they are natural components of the soil, which retains them in large quantities.

11.7.3 Nutrients extractions

It is necessary to know the fertility characteristics and nutrient levels in the soil, making the pertinent soil analysis, to schedule fertilization.

Normally, if the nutrient level is good, fertilization in practice is based on supplying the crop's uptake, corrected for the use efficiency, which allows for maintaining, after the crop cycle, a proper fertility and nutrient level. If the levels of any nutrient are high, or if the irrigation water contains it in sufficient amount, the inputs must be consequently corrected.

As a guide, Table 11.5 summarizes the approximate nutrient uptakes of some horticultural crops.

It is important to know the nutrient absorption dynamics, to adapt the inputs to the extraction rates, which vary through the cycle and are influenced by the climate conditions, especially by soil temperature and radiation.

When the availability in the soil of some nutrients is high, overconsumption may occur (in potassium) or it may negatively affect the quality (nitrogen) of the fruits, in extreme cases being possible to induce salinity, or even phytotoxicity. The availability of nutrients must be balanced and adapted to the plant requirements, to avoid antagonisms and possible restrictions to the nutrient absorption, which allow for optimum fertilization. Therefore, it is frequent to maintain predetermined relations between all or some of the nutrients.

11.7.4 Tolerance to salinity

The tolerance to salinity of the crops may be assessed in several ways. The most extensively used method (Ayers and Westcot, 1976) quantifies the tolerance (Table 11.4) by the percentage of the maximum yield that would be obtained for a certain level of electric conductivity of the saturated extract of the soil (EC_e) or the irrigation water (EC_w)

Table 11.5. Approximate nutrient uptake of some horticultural crops (compiled from very diverse sources).

		Plant uptake (kg ha^{-1})				
Crop	Yield (t ha^{-1})	N	P_2O_5	K_2O	CaO	MgO
Tomato	80	250	80	500	300	70
Pepper	40	180	60	180	160	50
Aubergine	50	250	40	300	150	25
Melon	60	230	80	400	300	70
Cucumber	200	320	160	600	250	100
Squash	40	70	70	390	–	–
Lettuce	40	100	50	250	50	12
Green bean	45	150	15	60	30	6

used. For greenhouse crops the tolerance to salinity can be quantified (Sonneveld, 1988) by means of the irrigation water salinity threshold below which there is no problem, and the percentage of yield decrease experienced by the crop per unit increase of salinity in the irrigation water, above the threshold value. This method is more useful for substrate crops. Table 11.6 summarizes the data in this respect. The specific growing conditions (cultivar, evaporative demand, management, microclimate) may affect these threshold values (Cohen, 2003). In Mediterranean greenhouses, Magán (2003) estimated the salinity threshold value of the nutrient solution to decrease the fresh weight tomato harvest between 4 and 5 dS m^{-1}.

The importance of the water quality to minimize the leaching fraction is enormous, influencing its environmental impact potential. Poor quality water will require considerable leaching and, as a consequence, will generate more negative impact than good quality water.

11.7.5 Fertigation

In HFLI, the restriction of root development, to only a part of the soil colonized by the roots, dictates the need to locate the nutrients in this soil to the volume occupied by the roots to ease their absorption. Therefore, the nutrients, normally dissolved in the irrigation water, are supplied by a localized irrigation system. This practice of joint application of irrigation and fertilization is known as fertigation.

Table 11.6. Tolerance of some vegetables to salinity in greenhouses. Threshold value of the irrigation water EC_w (dS m^{-1} at 25°C) below which there is no problem and percentage of yield decrease per unit increase of EC_w (Sonneveld, 1988).

	Threshold value EC_w (dS m^{-1})	Yield decrease by salinity (%)
Tomato	1.8	9
Pepper	0.5	17
Cucumber	1.5	15
Green bean	0.5	20
Lettuce	0.6	5

The control centre of a localized irrigation facility must have the necessary equipment to fertigate. This involves the use of soluble or liquid fertilizers, allowing for adjustable dosing and fractioning of the inputs which optimizes their use.

Criteria for fertigation

Traditionally, the fertigation criterion of supplying the nutrients as a function of the expected uptake by the plants prevailed (section 11.7.3).

Nowadays, the criterion of providing nutrients based on an ionically balanced physiological solution, used in soilless crops, is extending to conventional soil cultivation, when a suitable automated irrigation head is available.

In soilless cultivation the correction of the nutrient solution is performed based on its analysis. In soil cultivation, the classic method of analysing the saturated soil extract is being replaced by the use of suction probes, with which a sample of the soil solution is extracted for analysis. However, information on the ideal nutrient levels to use with this method is still scarce.

11.7.6 A practical example: a soil-grown tomato crop

Depending on each case's specific conditions (soil fertility, climate, irrigation type), there is notable variation in tomato fertilization (Castilla, 1995). Preliminary analysis of the soil is necessary.

In general, fertilizers are applied depending on the crop's estimated nutrient uptake. Although the variability in nutrient uptake is enormous (Zuang, 1982; Castilla *et al.*, 1990a), values that refer to harvest unit are in general lower, as estimated by different authors (Ward, 1964; Maher, 1976; Bar-Yosef *et al.*, 1980; Zuang, 1982; Castilla and Fereres, 1990):

- between 2.1 and 3.8 kg of N t^{-1} of harvest;
- between 0.3 and 0.7 kg of P t^{-1} of harvest;

- between 4.4 and 7.0 kg of K t^{-1} of harvest;
- between 1.2 and 3.2 kg of Ca t^{-1} of harvest; and
- between 0.3 and 1.1 kg of Mg t^{-1} of harvest.

The differences in nutrient uptakes are influenced by the type of pruning and, especially, by the timing of the removal of the axillary shoot. It is advisable to prune shoots as soon as possible to minimize the wasteful uptake of nutrients by the crop (Castilla, 1985).

The scheduling of fertilizer application must rely on the type of fertilizer used, on the irrigation technique and on the soil conditions, among other factors. In sandy soils, with low water storage capacity, supplies must be frequent with the irrigation (conventional), whereas in heavy soils it is only necessary to apply part of the nitrogen as a top dressing (Geisenberg and Stewart, 1986).

With surface irrigation, the most common practice is to apply the phosphorus with the pre-planting fertilization, for example when applying manure (around 30 t ha^{-1}), and at a time when half of the potassium is applied. The rest of the potassium and nitrogen are applied in alternate weeks after transplanting until 1 month before the end of the cycle (Nisen *et al.*, 1988). With drip irrigation, all fertilizers can be applied by fertigation, although it is common that at least part of the phosphorus is applied with the manure.

In drip irrigation, it is essential to know the absorption rhythm of the mineral elements in order to schedule fertilization (Zuang, 1982). In Mediterranean unheated greenhouse crops for autumn–spring cycles, fertilization rates higher than 0.3 g N m^{-2} day^{-1} do not seem advisable (Castilla, 1985). The fertilizer's content of the irrigation water is, in some cases, notable and must be taken into account for the fertilization schedule.

Nitrogen excesses negatively affect fruit quality, and maintaining an N:K ratio at 1:2 (or even 1:3) during the fruit enlargement stage, with drip irrigation, favours their quality (Geisenberg and Stewart, 1986). Equally, the balance between other nutrients, especially between calcium (when its supply is required) and potassium, and magnesium is necessary, as well as between the different forms of nitrogen (nitric/ammoniacal).

In drip irrigation, the amount of salts in the water must be limited, if possible, to 2 g l^{-1} (which is not feasible, in some cases, when saline water is used), to decrease possible dripper blockage problems. When using good quality water, it is a usual practice to add sodium chloride (common salt) to the water, up to the indicated limit, to improve tomato quality, because the soluble solids content increases with salinity which contributes to the improvement of its internal quality, although the fruit size is reduced.

A good irrigation efficiency is, logically, required for efficient fertigation and also contributes to a significant reduction in the environmental impact of fertilizer (nitrogen, especially) residues.

Foliar fertilization, in tomato crops, is usually limited to microelements, when deficiencies are forecasted or observed. Leaf analysis (of the limb, petiole or the whole leaf) is a good auxiliary index on which to base the scheduling of fertilization, being more common than sap analysis, as the latter displays a wider variability and requires more thorough sampling (Chapman, 1973; Van Eysinga and Snilde, 1981; Morard, 1984).

In greenhouse crops, low soil temperatures (15°C) in winter may limit absorption of nutrients, especially phosphorus (Wittwer, 1969) and nitrates (Cornillon, 1977). On the other hand, high temperatures favour nutrient absorption, although the nutrient uptake per harvest unit is not affected, as previously thought (Nisen *et al.*, 1988).

11.7.7 Fertigation of soilless crops

The nutrient cycle in soilless crops

Soilless crops, with free drainage, have similar problems to soil cultivation regarding the nutrient cycles. The management conditions (leaching percentage, characteristics of the nutrient solution) of the soilless crop (open system) will determine its environmental impact, which will be similar to that of crops grown in soil if the leachates are similar.

If leachates are recirculated (closed system) the salinity and pathology problems must be considered and the nutrient concentrations must be well monitored and controlled.

Preparation of the nutrient solution

In an ideal soilless growing system there are no mineral inputs from the substrate, and therefore, all nutrients must be supplied together with the water, in the nutrient solution.

The preparation of the nutrient solution requires prior analysis of the irrigation water, to allow for the formulation of the best nutrient solution depending on the crop to be grown. The preparation of this nutrient solution will also depend on the technical characteristics of the available fertigation hardware (and, possibly, software).

In the simplest case, one concentrated solution tank is available and another tank for the acid. Most facilities have two tanks for solutions (A and B), one of which already contains the acid. In this case, tank A has most of the acid to correct the pH (usually, nitric or phosphoric, and rarely sulfuric), the phosphates and the sulfates, as well as the microelements, except for iron. In this tank A, part of the potassium nitrate can be incorporated, but no calcium salts must be added, to avoid precipitates. Tank B contains the calcium nitrate and the potassium nitrate (all or only a part), as well as some nitric acid to regulate the pH and the iron chelates. The magnesium nitrate is usually added in tank B, but neither sulfates nor phosphates must ever be added, to avoid precipitates.

When three tanks are available, one of them is destined only for the acid that is usually nitrous acid, although sulfuric or phosphoric acids can be used. In sophisticated facilities, managed by means of a computer, several tanks are usually available, containing solutions of individual fertilizers.

The injection systems of concentrated solutions in the irrigation water flux are of such complexity or simplicity in agreement with the type of tanks used.

In mixing the fertilizers their solubility and compatibility must be taken into account, not forgetting that it depends on temperature (Tables 11.7 and 11.8). The literature on the preparation of simple solutions is extensive (e.g. Cadahía, 1998).

Table 11.7. Solid fertilizers most commonly used in fertigation: analysis and solubility at 20°C. (Source: Cadahía, 1998.)

Fertilizer	Analysis of N-P_2O_5-K_2O-others[a] (%)	Solubility (g l^{-1})
Calcium nitrate $4H_2O$	15.5-0-0-26.6 (CaO)	1200
Ammonium nitrate	33.5-0-0	1700[b]
Ammonium sulfate	21-0-0-22 (S)	500
Urea	46-0-0	500
Potassium nitrate	13-0-46	100–150
Potassium sulfate	0-0-50-18 (S)	110
Mono potassium phosphate	0-52-33	200
Mono ammonium phosphate	12-60-0	200
Magnesium sulfate $7H_2O$	16 (MgO)-13 (S)	700
Urea phosphate	17-44-0	150
Magnesium nitrate $6H_2O$	11-0-0-9.5 (Mg)	500

[a]The first three values in each entry refer to N-P_2O_5-K_2O. Where there is a fourth entry this refers to other compounds. The exception is the entry for magnesium sulfate which has no N-P_2O_5-K_2O and contains16 (MgO)-13 (S) as indicated.
[b]Steep water temperature decrease for concentrations above 250 g l^{-1}.

Table 11.8. Chemical compatibility of the mixture of some common fertilizers in fertigation: I, incompatible; C, compatible. (Source: Cadahía, 1998.)

	NO_3NH_4	Urea	$(NH_4)_2SO_4$	$(NH_4)_2HPO_4$	$(NH_4)H_2PO_4$	KCL	K_2SO_4	KNO_3	$Ca(NO_3)_2$
NO_3NH_4	–								
Urea	C	–							
$(NH_4)_2SO_4$	C	C	–						
$(NH_4)_2HPO_4$	C	C	C	–					
$(NH_4)H_2PO_4$	C	C	C	C	–				
KCL	C	C	C	C	C	–			
K_2SO_4	C	C	C	C	C	C	–		
KNO_3	C	C	C	C	C	C	C	–	
$Ca(NO_3)_2$	C	C	I	I	I	C	I	C	–

Parameters of fertigation with soilless crops

The proper management of fertigation requires the periodic analysis of the nutrient solution to assess its goodness of fit to the requirements of the crop and to perform necessary adjustments to its composition. In addition, it is necessary to frequently monitor (automated or manual) the pH and EC (electrical conductivity) of the nutrient solution and the leachate, to prevent any anomaly.

In practice, when a computer is available, a certain threshold of EC of the nutrient solution is fixed, for instance 2.5 dS m^{-1}, modulated as a function of the solar radiation, decreasing it by 0.1 dS m^{-1} for each 30 W m^{-2} of solar radiation that exceeds 400 W m^{-2} (Urban, 1997b). Obviously, these rules must adapt to the specific conditions of each operation.

Normally, the pH is not regulated but a certain value is fixed. The EC and pH sensors must be duplicated, at least, to prevent an eventual failure. In addition, the system must be fitted with alarms.

Other complementary analyses are carried out on the substrate solution (extracted with a syringe) and the drainage, to correct the nutrient solution.

Analysis of vegetable tissue and sap provide information on the nutrients that are really absorbed by the plants (Cadahía, 1998). The analysis of the conducting tissues is usually preferable to that of the leaves, whose composition varies slowly. In these analyses, the time variations are more relevant than the absolute values (Morard *et al.*, 1991).

Pathogens in the drainage waters

The recirculation of the drainage water requires the use of good quality water and its disinfection, to suppress pathogens (bacteria, fungi and virus) in the recirculation water.

The use of ozone, UV sterilization, thermal treatment and ultrafiltration are effective to a varying extent, but the last technique has the drawback that only 70–80% of the drainage water is recovered (Dasberg, 1999b). The use of bleach in recirculation water disinfection gives good results. Treatment with UV radiation is effective, but it is necessary to pre-filter the water so the radiation penetrates well (Dasberg, 1999b).

Automation

The use of computers, with several degrees of automation to manage the fertigation, is growing among greenhouse growers. It must be expected, in the future, that these systems will become integrated with climate control systems, in those greenhouses in which the technological level allows for it, for combined optimization (see Chapter 12).

11.8 Summary

- Water fulfils a number of basic functions for plants. Transpiration may involve up to a maximum of 98% of the total amount of water absorbed by cultivated plants in their life cycle.

- Transpiration requires energy for the water evaporation process, which mainly comes from the sun. Greenhouse heating also contributes to the evaporation process.
- The maximum transpiration values in greenhouse tomatoes in the Mediterranean area are of the order of 6 mm day^{-1}, although hourly rates of 1–1.5 mm may be possible.
- Soil water relations are characterized by its water retention capacity and by its water potential (energy status of the water in the soil).
- The water content in the soil is usually quantified by its moisture content (gravimetric or volumetric), by the readily available water and by its matrix potential or matrix tension.
- The most usual sensors used to measure the water in the soil are tensiometers, which measure the matrix tension (absolute value of the matrix potential) in the soil.
- Characterization of the water status of the plant (using methods based on measurements of the size of plant organs, of the sap flux, plant temperature, etc.) for irrigation scheduling is not common in practice.
- The guidelines to evaluate the irrigation water quality are based on its electric conductivity (EC), which depends on its dissolved salts content, and the SAR (sodium absorption ratio) (Table 11.9). Both indexes quantify the capacity of salination (EC) or alkalization (SAR) of the irrigation water.
- When a plant suffers water deficit or stress its growth slows down, possibly negatively affecting the yield, if the deficit is sufficient.
- Nowadays, the most common greenhouse irrigation systems are the high-frequency localized irrigation (HFLI) systems (drip and similar), which require different management to that of conventional irrigation systems.
- Irrigation scheduling involves giving a response to two basic questions: (i) When to irrigate?; and (ii) How much water to apply?
- The usual methods to schedule drip irrigation in greenhouse soil-grown crops are: (i) methods based on climate parameters to estimate the ET of the crop; (ii) methods based on soil parameters (tensiometers mainly); and (iii) methods based on plant parameters (seldom used).
- For soilless crops, the most usual methods are: (i) methods based on climate parameters to estimate transpiration; and (ii) methods based on substrate parameters (tensiometers, level sensors, weighing devices).
- The different irrigation scheduling methods, in both soil- and soilless-grown crops, can be complementary.
- The water use efficiency (WUE) defines the production (photosynthetic, biological or economic) per water unit (transpired, evapotranspiration or applied).
- In greenhouses, the WUE is, at least, double or triple that of the open field.
- Fertilization of horticultural crops must be based on replenishing the nutrient taken up by the crop, corrected as a function of the application efficiency, and avoiding excessive supplies that can generate aquifer contamination by leached nutrients and soil salinization.
- The salinity tolerance level of vegetable crops involves a decrease in yield when the tolerance thresholds are exceeded, in the water or in the soil.
- Fertigation, joint application of irrigation water and fertilizers (dissolved in the irrigation water), is compulsory if localized irrigation systems are used.
- Irrigation of soilless crops requires special management: supplying an excess of 20–30% of water, and, trying to maintain pre-fixed values of EC and pH in the nutrient solution. The analysis of the drainage solution (EC and composition) allows for adjusting

Table 11.9. Guidelines for the evaluation of the quality of irrigation water. (Source: Ayers and Westcot, 1987.)

	No problem	Growing problem	Serious problem
Salinity (EC_w in dS m^{-1})	<0.7	0.7–3.0	>3.0
Infiltration sodium absorption ratio (SAR)			
0–3 and EC_w	>0.7	0.7–0.2	<0.2
3–6 and EC_w	>1.2	1.2–0.3	<0.3
6–12 and EC_w	>1.9	1.9–0.5	<0.5
12–20 and EC_w	>2.9	2.9–1.3	<1.3
20–40 and EC_w	>5.0	5.0–2.9	<2.9
Specific ionic toxicity			
Sodium (me l^{-1})			
Surface irrigation	<3.0	3.0–9.0	>9.0
Sprinkler irrigation	<3.0	>3.0	
Chlorides (me l^{-1})			
Surface irrigation	<4.0	4.0–10.0	>10.0
Sprinkler irrigation	<3.0	>3.0	
Boron (mg l^{-1})[a]	<0.7	0.7–3.0	>3.0
Diverse effects			
Nitrogen (N in mg l^{-1})	<5.0	5.0–30.0	>30.0
Bicarbonates (me l^{-1}) sprinkler irrigation	<1.5	1.5–8.5	>8.5
pH		Normal range 6.5–8.4	

[a]Values for boron are the same for both surface irrigation and sprinkler irrigation.

the nutrient solution and the water applications.

- The recirculation of the drainage water in soilless crops (closed systems) requires the use of good quality water and can improve their WUE, but requires disinfection of the recycled water to prevent diseases.
- Nowadays, the use of fertigation management computers is expanding.

12

Regulation and Control Systems: Computer Climate Management – Mechanization

12.1 Regulation and Control Systems

12.1.1 Introduction

The regulation of a process, in a physical system, calls upon a set of dedicated techniques and materials to ensure that, under all circumstances, the physical quantity to be regulated equals a desired value, which is called the set point (Urban, 1997a).

The general principle of regulation is as follows:

1. The value of the parameter to be regulated is quantified, by direct measurement or by calculation. For instance, to regulate the opening of the vents as a function of the greenhouse air temperature, the air temperature is measured, which happens to be 25°C.
2. This measured value is compared with the parameter's set point, which can be predefined by the user or be the result of a calculation or the application of a pre-established rule. The air temperature set point value, in this example, can be fixed in 23°C, which is lower than the measured value (25°C).
3. Finally, by means of one or several actuators, one or more pieces of equipment begin to operate, to decrease the difference between the measured and the set point values of the parameter. In this example, the motors opening the vents would start operating, opening the vents to ventilate and approximate the air temperature of the greenhouse to the set point (23°C).

12.1.2 Input–output systems

In classical control, the processes to be regulated are considered as input–output systems. It may occur that there is more than one input or output (Fig. 12.1). The inputs can be of two types: (i) control inputs; and (ii) exogenous inputs (or disturbances).

Figure 12.1 schematizes an input–output system (Fig. 12.1a), an input–output system with a disturbance (Fig. 12.1b), and the inputs and outputs for the climate control of a greenhouse (Fig. 12.1c). In this last case, CO_2 enrichment, heat supply and vent opening for ventilation are considered as inputs of the system. The disturbances are the outside temperature, the outside wind direction and velocity, and the external global radiation, humidity and CO_2. The outputs are the inside temperature, humidity and CO_2. Considering radiation as a disturbance, even though it is essential for photosynthesis, is due to the fact that it is not a value that can be controlled by the user.

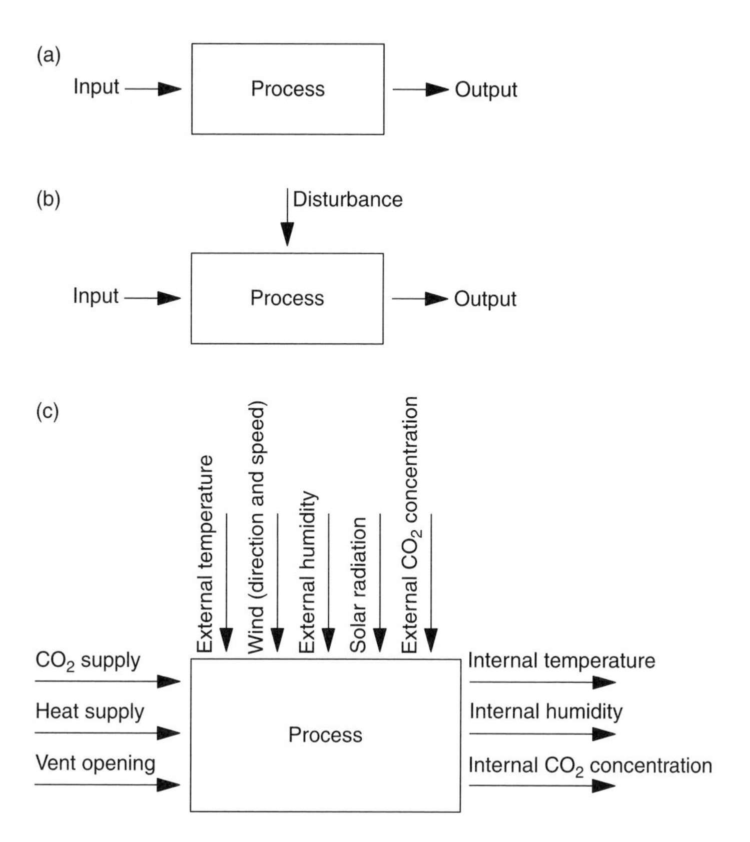

Fig. 12.1. Scheme of control systems: (a) input–output system; (b) input–output system with a disturbance; (c) greenhouse climate control system that details the inputs, outputs and the disturbances (external climate parameters) (adapted from Bakker *et al.*, 1995).

12.1.3 Regulation methods

There are two methods of regulation: manual and automatic. Manual regulation is not in use for the central boiler heating system and for fertigation, but it is still used for ventilation, shading and humidification. The disadvantages of manual regulation are: (i) that it is not possible without an operator; (ii) it is imperfect if there are no measurement instruments; and (iii) manual switches are imprecise (e.g. to set the duration of humidification, or the opening rate of the vents). However, manual regulation is essential when the climate conditions are exceptional (e.g. intense frost); its possible use must be foreseen.

Automatic regulation can be electromechanical or electronic. In electromechanical regulation, the parameter in question is regulated as a function of the set point value of the parameter. In electronic regulation, the parameter can be regulated as a function of the values of one or several parameters (i.e. night air temperature as a function of the previous day's radiation).

There are two types of regulation: 'closed loop' or 'open loop'. 'Closed loop' regulation takes into account only the average values of the parameter to be regulated. 'Open loop' regulation also considers the values of other parameters (e.g. to regulate the air temperature, it also considers the wind velocity or radiation).

12.1.4 Application to climate management

Climate management allows for simultaneously maintaining the set of climate factors (temperature, humidity, CO_2) close to pre-established set point values, respecting certain rules (absolute or conditional prohibitions, priorities, time delays) imposed by the user. The climate computer manages the climate.

We can distinguish several levels of climate management:

- *Level 1*: (base level) The time scale is very short (about 1 min). It excludes processing of the information. Most of the actual climate control computers employ this level of management nowadays.
- *Level 2*: The time scale is of the order of 1 h or a whole day. The objective is the management of the physiological functions involved in the growth and development of the plants in the short term (photosynthesis, transpiration). It involves the use of models.
- *Level 3*: The time scale is longer than 1 day. This level is the bio-economic optimization and strategic decision support. It allows solutions to be obtained that are close to the economic optimum, in each case. The realization of level 3 is ideal and has not yet been achieved in practice.

12.1.5 Types of controllers

The controllers can be classified depending on the type of regulation, which is the way in which the correction calculation can be made.

There are two main types of regulators: (i) *non-progressive controllers* that only regulate fixed positions of the controlled device; and (ii) *progressive controllers* that control any position.

Non-progressive controllers

In the 'on/off' type, the actuator can only take two positions: on or off. Mechanical ventilation, for instance, starts if the interior temperature exceeds 23°C and stops if it decreases below 20°C, pre-fixed set values. A set point value can be used for a 'dead zone', detailed later.

The 'on/off' mode is usual in dynamic ventilation, CO_2 injection, humidification, shading and hot air heating.

A disadvantage of the 'on/off' mode is the frequent starts and stops around the set point value. There are three ways to avoid it:

1. Using a time delay: After the equipment starts, it cannot stop until a certain minimum time has lapsed. It is used, for instance, in air heating. Once the equipment starts (for example, because the air temperature is at 14°C, lower than the fixed set point of 15°C) it has to operate, for instance, for 5 min before stopping, although the set point is reached again before this period ends. Therefore, the air temperature will exceed the set point value, delaying the next start.
2. Using a dead zone: Around the set point value a dead zone is fixed (x), so the regulated equipment for a certain set point (c) starts when the value ($c - x$) is reached and stops when ($c + x$) is exceeded. In this way frequent starts and stops can be avoided. In the previous example, we could fix a set point value (c) at 15°C, and the dead zone (x) at 1°C, so the system would start at 14°C and stop at 16°C.
3. Using average values: These are used for those parameters that can change a great deal, such as wind velocity or the light. As average values are used as set points, the variability is highly cushioned. For instance, as the wind velocity oscillates a lot because the wind is frequently gusty, the average value of the measurements of a set period is used instead of the last instantaneous air velocity measurement.

Progressive controllers

In the progressive type of controllers, the operation of the equipment is modulated to maintain the parameter values to be controlled inside the interval of pre-established set point values.

The most common progressive controllers are: (i) the proportional control (P); (ii) the proportional integrated control (PI); and (iii) the proportional integrated derivative control (PID).

Proportional regulation

In the ventilation process of high temperature or high humidity, when the temperature or the humidity reaches the set point the vent opening is operated. If this opening is operated proportionally to the measured temperature or humidity excesses, in relation to the threshold value, a proportional regulation is applied. Thus, if the difference is small, the vents will open a little and if the difference is very large they will open 100%.

The band of proportionality opening must be defined. If the band is 6°C and the set point temperature is 20°C, the vents will open 50% when the greenhouse temperature is 23°C, and will open 100% when it is 26°C. These band values may be increased or decreased, depending on the external temperature and wind velocity. Proportional regulation is also used in thermosiphon heating systems although in a more complex way.

Proportional integrated control (PI)

In a proportional controller (P) the amplitude of the action (the percentage opening of the vents in the previous ventilation example) is proportional to the difference between the average value and the set point value (temperature in the ventilation example).

In a PI the amplitude of the action is proportional to the integral of the differences between the average value and the set point. It eliminates the deviations step by step (Fig. 12.2).

In a similar way, the PI control is used in thermosyphon heating systems. If the interior temperature control is not linked to the external climate conditions (temperature and wind velocity), the control operates in feedback mode, so it only acts when the interior temperature changes and induces, after checking the set point, activation of the equipment if applicable. If, on the contrary, the alteration of the external climate parameters influences *a priori* the control (before it affects the internal temperature) the control operates in 'feedforward' mode. For example, a heating system by feedforward control can start or increase the energy supply induced by an increase in the wind velocity (which contributes to the cooling of the greenhouse), although the

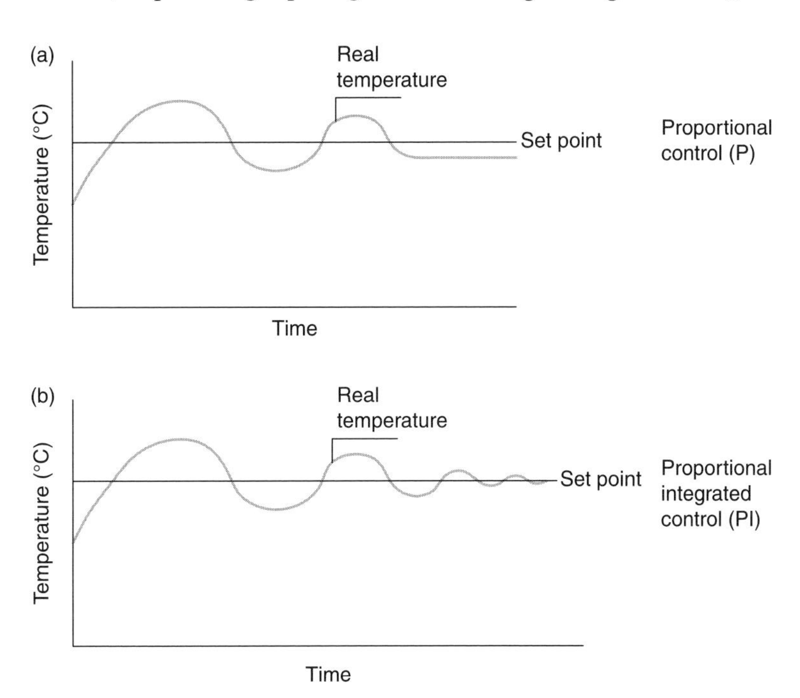

Fig. 12.2. Graphical representation of the performance of a good ventilation control system depending on the temperature. (a) Proportional control (P); (b) proportional integrated control (PI).

interior temperature has not yet decreased (which will happen after a certain time). In these cases, the rule or model that relates the wind velocity with the later predictable greenhouse air temperature changes must be pre-fixed.

The PID control improves the performance of the PI control.

In the control process of a certain parameter, the control system starts the actuators or equipment when it must correct the values of the parameter. The values of the parameter are still periodically measured and this information allows, if necessary, to correct the actions. In this way, the system can correct the control actions by means of 'feedback' control.

When a disturbance occurs, and its effect on the parameter being controlled is known, immediate action can take place, preventatively. This action is the 'feedforward' control, as already detailed.

Frequently, in the management of the process both feedback and feedforward intervene simultaneously. For instance, when the temperature of a greenhouse exceeds the set point, due to a radiation increase, the computer opens the vent (feedback). If later the wind velocity or the external temperature increases, the control system can adjust the vent opening in advance (feedforward).

Another type of control configuration is the cascade configuration that is used in complex processes.

12.1.6 Selection of the type of automatic control

In the simplest systems, non-progressive controllers (on/off) can be used. To select the type of progressive controller, it is recommended to choose the P type (simple proportional control) whenever possible. If the set points are not to be exceeded, the PI controllers must be used. When the speed of the process requires it, PID must be used.

Once the type of controller has been chosen, the management value of the parameters to control must be selected.

12.1.7 Models

Introduction

A model is a simplified representation of a system or one of its parts. The greenhouse, the crop and its management constitute a system. Normally, a model is represented by a number of mathematical equations.

There are a large number of model types. A *static model* is a set of equations that relate several aspects such as, for instance, heat losses, or ventilation, that occur at a time when, essentially, the system is balanced; thus, a static model can be considered a *steady-state model*. In these models the equations are based on physical laws, so they are called *mechanistic models*.

A *dynamic model* incorporates the time variable. These models are necessary when a process whose response is slow is represented, such as the heating of the soil. They are *stochastic models*.

The term *heuristic* or *stochastic* refers to the mediums used in the resolution of the models, thus, heuristic models are solved by exploration or by means of trial and error, whereas stochastic models are solved using statistical methods.

The feedforward control systems use models, which determine the predictable effects of a disturbance in the regulated process and, preventatively, adjust the set points to this new situation.

In greenhouses, two groups of models can be distinguished: (i) *physical* models, which focus on the greenhouse microclimate as a function of the external climate; and (ii) *physiological* models that focus on the plants and their relations with the greenhouse microclimate.

Simulation models can be of any kind, from the simplest to the most complex, and are of great use, if they are well conceived and validated, to simulate several real situations at a low cost. A very simple example, in Mediterranean greenhouses, is the simulation model of transmissivity to solar radiation (Soriano *et al.*, 2004b). This has been of great use in designing new low-cost greenhouse structures that are more efficient in capturing solar radiation at a low cost. At a more complex level, there is a diversity of simulation models, both for energy balance and for crop growth and production.

Use of models

Models have constituted a very useful tool for research of the greenhouse physical environment and the crop's growth and production (Challa, 2001).

At the beginning, obviously, simple stationary models were used. The use of models in the design and management of control systems has been widespread and very positive, but its application on a commercial scale has been limited and restricted to well-equipped greenhouses (Gary, 1999).

The simplest models, such as the rule of thumb, and scale models, have been used in Mediterranean greenhouses (Soriano *et al.*, 2004b). In these greenhouses, at different levels of complexity and from the practical point of view, the models which have attracted the most interest are those of irrigation control and analysis of yield potential of the crops. Nowadays, in well-equipped greenhouses, the control of the air temperature (that is regulated depending on the available light) is widely used and this is based on a simple model.

However, there are reservations in using models, from the user's side, and these stem from the need for simple, robust and universal models (Bailey, 1999). In addition, previous work gathering relevant information (for instance, on assimilates distribution) needs to be done prior to the application of a model, and in many cases this information is not available (Gómez *et al.*, 2003).

A primary aspect to be considered in greenhouse climate control models is that the grower's goal is to maximize the profit (Bailey, 1998; Fig. 12.3). Therefore, and given the normally existing variability in a crop, the grower/user must be the one who finally makes the decisions.

In Appendix 1 section A.9.1 the solar radiation transmissivity models most commonly used in greenhouses are listed.

12.2 Computer Climate Management

12.2.1 Controls performed by greenhouse management systems

In heating, the primary goal of the control systems is to adapt the heat supply to the crop requirements. The secondary goal is to dehumidify the air.

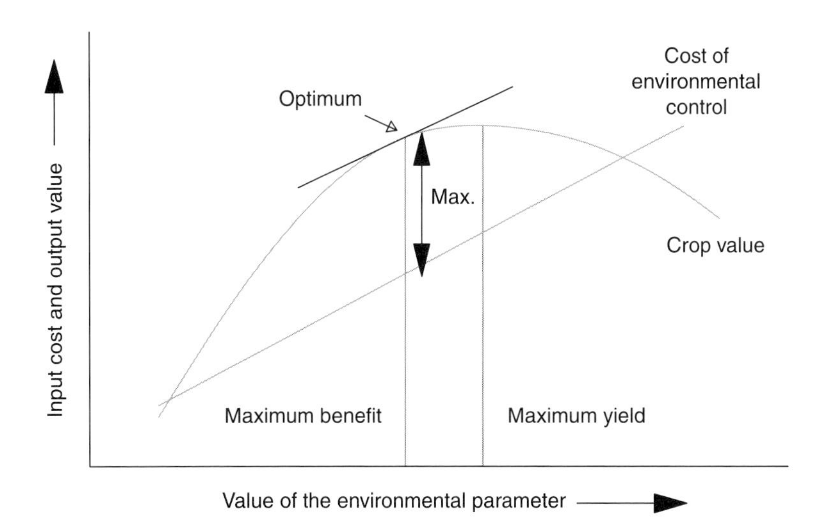

Fig. 12.3. Graphical representation of the economic optimization of greenhouse climate control (adapted from Bailey, 1998).

The main goal of ventilation regulation systems is to avoid the interior air temperature exceeding the fixed threshold. Secondary objectives are to dehumidify and favour the input of CO_2. Temperature, humidity and CO_2 sensors are needed for their management. They may be limited by the rain or the wind.

The only function of the shading control system is to decrease radiation, normally to reduce the temperature at times of high radiation load.

The supply of CO_2 is only practised during the daytime, with intervention of the CO_2, radiation and vent-opening sensors.

Humidification tends to maintain the hygrometry, using humidity and air temperature sensors.

For the regulation of all these systems (thermal screens, dehumidification systems) several sensors can be used. The simplest control systems use clocks.

12.2.2 Digital control systems

Systems developed during the Second World War enabled analogue technology, which used electrical circuits to obtain inputs (measurement of environmental parameters) and calculate, automatically, outputs, to control mechanisms and equipment. The arrival of digital control systems, which could manage more complex systems at lower cost, has helped digital control systems to supersede analogue control systems.

A digital control system is basically composed of: (i) the controller, that is, the climate computer; (ii) the correction equipment (heating, ventilation, etc.); and (iii) sensors, to measure the different parameters to be regulated.

12.2.3 The climate control computer

The climate control computer controls different processes to regulate, mainly, temperature, humidity, light, CO_2 and air circulation. Its functions are to measure different parameters, perform calculations with resident programs, and give activation orders to existing equipment, to maintain the regulated parameters within the desired values (set points).

When sensors are monitored using analogue technology, the signals must be converted into digital information before they can be interpreted by the computer. For this, an analogue–digital converter (ADC) is used. When a sensor generates a measurement signal that is not interpretable by the ADC, an interface that adapts the signal (to make it readable by the ADC) is used. For instance, a solar radiation sensor generates a potential difference, which is proportional to the incoming radiation, in the form of an analogue signal that is converted into a digital signal by the ADC converter in order to be interpreted by the computer.

The activation orders of the computer or output signals, at low tension (24 volts), activate relays that operate the different correction equipment.

Until now, different computers performed the fertigation and climate management. Nowadays, the trend is to integrate them, which allows for a better joint management.

12.2.4 Functions of climate control computers

It is impossible to provide a full list of all the possible functions of climate control computers, because each user has specific requirements. The ones detailed below are the minimum required for a well-equipped greenhouse.

The set points are generally different during the day and the night, and can vary even during the same period (day or night). A clock can perform the day/night changes, or changes can be triggered by measurements of the radiation or by calculations of sunrise and sunset (depending on the latitude and date; i.e. the astronomical clock).

Temperature control

In the simplest systems, the user normally fixes a temperature below which the heating system is activated (heating set point), being common to use different set points during the day and the night. In addition, the user indicates a maximum temperature, above which the vents open (ventilation set point). Equally, the system can control water evaporation equipment (fog, pad and fan) or a shading screen. Nowadays, most systems can adjust the set points for several independent periods or recalculate them periodically.

The temperature set point can be modulated as a function of other parameters, such as radiation, increasing the set points as radiation increases. It is usual to fix the night set point temperature as a function of the radiation of the previous day.

In thermosyphon heating systems, the temperature of the heating pipes is usually controlled independently from that of the air, it being usual to maintain a minimum pipe temperature, to achieve a leaf temperature higher than that of the air with the aim of avoiding *Botrytis* (induced by water condensation).

The presence of a thermal screen affects the temperature set points. If soil or substrate heating is available, besides air heating, they must be controlled in coordination with each other. The management of the screens can be done with a clock, by radiation or by temperature. The opening of screens must be gradual.

In exceptional cases (heavy frost) the set points can be unreachable due to insufficient capacity of the existing heating system. In these cases survival temperature set points, lower than the usual are used.

Hot-water heating systems have considerable thermal inertia. Therefore, their operation should be scheduled in advance, using a proportional controller. In air heating systems the on/off control is used.

A typical example of the changes in heating and ventilation set point temperatures is represented in Fig. 7.12, in thermosyphon heating systems. The set point temperatures change before sunrise. At sunrise, transpiration increases quickly raising the humidity whereas temperature increases more slowly, causing condensation on the plants that favours fungal attacks. To avoid this situation the heating set point must be progressively increased before sunrise (point A, Fig. 7.12). At sunset a similar procedure is followed, for energy-saving purposes, progressively decreasing the heating set point from point C (Fig. 7.12).

In case of contradiction between the ventilation control orders as a function of temperature and humidity, priority control by humidity is usually established.

For control of high temperature by ventilation a proportional controller is normally used. As the ventilation rate is difficult to measure, a temperature set point is used instead, corrected by the wind velocity and the interior–exterior temperature difference (Fig. 12.4). The vent opening percentage is measured or estimated. Knowledge of the wind direction allows for choosing which vents to be opened: that is, the windward vents (facing the wind) or the leeward vents (opposite to the wind). Normally, the leeward vents open first. Maximum and minimum vent openings (in degrees) must be pre-established, in case of storm, frost or rain.

Hygrometry control

If the greenhouse has a de-humidification system, which is quite infrequent, it can be activated with a humidity set point, to decrease the humidity.

If the greenhouse does not have a heating system the only way to limit the humidity is to ventilate. When heating is available, the humidity excesses can be avoided by heating and ventilating, although at a high energy cost. The humidity set points are different during the day and the night.

In some crops, such as tomato, this simultaneous heating and ventilation is performed every morning to decrease the RH and avoid condensation on the plants.

The fog or pad and fan systems can be activated when the hygrometry is low, normally during the daytime.

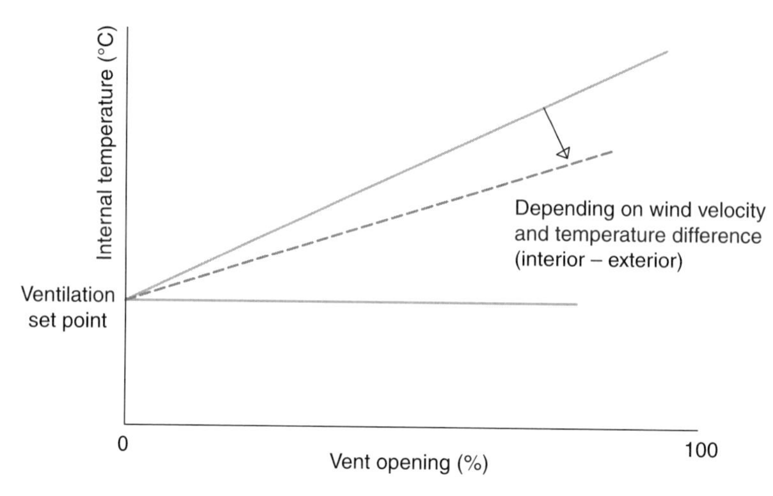

Fig. 12.4. Scheme of the proportional control of the vent opening to ventilate, depending on the internal temperature. The slope of the line depends on the wind velocity and of the internal–external temperature difference (adapted from Bakker *et al.*, 1995).

Light control

Photoperiodic illumination and darkening screens are managed with clocks. Shading screens are controlled by means of maximum radiation or temperature thresholds. It is usual to maintain openings or slits in the screen, to avoid affecting the ventilation and to maintain the 'chimney effect' of ventilation. As the screens are sensitive to the wind, screens cannot be deployed when the wind is above a certain speed (see Chapter 9).

Artificial lighting, for photosynthesis, is activated by a clock or by temperature set points.

CO_2 management

The CO_2 set point can be modulated as a function of the temperature and the radiation. In practice, it can be modified depending on the wind velocity and direction and the degree of vent opening (see Chapter 9).

Screens control

Thermal screens are deployed during the night and gathered in during the day. The deployment is done when the temperature difference (between the greenhouse and the exterior) exceeds a pre-fixed value.

The opening of the screen is done, at the pre-fixed time or depending of the light level, gradually to avoid a sudden fall of the cold air mass over the crop. When humidity is excessive the screen can slightly open to evacuate the excess humidity.

Shading screens are managed with two light levels, one to deploy and one to retract. When the interior temperature is excessive, an opening must be left to avoid excessive blockage of ventilation.

Darkening screens are controlled by clock and must perfectly block the light in order to effectively shorten the day length.

Alarms control

Alarms are essential to prevent crop and property damage. The control systems usually incorporate a series of security functions, whose thresholds are fixed by the user.

The most important climate alarms are the ones announcing a violation of the minimum and maximum temperature set points. Other alarms relate to the humidity, CO_2 and screens set points.

The existence of alarms does not eliminate the need for preventative maintenance (verifying probes, circuits, etc.). The system must be protected against electromagnetic and electric disturbances, especially in zones where storms are common (e.g. to avoid damage by lightning).

Communication with the user

The user can change the program set points and also identify the factors that must be considered to respect the set points, intervention priorities, and intervention delays. Communication with the user allows for remote control of equipment when the appropriate communication interface is available. The most commonly used systems are wireless communication (by radiofrequency), and phone and wire (with conventional cable) communication.

An essential aspect to take into account in communication with the user is the alarm notification in case of serious failure.

All the data registered during the day can be stored. In addition, the system may have the usual performance of a personal computer, providing it with the required elements (data visualization screens, data downloading, printer, etc.).

12.2.5 Towards integrated control

Integrated management of fertigation and climate control is already used in some modern commercial greenhouses.

In the future, in order to optimize control in well-equipped greenhouses it will be necessary to also integrate plant growth management and economic aspects of production with the climate control and fertigation, so that the new generation of growers will have to be experts in interpreting and using technical information in decision making (Papadopoulos and Hao, 1997a, b). Before this happens, it will be necessary to generate the required information about plant growth and other non-documented aspects of the local conditions.

12.3 Mechanization

12.3.1 Introduction

In the conception of the general arrangement of a greenhouse, besides the structural design and energy considerations (see Chapter 4), other aspects that must be taken into account include: (i) accessibility to all points of the greenhouse; (ii) efficient use of the available machinery; and (iii) for the labour force, ergonomic efficiency.

Labour costs, which constitute the main variable cost in most greenhouses (see Chapter 14), can be decreased with: (i) aids that assist with transport and management (e.g. trolleys, vehicles; Photo 12.1); (ii) better

Photo 12.1. Mechanized transport, especially for harvest, is important for its impact on production costs.

organization of work; and (iii) the automation of operations, such as fertigation.

A well-conceived circuit of passages and roads which allows for easy removal of the products from the greenhouse is of primary importance, because harvesting is one of the most labour-intensive operations in greenhouse vegetable production. Any point inside the greenhouse must be close to an accessible passage for a vehicle or trolley; this distance must not exceed 20 or 30 m.

The situation of the different buildings annexed to the greenhouse must ease and minimize the traffic.

The greenhouses must have double access (interlocking) doors, to limit entrance of insects, wide enough to allow for the machinery to pass through and with simple opening–closing systems, allowing for good insulation and airtightness.

12.3.2 Mechanization of operations

The operations to prepare the soil can be mechanized, because there are several rotovators (one-wheeled tractor) and small articulated tractors, with a wide range of implements for these operations. Sowing or transplanting is usually performed by hand, although it can be mechanized for large areas.

Irrigation and fertilization, given the general use of HFLI (normally drip irrigation), are automated, to a varying degree depending on the specific characteristics of the fertigation head used.

Pesticide treatments are usually performed by means of spraying. In small greenhouses, they can be done with light equipment such as backpacks or with movable sprayers equipped with a small tank, pump and hoses, powered by a tractor, a dedicated petrol engine or a small electric motor. In larger greenhouses, a fixed-pipe network is usually installed inside the greenhouse to which the hoses are connected, the tank and the pumping system remaining in the service areas outside the greenhouse (Aranda, 1994). For ultra-low volume treatments appropriate sophisticated equipment is available on the market (Photo 12.2).

In crops trained to grow tall, the diversity of work operations (pruning, defoliation, harvest) that must be performed at a certain height, induced the development of different systems to lift the workers from the ground level, such as lifting platforms. In other cases, in crops like tomato, different plant lowering and training techniques allow for working at the soil level.

Handling the harvested product is a highly labour intensive operation, if it is not well thought out and organized. The efficiency in the use of several types of trolleys and transport elements will depend on the specific characteristics of each greenhouse. The use of guiding rails, which act simultaneously as heating pipes, is widespread (Photo 12.3). Some sophisticated greenhouses used as nurseries or for ornamental crops also have aerial transportation systems.

Photo 12.2. Mobile equipment for greenhouse pesticide treatments.

Photo 12.3. Use of rails (heating pipes) to move the lifting platform.

The use of small channels in which water circulates, activated by a pump, in which the harvested fruits are transported floating to the packing house is another alternative. This is used in sophisticated greenhouses in northern countries, but has not extended to other parts of the world.

Robots employed in greenhouse operations are already used in nurseries and pot plant production (Giacomelli *et al.*, 2008; Montero *et al.*, 2010). Several automatic systems have allowed for the maximization of the use of the greenhouse area (see next section) minimizing labour (Plate 24).

12.3.3 Occupancy of the greenhouse

The rate of occupancy of a greenhouse describes the proportion of the total surface which is covered by the crop. When the crops are in the soil or in substrate at soil level, the normal arrangement is one or two main passages such as indicated in Fig. 12.5. This arrangement is used for low-density species, such as vegetables, because the workers can access, by means of secondary passages, any point in the greenhouse. It is preferable to have a single main passage on the north side (in the northern hemisphere), as this is the coldest side. The secondary passages allow access to the different zones to perform different crop-care tasks (Fig. 12.5). In tall species, the secondary passages are usually covered or almost completely covered by vegetation. In high density crops, such as carnation, that prevent workers from passing between the plants, it is normal to adopt a longitudinal arrangement (Fig. 12.6).

However, the important point from the productive point of view is the interception of solar radiation by the vegetation. The interception really determines the real space occupancy coefficient, and depends on the size of the species, as well the pruning and training techniques used, as well as the density and the arrangement of the plants.

In practice and in the majority of tall crops the secondary passages (Fig. 12.5) remain shadowed by the vegetation for most of the day. In species that have more-or-less horizontal leaves, radiation is admitted to be conveniently intercepted by several leaf strata when the leaf area index (LAI) reaches a value of 3.0 (Berninger, 1989).

The orientation of the crop rows in a north–south direction homogenizes the radiation conditions between the plants, which is especially relevant in some ornamental crops (phototropic plants).

When growing ornamental pot plants or in nurseries, plants can be elevated on tables or bedplates, of 0.8 m height, to make tending the crop easier. In these cases, the traditional arrangements of passages are longitudinal and peninsular when the bedplates are fixed (Fig. 12.6). The occupancy rate in the longitudinal arrangement is slightly lower (around 65%) than

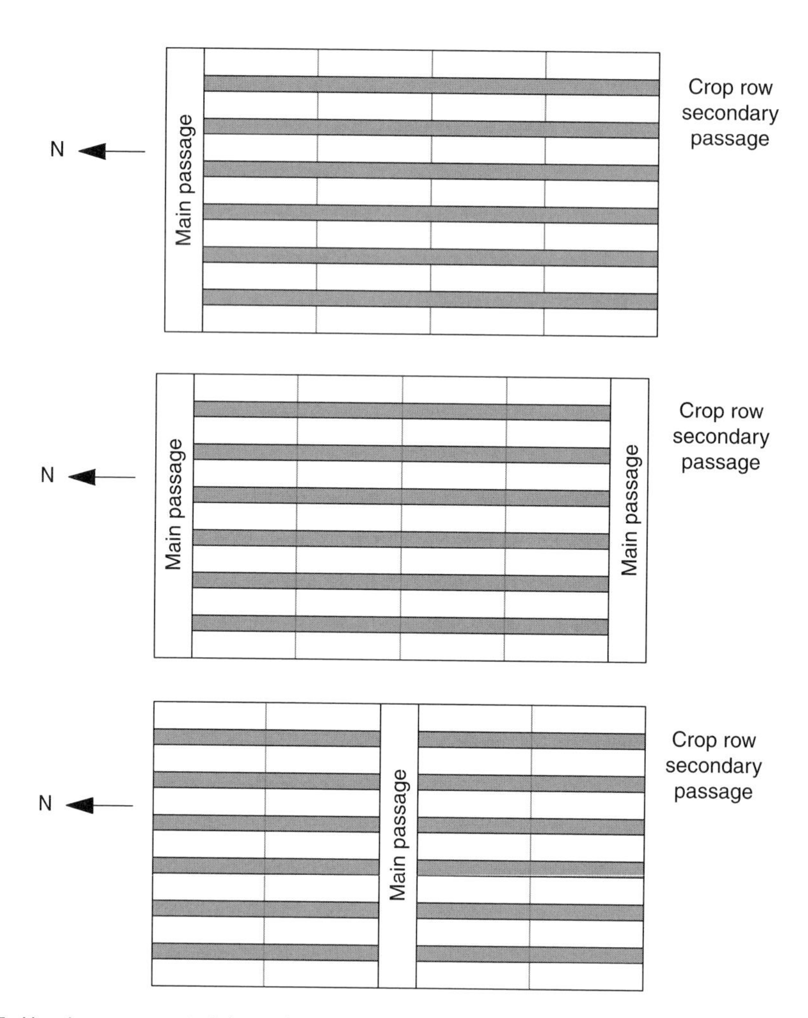

Fig. 12.5. Usual arrangement of the main passages in a greenhouse.

in the peninsular type (around 75%). Another arrangement, which is more expensive, is that of semi-mobile rolling trays which can be moved (Fig. 12.7), but allow for a greenhouse occupancy rate from 85 to 90%.

The passages are usually between 40 and 50 cm wide and the bedplates or trays 1.6–2.0 m wide, accessible from both sides.

The use of crop frames, with the aim of increasing the occupancy rate, generates different microclimate conditions between plants, possibly limiting the PAR available to the plants (Photo 12.4); the profitability of its use is questionable.

In ornamental crops, the location of several lines of pots at some height allows for increasing the occupancy rate by 10–20%, although care must be taken not to cause excess shade (Berninger, 1989). There are other mobile or fixed devices, which allow for placing a higher number of pots inside the greenhouse but which are only of interest in species of low light requirements (Photo 12.5).

An important aspect to be considered in non-conventional arrangements, is the notable differences in microclimate that are generated, not only in light but also in temperature.

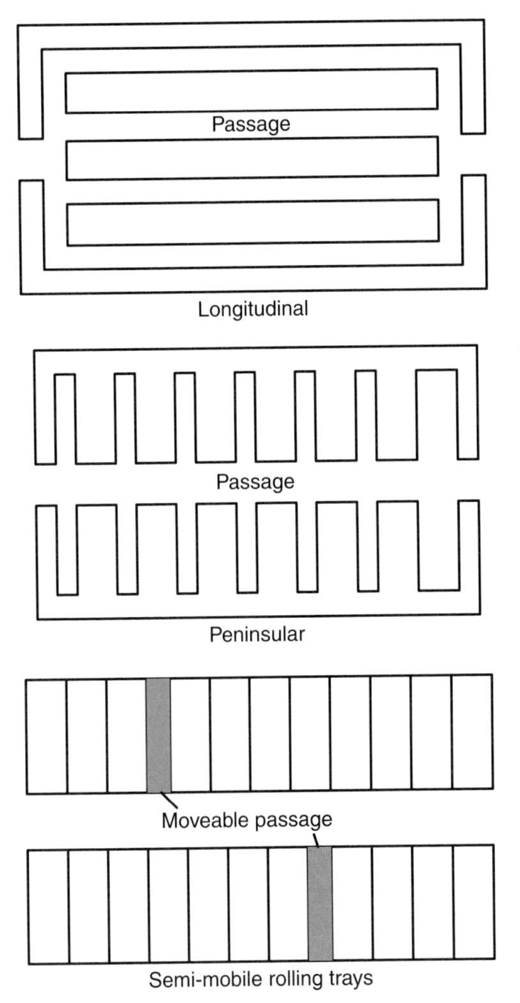

Fig. 12.6. Usual arrangement of the passages in a greenhouse, for crops planted at high density or grown on elevated tables (longitudinal or peninsular arrangements). The semi-mobile rolling trays arrangement is used on elevated tables.

12.4 Summary

- The regulation of any parameter (temperature, humidity, CO_2) consists of comparing the measured value of the parameter with a threshold value and, if there is a difference between them, operating equipment (heating, ventilation) to decrease this difference.
- Regulation can be manual or automatic. Depending on how the correction to be made is calculated, the controllers can be progressive or non-progressive. In the non-progressive controllers, only two options are available, on/off. Progressive controllers can be of several types: proportional control (P), proportional integrated control (PI) and proportional integrated derivative control (PID), depending on the type of action on the equipment that is regulated.
- Greenhouse control systems use the different types of regulators mentioned. The control can be feedforward, when it acts preventing the future values of the parameter regulated, or by feedback, when it only acts in response to the measured values. In the control of a process both can intervene (feedforward and feedback).
- There are different types of models, which are simplified representations of a system or of a part of a system. In greenhouses, we may distinguish two types of models: (i) physical models, related to the microclimate of the greenhouse depending on the outside climate; and (ii) physiological models, which focus on the plants and their relations with the microclimate.
- Simulation models are of great use, if they are well conceived and validated, to simulate several real situations at a low cost. The models constitute a useful tool in the research of the physical medium of the greenhouse and the growth and production of the crop.
- Nowadays, greenhouse climate control based on simple models is usual in well-equipped greenhouses.
- The use of climate control computers in greenhouses is more and more usual, and they may be independent or coordinated with the fertigation control.
- A greenhouse climate digital control system is composed, basically, of: (i) a controller (the climate computer); (ii) correction equipment (heating, ventilation); and (iii) sensors to measure the parameters to be regulated.
- Each greenhouse has specific control requirements, depending on its equipment

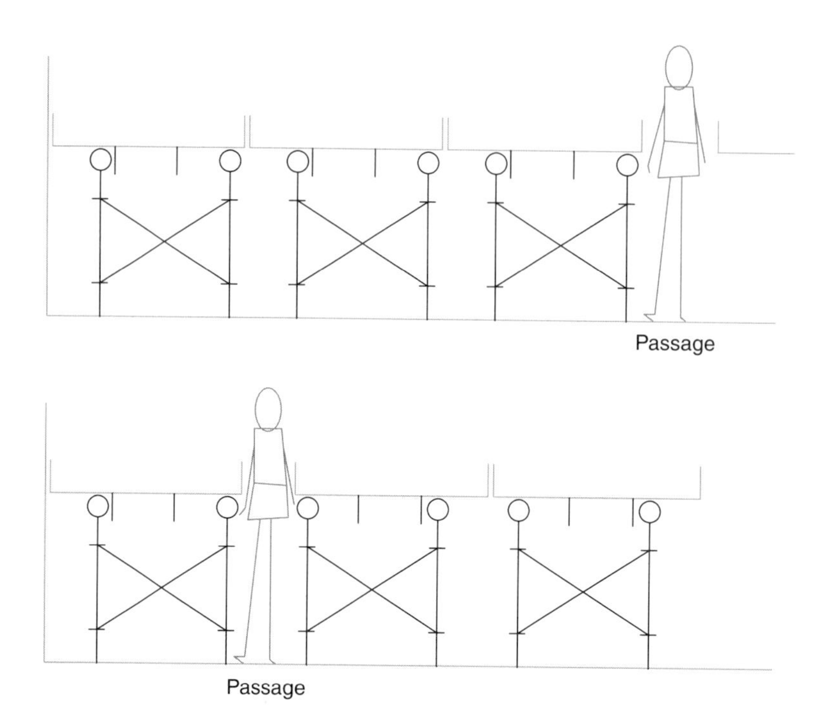

Fig. 12.7. Functioning scheme of the semi-mobile rolling trays (adapted from Urban, 1997b).

Photo 12.4. Greenhouse training framework.

and its technological level. The climate control system manages different processes to be regulated, mainly temperature, humidity, CO_2, light and air circulation.

- In the conception of the general arrangement of a greenhouse aspects that must be taken into account include: (i) accessibility to all points of the greenhouse; (ii) efficient use of the

Photo 12.5. Elevated ornamental pot plants.

available machinery; and (iii) occupancy of the space compatible with ergonomic efficiency for the labour force.

- A well-conceived system of passages, foreseeing aids to management and transport (trolleys, vehicles), well-organized work and the automation of operations (such as fertigation) are essential to minimize labour costs, which are the main production costs.
- In order to optimize the rate of occupancy of a greenhouse (proportion of the total area that is occupied by the crop) there are several possible arrangements, depending on the crop characteristics (density, size, training), but the most important, from the productivity point of view, is the interception of solar radiation by the crop.
- For nurseries or in greenhouses growing pot plants, tables or trays can be used to place the plants above the soil and make the crop cultural practices easier to carry out. In these cases several options are possible, fixed or semi-mobile, to maximize the occupancy rate of the greenhouse.

13

Plant Protection

13.1 Introduction

In protected cultivation, pests and diseases find more favourable conditions for their development than in open field cultivation. The mortality of insects due to abiotic factors (rain, wind, cold temperatures) is enormously decreased and the climatic conditions (high humidity, higher temperatures) favour the development of diseases (Elad, 1999).

This chapter includes some brief notes on the phytosanitary controls used in cultivating greenhouse vegetables, a subject widely covered in specialized books (Fletcher, 1984; García-Marí *et al.*, 1994; Messiaen *et al.*, 1995; De Liñán, 1998).

13.2 Chemical Control

13.2.1 Main aspects

The preventative methods of pest and disease control include the use of healthy transplants, of tolerant or resistant cultivars and rootstocks, of non-contaminated substrates, soil disinfection, and environmental control to avoid favourable conditions for the development of pests and diseases and the application of pesticides as means of protection (Elad, 1999).

From the middle of the last century, the control of insects and diseases of crops has been based on the use of chemical pesticides. The ability of some insects and pathogens to develop resistance to pesticides, the general awareness of the need to conserve the environment and for health and food safety have been the main causes for the displacement of dependence on chemical control in favour of integrated pest management (IPM; see section 13.4).

The fast growth of plants in the greenhouse forces frequent chemical disease protection treatments (preventive), because the systemic fungicides can only partially reach some organs, such as flowers or fruits, as their translocation is induced by transpiration (Elad, 1999). The use of tensio-active substances can improve the application efficiency.

There has been a general development of resistance to pesticides in various insect populations. Among the pathogens that have developed resistance to fungicides we may highlight *Botrytis cinerea, Fusarium oxysporum, Pseudoperonospora cubensis* and *Sphaerotheca fusca* (Elad, 1999).

The timing of when to perform the preventative treatment is critical for the achievement of the desired effect, because the pest or pathogen's sensitivity depends on its developmental stage.

The safety conditions in the application of pesticides are an essential aspect for the hygiene and safety of the staff working in the greenhouses. The increasing interest of consumers in health and food safety has contributed to improved risk prevention mechanisms in the use of pesticides.

13.2.2 Treatment equipment

Pesticide application methods have a great influence, both on their efficiency and on the labour costs. The application methods depend on the vehicle used to distribute the pesticide: (i) dust, when a solid is used (e.g. talc); (ii) spraying and fogging, when a liquid is used (usually in water); and (iii) fumigation when a gas is used.

The majority of horticultural treatments consist of spraying solutions or suspensions of the active materials. The most usual greenhouse sprayings are performed with a motor, provided with hoses and pistols to cover all areas of the greenhouse and direct the spray to the desired points. The service pressure must be sufficient to achieve a very small droplet size and to generate turbulence that helps the product cover the whole canopy (Aranda, 1994).

The ultra-low volume systems use a fan and a fog mechanical generator in association with each other, to obtain a very small droplet size and great canopy penetration, which allows for treating large areas from a fixed point. If located over a mobile trolley they can be operated in pre-fixed locations (as a semi-fixed system).

The thermo-foggers cause a very fine spray (of less than 100 μm diameter) by the explosion of a mix of fuel and air (Urban, 1997a).

Among the mobile automatic systems the most common is a treatment trolley that moves automatically along guiding pipes (also used for heating) between the crop rows.

13.3 Biological Control

Biological control is based on the use of natural enemies of the pests (parasitoids, predators and pathogens) to maintain their infestation below an economic damage threshold. Although biological control has been known about for more than half a century, its use had not expanded until a few decades ago (Blom, 2002).

The biological control of diseases is not widely used in practice. Techniques that may be highlighted include: (i) crossed protection techniques (where the organism that arrives first to an infection point acts against the pathogen that arrives later; this is used for the control of tobacco mosaic virus (TMV) by inoculating the plant with an innocuous form of the virus); (ii) induced resistance (the organism arriving first induces a defence reaction in the host); (iii) passive occupation (previous occupation of the infection point by an innocuous organism); and (iv) hyper-parasitism (*Trichoderma*). However, it should be pointed out that there are many other techniques (alelopathy, antibiosis) of possible use (Jarvis, 1997).

The use of biocontrol agents has been efficient against some diseases such as the use of *Trichoderma* against *Pythium, Fusarium* and *Rhizoctonia* (Elad, 1999).

The pest's natural enemies, besides not competing for the resources, carry out a predatory activity, feeding on the pest species. Parasitoid insects carry out a particular parasitic activity (external or internal oviposition of one egg on the host, from which a larva emerges that eats the animal as it develops). Parasitoids are more specific (monophagous in many cases), whereas predators are usually polyphagous (García, 1994).

The use of entomopathogens in biological control in the greenhouse, as an insulated enclosure, is of special interest. The massive use of *Bacillus thuringiensis* for the control of Lepidoptera has expanded, whereas the use of other pathogens, such as *Verticillium lecanii* (limited by the optimum temperatures range, Photo 13.1) or *Archensonia* (for white fly) and *Beauveria* or *Paecilomyces* (for white fly, aphids or thrips) is not so widespread (Parrella, 1999). At the time of writing this text, in Spain only the use of *Beauveria* is authorized, although the registration process for *Verticillium* and *Paecilomyces* is in progress (J.V.D. Blom, 2007, personal communication).

Table 13.1 summarizes the main natural enemies of the most common greenhouse pest insects in Spain (Photos 13.2 and 13.3).

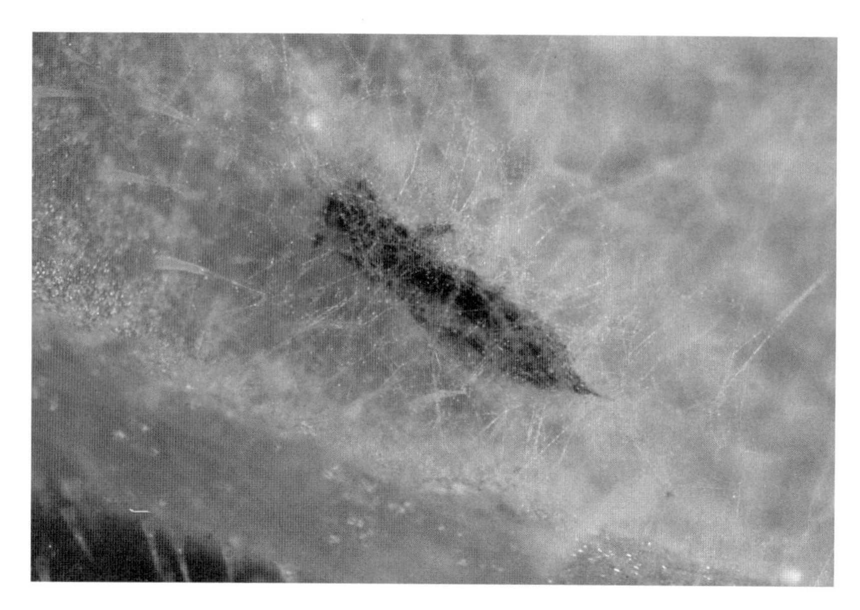

Photo 13.1. Adult thrip covered with *Verticillium lecanii* hyphae.

The use of *Encarsia formosa*, for the control of white fly (*Trialeurodes*), as well as the use of *Phytoseiulus persimilis* for the red spider mite, are widespread in protected cultivation all over the world (Parrella, 1999). In Spain, the most common control for white fly is *Eretmocerus mundus* (J.V.D. Blom, 2007, personal communication).

13.4 Integrated Pest Management

Integrated pest management (IPM) is a different approach to the control of pests and diseases, which tolerates the presence of pests and diseases at a low level of incidence that does not cause economic losses, and gives preference to other types of control (biological, genetic, cultural) over chemical control, which is only used as a last resort. All this allows for a notable decrease in the intensity of the phytosanitary treatments.

The integration of several types of control in a common strategy, considering economic, ecologic and toxicological criteria, forms the basis of the integrated management (Moreno, 1994).

In the literature, the acronym IPM and more recently the acronym IPP (integrated production and protection) proposed by FAO (Hanafi, 2003) can be found, highlighting the relevance of cultural practices and management in integrated control.

IPM is based on fixing the 'economic damage thresholds' of pests and diseases for each crop, which define the population density of a pest above which economic losses occur (Moreno, 1994; Bielza, 2000). The IPM strategy is established with the aim of maintaining the populations below the economic damage threshold.

Among cultural practices of interest in IPM we may highlight: (i) the use of healthy seed; (ii) the quality of the substrate; (iii) planting density; (iv) the type of pruning and training, grafting and soil management practices (especially solarization; Katan, 1981); and (v) greenhouse climate control (Garijo and Frapolli, 1994; Bielza, 2000). The use of methyl bromide in the suppression of soil-borne diseases has been prohibited due to its effect on the ozone layer, so its use has been substituted by solarization, when enough solar radiation is available, as is the case in the Mediterranean Basin. The irrigation and fertilization schedule greatly affects the development of pests and the elimination of weeds contributes to the efficacy of IPM.

The development of resistant or tolerant cultivars to certain viruses, insects, fungi or nematodes has become an efficient indirect

Table 13.1. Main pests and natural enemies used in greenhouse biological control in Spain. (Source: Belda and Cabello, 1994; Cabello and Belda, 1994; Cabello and Benítez, 1994; Rodríguez, 1994; Aparicio *et al.*, 1998.)

Pest	Natural enemy
Spider mite	*Phytoseiulus persimilis*
	Metaseiulus occidentalis
	Therodiplopsis persicae
White fly	*Encarsia formosa*
	Macrolophus caliginosus
	Eretmocerus mundus
	Delphastus pusillus
	Verticillium lecanii
	Aschersonia aleyrodis
	Beauveria bassiana
	Paecilomyces fumosoroseus
Aphids	*Aphidius matricariae*
	Aphelinus abdominalis
	Aphidius colemani
	Aphidoletes aphidimyza
	Chrysoperla carnea
	Hippodamia convergens
	V. lecanii
Leaf miners	*Diglyphus isaea*
	Dacnusa sibirica
Thrips	*Amblyseius cucumeris*
	Orius laevigatus
	Orius albidipennis
	V. lecanii
Noctuids	*Bacillus thuringiensis*
(lepidopteran larvae)	
	Trichogramma evanescens
Soil insects	*Steinernema feltiae*
	Hypoaspis miles

Photo 13.2. White flies are one of the main pests of greenhouse vegetables.

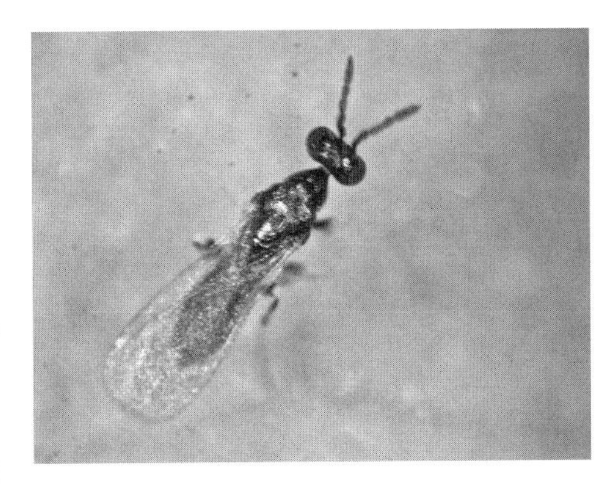

Photo 13.3. *Diglyphus isaea*, parasitoid of leaf miners (*Liriomyza* spp.).

method of combating the problem caused by pests and diseases. In general, fewer plants have been developed, by classical breeding methods, that have resistance to pests than those that are resistant to diseases. This is possibly due to the wrong strategy being adopted, because it could be of greater interest to develop plants that improve the predating efficiency of the natural enemy, at least in some cases, than to try to obtain plants resistant to a certain pest (Parrella, 1999). Nowadays, biotechnology opens new possibilities of incorporating resistance to pests, such as the development of transgenic *Bt* (*Bacillus thuringiensis*) plants (T. Cabello, 2007, personal communication).

Other means of physical control have proved to be efficient to decrease the populations of pests, such as: (i) the use of insect-proof screens (even impregnated with pesticides) that prevent the access of pests through the greenhouse vents; (ii) the use of chromatic sticky traps to attract insects; and (iii) the use of trap-lamps or pheromones. In some Spanish regions, such as in Andalusia, the use of very fine screens (10×20 threads cm^{-2}) in the vents is compulsory, as long as proper ventilation is not prevented. The recent introduction to the market of plastic anti-pest films for covering the greenhouse that restrict the mobility of some insects (affecting their vision) and that hinder the development of diseases (e.g. *Botrytis*) are, also, complementary measures of interest in IPM (see Chapter 4).

The use of pollinating insects (bees and bumblebees) inside the greenhouse (Plates 25 and 26) has contributed to the rationalization of integrated phytosanitary control, at the grower's level, avoiding the use of harmful pesticides (Meneses and Castilla, 2009).

The transfer of the greenhouse structure to another location, which was common for the tunnel greenhouse years ago and that constituted a peculiar variant of crop rotation, was efficient in the fight against soil-borne diseases. Nowadays, soilless cultivation can be an easier alternative.

13.5 Climate Control and IPM

Climate factors directly affect the development of pests and diseases of crops, and so an appropriate greenhouse climate control can help to decrease their development.

Avoiding water condensation on the plants avoids the proliferation of *Botrytis*. The ambient humidification, to increase the RH, contributes to the fight against *Oidium* and hinders the development of spider mites and thrips.

On the other hand, the improvement of the thermal conditions in greenhouses, the absence of rain (inside greenhouses) that washes the phytosanitary products (in the open air) and the filtering of UV light of some plastic covering materials contribute to an increase in the efficiency of the treatments although they could also increase the risk of phytotoxicity (Urban, 1997a).

Alteration of the solar radiation inside the greenhouse can influence the development of diseases, not just due to the effect on the spectral composition of the radiation that affects the formation of fungal spores, but also due to the higher sensitivity of plants to diseases at reduced light levels, as is usually the case in dense crops (Louvet, 1984).

Inside greenhouses, the temperatures are usually higher than outside, limiting the development of some diseases (*Cladosporium, Peronospora*) or favouring the spread of others (*Pythium, Phytophtora, Alternaria*). Soil-borne diseases, in general, are favoured by high temperatures (Louvet, 1984).

The usually high humidity inside greenhouses, at least at certain periods of time, favours the formation of fungal spores, accelerating their development, as well as the growth of bacterial colonies, especially if temperatures are favourable. These risks become higher when water vapour condenses on the plants. Therefore, indirect prevention methods include procedures that decrease the environmental humidity, such as the use of drip irrigation and mulching the soil.

The wind is limited inside the greenhouse, except in greenhouses with mechanical ventilation, which limits the dispersal of disease inocula.

13.6 Most Common Greenhouse Diseases

Viruses, bacteria, fungi and nematodes can cause diseases in greenhouse crops. Fungal pathogens are the main cause of disease, most frequently species in the genera *Fusarium, Pythium* and *Phytophthora*, among those of soil origin (Gómez, 1994). Nematodes or bacteria cause other soil-borne diseases. Bacterial diseases are difficult to control, because there are very few efficient products, so it is essential to prevent these by means of disinfecting the root medium, among other measures. Among diseases that affect the aerial organs of plants we may highlight *Botrytis* (Photo 13.4), *Phytophtora, Pseudoperonospora* and *Sclerotinia*. Table 13.2 summarizes the most common diseases.

Photo 13.4. *Botrytis* on a tomato fruit.

Table 13.2. Most common greenhouse diseases. (Source: Cuadrado, 1994; Gómez, 1994; Moreno, 1994; Aparicio *et al.*, 1995, 1998; Cuadrado *et al.*, 2001; Verhoeven *et al.*, 2003; Segundo *et al.*, 2004.)

(a) Mycoses (diseases caused by fungi) and bacterial diseases.

Diseases caused by	Disease (pathogen)
Fungi	Damping-off (*Pythium, Phytophthora, Rhizoctonia, Botrytis, Sclerotinia*)
	Black root rot (*Chalara elegans*)
	Foot and root rots (*Pythium*)
	Foot and root rots (*Rhizoctonia solani*)
	Basal stem rot (*Phytophthora capsici*)
	Buckeye rot (*Phytophthora nicotianae*)
	Root rots (*Fusarium*)
	Fusarium oxysporum f. sp *radicis lycopersici*
	Verticillium wilt (*Verticillium dahliae*)
	Fusarium wilt (several specialized forms of *Fusarium oxysporum*)
	Black stem rot (*Didymella bryoniae*)
	Grey mould (*Botrytis cinerea*)
	Sclerotinia (*Sclerotinia sclerotiorum*)
	Tomato blight (*Phytophthora infestans*)
	Downy mildew (*Pseudoperonospora cubensis*)
	Powdery mildew of the *Cucurbitacae* (*Podosphaera fusca*)
	Powdery mildew of the *Solanaceae* (*Leveillula taurica*)
	Tomato alternaria canker (*Alternaria dauci*)
Bacteria	Tomato bacterial speck (*Pseudomonas syringae* pv. *tomato*)
	Tomato pith necrosis (*Pseudomonas corrugata*)
	Leaf spot (*Xanthomonas campestris*)
	Angular leaf spot of the *Cucurbitacae* (*Pseudomonas syringae* pv. *lacrimans*)
	Bean bacterial speck (*Pseudomonas syringae* pv. *phaseolicola*)
	Fruit scab (*Xanthomonas campestris*)
	Soft rots of stems, leaves and fruits (*Erwinia carotovora, Pseudomonas fluorescens*)

(b) Viral diseases

Viral diseases in	Transmitted by	Name of virus
Solanaceae	Thrips	Tomato spotted wilt virus (TSWV)
	Aphids	Cucumber mosaic virus (CMV)
		Potato virus Y (PVY)
	White fly	Tomato yellow leaf curl virus (TYLCV)
		Tomato chlorosis virus (ToCV)
	Seeds and direct contact	Tomato mosaic virus (ToMV)
		Pepper mild mottle virus (PMMoV)
	Soil	Tomato bushy stunt virus (TBSV)
Cucurbitaceae	Aphids	CMV
		Watermelon mosaic virus-2 (WMV-2)
		Zucchini yellow mosaic virus (ZYMV)
	Fungi	Melon necrotic spot virus (MNSV)
	Seeds and direct contact	Squash mosaic virus (SqMV)
	White fly	Cucumber vein yellowing virus (CVYV)
		Beet pseudo-yellows virus (BPYV)
		Cucurbit yellow stunting disorder virus (CYSDV)
Beans	Thrips	Tomato spotted wilt virus (TSWV)
	Aphids	Bean common mosaic virus (BCMV)
		Bean yellow mosaic virus (BYMV)
	Seeds or direct contact	Southern bean mosaic virus (SBMV)
	White fly	Bean yellow disorder virus (BnYDV)

Viral diseases are worth a special mention as their devastating effects have been frequently felt. The only control measure against these diseases is prevention, unless tolerant or resistant plant material is available. Where these diseases are introduced by a vector there is a need for strict control of the vector's populations.

13.7 Most Common Greenhouse Pests

Due to the non-existence of a seasonal climatic break in the greenhouse, such as it happens in cold winter areas with open field crops, many pests are permanent, helped by the long occupation of the greenhouses and due to the existence of pest reservoirs in open-air host weeds or crops, when the greenhouse is empty.

In greenhouses, in decreasing order of importance the Aleurodidae (white flies), Noctuidae, thrips, leaf miners, mites and aphids are the most common greenhouse pests in the Mediterranean area (Aparicio *et al.*, 1998). At the end of the first decade of the 21st century, the appearance in the Mediterranean area of the previously unknown lepidopteran 'tomato moth' (*Tuta absoluta*) is causing severe problems in the tomato crop.

In general, it is easier to avoid the growth of pest populations than to decrease their number. Therefore, prevention is essential.

13.8 Prophylaxis

Disinfection of the soil can be necessary in some cases, especially in nurseries or for pot plant production. After disinfection, it is essential to prevent contamination by using the following measures: (i) using footbaths at points of access to the greenhouse; (ii) limiting the movement and transport of soil particles (with plastic mulch); (iii) disinfecting the tools; (iv) eliminating weeds around the greenhouse; (v) ensuring the sanitary state of the plant material; and (vi) using biologically inert substrates.

13.9 Other Aspects

Phytosanitary control costs are very variable, depending on the cultivated species and the general growing conditions, especially the machinery and labour used (see Chapter 14).

For other aspects, such as control of other pests (rodents, birds) and weeds, the prevention of pathophysiologies, etc. the reader should refer to the specialized literature on these subjects.

13.10 Summary

- In protected cultivation, pests and diseases find favourable conditions for their development.
- The control of pests and diseases has been based, until recently, on the use of chemicals.
- Biological control is based on the use of natural enemies of the pests and pathogens, to maintain their populations below the 'economic damage threshold'.
- The development of resistance to pesticides combined with general concerns for the environment and for food safety and health have been the main causes for the recent move away from sole dependence on chemical control to an integrated pest management (IPM) approach.
- IPM constitutes a different way of understanding pest and disease control, which tolerates the presence of pests and diseases below an 'economic damage threshold' and gives preference to other types of control (biological, cultural, genetic) rather than to chemical control, which is used only as a last resort. This allows for a notable decrease in phytosanitary treatments.
- Climatic factors directly influence the development of pests and diseases of crops. Therefore, proper greenhouse climate control can help to decrease their development.

- Fungal diseases of the soil and aerial parts of the plant, together with viral diseases are the main diseases of greenhouse crops. Normally, bacteria and nematodes are of less importance.
- The main pests of greenhouse crops are the aleourodids (white flies), noctuids, thrips, leaf miners, mites and aphids.
- Biological control is widely used for pest control, especially against white fly, noctuids, leaf miners, apids and spider mites.
- Prevention is essential in greenhouse phytosanitary control.

14

Economic and Environmental Analysis

14.1 Economic Analysis

14.1.1 Introduction

Three basic factors determine the selection of the greenhouse cultivation area: (i) the production costs; (ii) the quality of the produce; and (iii) the transport costs (Nelson, 1985). This is because these factors determine the ability to compete in an increasingly globalized market. Obviously, the cost and quality of production will depend on the local climatic conditions (see Chapter 3), and both will determine the technological level of the greenhouse enterprise (degree of sophistication required in the greenhouses and climate control equipment) as well as its management. The possibilities for long-distance transport have evolved enormously during recent decades, allowing production areas to be located far away from the large consumption centres, which was an unfeasible situation just half a century ago, and this has permitted the development of large greenhouse industries in the coastal areas of the Mediterranean Basin (e.g. in Italy, France, Spain and Morocco; see Chapter 16).

Despite the increasing negative impact of the commercialization costs (classifying, packaging, etc.) to the final cost of the product in the market, these costs are usually less important than the production and transport costs when the products, coming from different origins, meet a certain minimum quality, and compete successfully in a specific market (Castilla *et al.*, 2004).

Across Europe, we can distinguish between two greenhouse production agrosystems: (i) the 'northern or Dutch agrosystem', typical of Northern Europe, which requires a large initial investment (in the greenhouse structure and its equipment) and which is characterized by large energy use (Photo 14.1); and (ii) the 'Mediterranean greenhouse agrosystem', characterized by low investment and lower energy consumption, which is common in countries in the Mediterranean Basin. Between these two technological extremes, there are different gradations.

In the 'Dutch agrosystem' the crop production strategy has been to optimize the greenhouse microclimate, whereas in the Mediterranean countries the prevailing strategy has been to adapt the crops to suboptimal climates, which has meant less yield and in some cases limited quality, but also lower production costs (Castilla, 1994).

The need to improve the product quality in Mediterranean greenhouses by means of proper climate control involves an

Photo 14.1. Venlo-type glasshouse, characteristic of production in the 'Dutch agrosystem'.

increase in their technological level using well-equipped greenhouses. To find a good economic compromise between high investments in greenhouse structures and equipment and their productive performance requires different approaches, according to the local technical and socio-economic characteristics, to produce high quality vegetables competitively (Castilla, 2002).

14.1.2 Greenhouse structures and equipment

The choice of a certain greenhouse structure carries with it the choice of technological equipment in agreement with the chosen structure, so that the 'technological package' (the set of greenhouse and equipment) allows for achieving the right agronomic performance at a given cost.

Table 14.1 presents the average construction costs of the most usual structures in southern warm climates. The cost of the multi-tunnel-type greenhouse structure with a curved-shape cover duplicates that of a low-sloped roof low-cost greenhouse (common in Spain).

Among the different equipment choices for these structures, Table 14.2 presents some of the possibilities, incorporating in the case of the improved low-cost greenhouse a set of de-stratification fans and the cheapest fog and heating systems. In the more expensive multi-tunnel greenhouse structure (Photo 14.2), better fog and heating systems are implemented, as well as thermal and shading screens. Obviously, the specific choice of the equipment depends on the prevailing conditions; the example in Table 14.2 only aims to illustrate the level of investment for a certain 'technological package'. Therefore, the simplest case of low-cost 'parral-type' greenhouse

Table 14.1. Average construction cost of the greenhouse structures in the south of Spain (excluding taxes). Values calculated for a minimum surface of 1 ha, including the installation of a plastic film cover (year 2007).

Greenhouse type	Cost (€ m^{-2})
Low-cost parral-type greenhouse (motorized sidewall and roof ventilation)	
Low roof slope (old type)	10.0
High roof slope (27°/27°)	14.0
Multi-tunnel greenhouse, with motorized sidewall and roof ventilation	19.0

(old type) without auxiliary equipment, which is common nowadays in Spain, costs around €10 m^{-2} (Table 14.1), whereas a well-equipped curved-type multi-tunnel would cost around €43.1 m^{-2}, which is less than half of the cost of a completely equipped modern glasshouse (Castilla *et al.*, 2004). The cost of the improved low-cost greenhouse with high roof slope (Photo 14.3) is intermediate (Table 14.1).

Table 14.2. Average construction and equipment costs of Mediterranean greenhouses equipped with motorized sidewall and roof ventilation (excluding taxes) including the installation of the plastic film cover (year 2007). Values calculated for a minimum surface of 1 ha. Climate control and fertigation computer are excluded. (Source: Hernández and Castilla, 2000; Castilla *et al.*, 2004; updated to 2007.)

Type of greenhouse	Cost (€ m^{-2})
Improved low-cost parral-type greenhouse, roof slope 27°/27° (symmetric)	
Structure	14.0
Fans (de-stratification)	1.3
Fog systems (low pressure)	1.3
Heating (hot air)	3.0
Total	**19.6**
Curved-shape multi-span (multi-tunnel) greenhouse	
Structure	19.0
Fans (de-stratification)	1.3
Fog system (high pressure)	2.8
Heating (metal pipes) and CO_2 enrichment	15.0
Double-use screen (thermal and shading)	5.0
Total	**43.1**

These figures for the initial investment needed for greenhouse cultivation must be increased to take account of the cost of, for example, preparing the land, water and electricity supply, etc. (Table 14.3), excluding the value of the land and taxes.

The level of sophistication of the chosen 'greenhouse technological package' will determine the possibilities for climate control and, therefore, the productive performances and profitability of the chosen package.

14.1.3 The Spanish greenhouse horticultural farm

Although there are farms of different sizes in Spanish protected horticulture, it is generally characterized by small size farms that, for the specific case of Almeria, range between 1.5 and 2.4 ha, and most exploit family labour (Martínez-Carrasco, 2001; Pérez *et al.*, 2002). Their technological level is low, with a predominance of low-tech

Photo 14.2. Arch-shaped plastic multi-span.

Photo 14.3. Improved high roof slope low-cost parral-type greenhouse, of metallic structure (steel and wire).

Table 14.3. Average initial investment in a greenhouse, in addition to the structure and equipment, in the south of Spain (year 2004), excluding taxes and value of the land (p.p.: proportional part).

Concept	Cost (€ m^{-2})
Preparation of the land	3.5
Water reservoir (p.p.)	1.2
Store (p.p.)	1.1
Electricity (p.p.)	0.7
Preparation of the soil	1.5
Irrigation system	2.3
Substrates	1.1
Total	**11.4**

greenhouses, without any auxiliary climate control equipment (Photo 14.4).

The range of species that are cultivated is enormous. Focusing on vegetables, although the relative importance changes every growing season, pepper, tomato and melon are the most important, followed by watermelon and cucumber, and with lower growing areas devoted to courgette, climbing bean, aubergine, Chinese cabbage and dwarf bean, among others.

Land is held predominately by growers who own their property (more than 90%). During recent years there has been a decrease in the number of growers involved in partnerships (sharecropping) and those who rent the land (Martínez-Carrasco, 2001).

Production planning, by growers and the marketing companies, is limited; the main criteria for choosing the species to be cultivated being the degree of knowledge and specialization of the grower with regards to a certain crop and the availability of advice from technicians working for the commercialization companies (Martínez-Carrasco, 2001).

Summarizing, the productive infrastructure in Almeria (the centre of the Spanish greenhouse industry) is based on the low-cost greenhouse, both the symmetric multi-span type (52% of the total) as well as the flat-roof type (40% of the total). The majority of the cultivation (80% of the total) is in soil although an increasing area is in substrates (20% of the total). Irrigation is almost completely by drip systems (99.7%), with a limited use of heating (air heating or low-temperature water heating), but wide use of natural ventilation and little use of fog systems (Pérez *et al.*, 2002).

14.1.4 Production costs

The fixed production costs depend on the chosen technological package, but these

Photo 14.4. Conventional flat-roofed low-cost parral-type greenhouse.

costs can vary greatly due to the differing levels of investment and the expected useful life of the technological package.

For a low-cost parral-type greenhouse, without auxiliary climate control equipment, the fixed costs in the year 2000 are of the order of €0.91 m^{-2} $year^{-1}$; the main components comprising depreciations of the structure (with 38.3%) and the cost of the plastic film (27.1%) (Aznar, 2000). In these greenhouses, the fixed costs represent about 21% of the average total cost of production (Aznar, 2000).

In higher technology Mediterranean greenhouses, these fixed costs can increase significantly, even reaching 40% of the total cost for a pepper crop (Caballero and Miguel, 2002).

There is a wide range in variable costs as these vary depending on the horticultural species being considered, and also on the yield, which is determined by the cultivar and growing cycle (Tables 14.4–14.8), as well as on the technological level of the greenhouse. Market conditions sometimes limit the length of the production cycle and, therefore, the yield, modifying the variable costs. In general, there is a lot of variability in the described costs (Tables 14.4–14.8).

In low-tech greenhouses, without heating, the main variable cost (per unit of harvested product) is labour (salaries), especially in green bean, where it involves 78.3% of the total variable costs, followed by cherry-type tomato and short cucumber (spring cycle) (Tables 14.4–14.8). The differences between labour costs within the same species, besides depending on the cultivar or commercial type grown in the case of tomato, conventional type or cherry-type (Table 14.4), are influenced by the harvesting method (one by one or trusses), as has been described by Berenguer *et al.* (2003). The following variable costs in order of importance, after the salaries, are fertilizers, seeds or seedlings and pesticides, for the majority of the mentioned crops (Tables 14.4–14.8). Similar results have been quantified in other cost studies (Cañero *et al.*, 1994; Caballero and Miguel, 2002).

Water has little influence on the total variable costs, not exceeding 7.3% in any case (Tables 14.4–14.8), so there is no incentive for the grower in terms of savings (Martínez-Paz and Calatrava-Requena, 2001; Colino and Martinez-Paz, 2002).

The costs structure, both fixed and variable, for greenhouse cultivation is altered when conventional soil cultivation

Table 14.4. Variable production costs and yield of the tomato crop in an unheated low-cost greenhouse in Almeria. (Source: Calatrava-Requena *et al.*, 2001 and personal communication.)

	Tomato[a]		Cherry-type tomato[b]	
Variable cost	Cost (€ kg^{-1})	Percentage of total variable costs	Cost (€ kg^{-1})	Percentage of total variable costs
Fertilizers	0.035	23.3	0.056	12.5
Pesticides	0.013	8.7	0.025	5.6
Seedlings/seeds	0.012	8.0	0.055	12.2
Water	0.011	7.3	0.015	3.3
Salaries	0.070	46.7	0.275	61.1
Others	0.009	6.0	0.024	5.3
Total	**0.150**	**100.0**	**0.450**	**100.0**
Yield (kg m^{-2})	13.57		6.93	

[a]Tomato data: seasons 1996–1997 and 1997–1998 from 16 tomato farms.
[b]Cherry-type tomato data: growing season 2000–2001 from12 cherry-type tomato farms.

Table 14.5. Variable production costs and yield of melon and watermelon crops in an unheated low-cost greenhouse in Almeria. (Source: J. Calatrava-Requena, unpublished data.)

	Melon[a]		Watermelon[b]	
Variable cost	Cost (€ kg^{-1})	Percentage of total variable costs	Cost (€ kg^{-1})	Percentage of total variable costs
Fertilizers	0.062	27.7	0.024	17.8
Pesticides	0.045	20.1	0.020	14.8
Seedlings/seeds	0.039	17.4	0.026	19.3
Water	0.011	4.9	0.008	5.9
Salaries	0.043	19.2	0.029	21.5
Others	0.024	10.7	0.028	20.7
Total	**0.224**	**100.0**	**0.135**	**100.0**
Yield (kg m^{-2})	3.57		5.76	

[a]Melon data: growing season 2000–2001 from 12 melon farms.
[b]Watermelon data: growing season 2000–2001 from 20 watermelon farms.

Table 14.6. Variable production costs and yield of pepper and green bean crops in an unheated low-cost greenhouse in Almeria. (Source: J. Calatrava-Requena, unpublished data.)

	Pepper[a]		Green bean[b]	
Variable cost	Cost (€ kg^{-1})	Percentage of total variable costs	Cost (€ kg^{-1})	Percentage of total variable costs
Fertilizers	0.039	12.2	0.058	8.3
Pesticides	0.047	14.7	0.057	8.1
Seedlings/seeds	0.056	17.5	0.012	1.7
Water	0.012	3.8	0.011	1.6
Salaries	0.150	46.8	0.548	78.3
Others	0.016	5.0	0.014	2.0
Total	**0.320**	**100.0**	**0.700**	**100.0**
Yield (kg m^{-2})	5.8		2.18	

[a]Pepper data: growing season 2000–2001 from 20 pepper farms, of which 13 were of the thick-wall type and the rest of thin-wall type.
[b]Green bean data: growing season 2000–2001 from 13 green bean farms.

Table 14.7. Variable production costs and yield of cucumber and short cucumber crops in an unheated low-cost greenhouse in Almeria. (Source: J. Calatrava-Requena, unpublished data.)

	Cucumber (autumn)[a]		Short cucumber (spring)[b]	
Variable cost	Cost (€ kg^{-1})	Percentage of total variable costs	Cost (€ kg^{-1})	Percentage of total variable costs
Fertilizers	0.037	14.0	0.036	15.4
Pesticides	0.038	14.3	0.010	4.3
Seedlings/seeds	0.049	18.5	0.033	14.1
Water	0.006	2.3	0.007	3.0
Salaries	0.129	48.6	0.143	61.1
Others	0.006	2.3	0.005	2.1
Total	**0.265**	**100.0**	**0.234**	**100.0**
Yield (kg m^{-2})	7.22		5.38	

[a]Cucumber data: growing season 2000–2001 from 12 autumn-cycle cucumber farms.
[b]Short cucumber data: growing season 2000–2001 from 14 spring-cycle cucumber farms.

Table 14.8. Variable production costs and yield of autumn and spring courgette crops in an unheated low-cost greenhouse in Almeria. (Source: J. Calatrava-Requena, unpublished data.)

	Courgette (autumn)[a]		Courgette (spring)[b]	
Variable cost	Cost (€ kg^{-1})	Percentage of total variable costs	Cost (€ kg^{-1})	Percentage of total variable costs
Fertilizers	0.057	18.3	0.016	9.1
Pesticides	0.032	10.4	0.015	8.5
Seedlings/seeds	0.049	15.9	0.023	13.0
Water	0.017	5.5	0.013	7.3
Salaries	0.138	44.7	0.099	55.9
Others	0.016	5.2	0.011	6.2
Total	**0.309**	**100.0**	**0.177**	**100.0**
Yield (kg m^{-2})	3.42		6.16	

[a]Courgette (autumn) data: growing season 2000–2001 from four autumn-cycle courgette farms.
[b]Courgette (spring) data: growing season 2000–2001 from five spring-cycle courgette farms.

is compared with soilless cultivation (Caballero and Miguel, 2002; D'Amico *et al.*, 2003).

Growing crops organically also modifies the variable costs structure in greenhouse cultivation (Engindeniz and Tuzel, 2003).

Table 14.9 summarizes the composition of the variable costs of greenhouse tomato production in the two agrosystems (Mediterranean and Dutch or northern). In both cases, we can highlight the importance of labour, which reaches similar proportions. The greatest difference is in the energy costs (natural gas), which involves 35% of the total variable costs in the northern agrosystem (Table 14.9). In this system there is no cost for irrigation water, as previously stored rainwater is used.

The presence of the grower permanently heading the farm management, as well as the use of better fertigation systems, are factors that clearly increase the production efficiency in Mediterranean greenhouses, without having detected scale economies (Cañero and Calatrava-Leyva, 2001).

14.1.5 Other aspects of interest

Commercial greenhouse cultivation systems, besides their direct economic impact, generate multiple external economies, of

Table 14.9. Composition of the variable costs of conventional greenhouse tomato production in Spain and Belgium with yields of 14 kg m^{-2} in Spain (unheated plastic greenhouse) and 55 kg m^{-2} in Belgium (glasshouse with climate control). (Source: Benoit, 1990 and personal communication; Calatrava-Requena *et al.*, 2001; Castilla *et al.*, 2004.)

	Spain		Belgium	
Variable costs	Cost (€ kg^{-1})	Percentage of total variable costs	Cost (€ kg^{-1})	Percentage of total variable costs
Natural gas (heating + CO_2)	–	–	8.80	35.0
Salaries	0.92	46.0	10.91	43.4
Plant	0.17	8.5	2.00	8.0
Fertilizers and pesticides	0.65	32.5	1.37	5.4
Irrigation	0.13	6.5	–[a]	–
Others	0.13	6.5	2.06	82
Total	**2.00**	**100.0**	**25.14**	**100.0**

[a]Water for greenhouse irrigation in Belgium comes from previously stored rainwater.

which the following have notable economic impact: (i) companies that supply production inputs; (ii) commercialization companies; and (iii) banks and credit cooperatives, etc. The multiplying effect of protected horticultural systems on the surrounding economic system is very important.

14.2 Environmental Analysis

14.2.1 Introduction

Nowadays, it is becoming increasingly clear that any economic activity must be at the service of sustainable development, in general, and the protection of the environment, in particular. Sustainable development must be understood as one that satisfies present requirements without compromising the possibilities of future generations to satisfy their needs.

High quality vegetable production is nowadays conceived in a broad sense in such a way that, besides providing nutritive value, health guarantees, etc., production must also have been achieved with minimum environmental impact (Galdeano, 2002). 'Minimal environmental impact' must include a better use of natural resources and the reduction of the residues generated.

Important research and development efforts during recent decades have allowed for improvements in the efficient use of resources like water (see Chapter 11), so that water management, which up until now was oriented towards making the best use of what was available, is now more oriented towards the management of what the crop demands (Scoullos, 2003).

In general, greenhouse cultivation generates internal type residues, present in any kind of agricultural activity, which affect soil and water, and external type of residues, such as plastics, plant waste and other residues (e.g. substrate waste, greenhouse wires, supports, etc.).

Although plant waste was not traditionally considered as a residue, the vast accumulation of plant waste in areas where there is a high density of greenhouses has forced many to consider it as a residue (Escobar, 1998), even in legal terms (Parra *et al.*, 2001).

Greenhouse horticultural systems have a great multiplying effect on the surrounding economic system. However, these systems also generate several negative externalities mainly of the environmental kind, which constitute the basis of their most negative aspects. Although several research studies have been initiated on this subject, it would be desirable to investigate this further to identify and quantify the environmental impacts of these systems, so that there was greater understanding of their total economic value and their social productivity.

14.2.2 Most important residues

The environmental impact of irrigation water leachates in agricultural systems has already been described (Tanji, 1980; Stewart and Nielsen, 1990). In the specific case of greenhouses, substrate cultivation allows for the use of closed systems, which recirculates the drainage water reusing it for irrigation and avoiding the environmental impact associated with the leachates and subsequent pollution of the underground aquifers.

Sustainable management of the greenhouse soil is more complex than in conventional agriculture, mainly because of the lack of crop rotation and due to crop intensity, which enhances the possibility of salinity problems and, in the case of low quality water, increases the danger of salinization and alkalinization of the soil.

Plastic residues consist mainly of greenhouse cover films, which amount to between 2000 and 2260 kg ha^{-1} of PE, every 2 or 3 years in the south of Spain (Escobar, 1998), and whose recycling is nowadays almost universal (Photo 14.5). Other plastic materials, of much less importance, are the films used for double roofs and mulching as well as the polypropylene strings and waste irrigation dripper lines.

Plant waste is of enormous importance in those areas where the density of greenhouses is high. Its removal at the end of the growing cycles (Photo 14.6) generates waste volumes ranging from 40 to 60 m^3 ha^{-1} for a watermelon crop and from 130 to 150 m^3 ha^{-1} for a tomato crop (Escobar, 1998). If expressed in fresh weight, the estimation of the residues may reach 55 t ha^{-1} for pepper or 40 t ha^{-1} for tomato and aubergine, with initial moisture of 85% (Parra *et al.*, 2001). To eliminate plant waste, composting and controlled combustion (with cogeneration of electric energy, for instance) are the systems of most interest (Parra *et al.*, 2001). Recently another option has been introduced that of biofumigation.

Substrate cultivation generates residues in the order of 75 m^3 ha^{-1} of rockwool or 128 m^3 ha^{-1} of perlite, as an average, every 2 or 3 years, depending on their shelf life (Escobar, 1998).

Photo 14.5. The recycling of plastic greenhouse cover film residues is common in the south of Spain.

Photo 14.6. The collection of plant waste is common in the Spanish greenhouse industry.

Other residues, such as the obsolete elements of greenhouse structures, are of a much less frequent and relevant nature.

14.2.3 Environmental impact assessment

Proper assessment of the environmental impact of greenhouse cultivation requires methods that integrate its different aspects from a global perspective (Van Uffelen *et al.*, 2000). One of the most widespread methods of environmental impact assessment, in different sectors of economic activity, is life cycle assessment also known as life cycle analysis (LCA) (Antón and Montero, 2003).

LCA is a methodology that allows the environmental damage attributable to a certain product or activity throughout its life cycle to be assessed, that is, from its origin as raw matter until its end as a residue (waste) (Antón and Montero, 2003). The LCA methodology considers not only the environmental effects derived from the productive process (for instance, underground water contamination) in the case of greenhouse cultivation, but also takes into account all the other aspects with a potential to impact on the environment throughout its life cycle (e.g. environmental impact of the manufacture of the greenhouse structure or impact of the waste greenhouse).

The European Union published, in regulation ISO-14040, the LCA methodology, which besides defining its objective and scope, provides an inventory of inputs–outputs, the analysis of the impact linked to these inputs–outputs and the interpretation of the results. The LCA is, summarizing, an eco-balance that quantifies the environmental impact in categories such as: energy demand, water use, land use, climate change, ozone layer depletion (Schnitzler, 2003; Antón and Montero, 2003).

Methodologies such as LCA have allowed the improvement of the existing 'quality systems' incorporating advice to the growers (see Chapter 16) on good agricultural practices (GAP) to be refined in order to minimize the environmental impact.

There are only a limited number of existing LCA studies on Mediterranean greenhouses but these highlight the interest of reusing the drainage waters in soilless cultivation, and underline the low impact of Mediterranean greenhouses, in terms of energy use, compared with greenhouses in Northern Europe (Ammerlaan *et al.*, 2000; Antón *et al.*, 2003).

Many people do not respect the seriousness of environmental issues. Perhaps such opinions are not based on solid data, and as a consequence intensive production systems such as greenhouse horticulture are perceived as artificial processes which must be considered as highly polluting. But quantitative environmental assessments do not always agree with this point of view. For instance, Muñoz *et al.* (2008) conducted an LCA to compare the environmental impacts of greenhouse tomato production with that of open-field tomato production in the Mediterranean region. They determined that greenhouse production, *when properly managed*, has a smaller environmental impact than open field crops in most of the evaluation categories. The great advantage that could be gained by reducing the water consumption in greenhouse systems located in semi-arid regions, by Muñoz's comparative study, was that the water consumption to produce 1 kg of tomato was 24.2 l for greenhouse production and 42.8 l for open field production.

This is not meant to say the greenhouses do not have any negative impact on the environment. For instance, large areas filled with greenhouses create a big visual impact, a factor which is especially important in the tourism areas of the Mediterranean coast. Muñoz *et al.* (2008) observed that the greenhouse structure itself had the greatest influence in the global warming category, due to the energy and emissions during the production of the steel and concrete structure.

Other LCA studies on unheated, naturally ventilated greenhouses (Antón, 2004) showed that the greenhouse structure and auxiliary equipment (irrigation pipes, plastics for mulching, crop supports, etc.) accounted for 51% of the total gas emissions of the whole production process (greenhouse construction and operation). Within the structure itself, the foundations and perimeter walls made of concrete were responsible for most of the emissions. Currently efforts to redesign the foundation system and to use recyclable concrete are in progress to reduce the energy of construction. Antón's study also concluded that fertilizer production and use is the main factor that influences the environmental burden associated with the increase of chemical nutrients in water (eutrophication) and the emission of sulfur and nitrogen compounds that generate acid rain.

Good agricultural practices (GAP), especially regarding irrigation and fertilization programmes, are very important to reduce emissions. However, waste management by composting the plant biomass and recycling of operational materials is another obligation for future sustainable greenhouses. Recycling of waste has become very important, as the reuse of plastic is contributing to increases in productivity and energy efficiency. In the Almeria area of Spain, practically all the greenhouse plastic cover residues are recycled, as well as the crop residues.

The lower energy inputs of simple climate control methods in Mediterranean greenhouses, for example whitewashing to provide shading as compared with mechanical ventilation, contribute to reduce their environmental impact (Antón *et al.*, 2003). Recent data, comparing the sustainability of the greenhouse produce in Spain and The Netherlands, show that primary fuel consumption for cultivation and transport purposes per kilogram of tomato, sweet pepper and cucumber is estimated to be 13, 14–17 and 9 times greater, respectively, in The Netherlands (Van der Velden *et al.*, 2004). Recent quantitative analysis has shown that greenhouse production under passive low technology greenhouses is not a highly polluting process, provided that GAP are followed, and significant input reductions are achieved (Antón, 2004).

In some municipalities of the south of Spain, motivated mainly for landscape reasons or to avoid large concentrations of greenhouses, regulations have been introduced that restrict the development of greenhouses.

Organic horticultural production involves important restrictions in the use of fertilizers and pesticides that may have harmful effects on the environment or can give rise to the presence of residues in the horticultural products. Organic production enjoys an incipient and increasing demand in the

European markets, and it may grow further in the future because of its profitability in some cases (Engindeniz and Tuzel, 2003); however, it is not always economically feasible. The integrated production method (see Chapter 13) is, nowadays, the most feasible alternative.

14.3 Summary

- For greenhouse horticultural products, the production costs, the quality of the product and the transport costs to the markets are determinants in the selection of production sites.
- Climate conditions determine the technological level of the investment in the structure and equipment of the greenhouses (technological package), and this has a major impact on production costs.
- In Spain, a typical 'greenhouse horticultural company' is characterized by its small average size and family structure, with some exceptions. Its technological level is low, based mainly on the low-cost parral-type greenhouse, with limited use of active climate control equipment.
- The fixed production costs of these greenhouses are low, the main items being the depreciation of the greenhouse structure and the cost of the plastic covers.
- The variable costs of production vary depending on the horticultural species to be cultivated, affected further by the chosen cultivar and growing cycle, as well as by the technological level of the greenhouse.
- The fixed greenhouse production costs vary depending, mainly, on the type of structure and equipment. The variable costs are linked to the management and, therefore, have a great variability.
- In an unheated low-cost greenhouse, the main variable cost is the labour (salaries), reaching its maximum values for green bean and cherry-type tomato crops, followed in order of importance by fertilizers, seeds or seedlings and pesticides.
- Water has a limited impact on the total variable costs.
- The highest percentage difference between the variable costs of production in the 'Mediterranean greenhouse agrosystem' and the 'Dutch greenhouse agrosystem' lie in the energy cost (natural gas).
- High quality vegetable production is nowadays conceived in a broad sense in such a way that, besides providing nutritive value, health guarantees, etc., it must also ensure a minimum environmental impact.
- Greenhouse growing systems generate internal type of residues, present in any kind of agricultural activity, which affect the soil and the water, and external type residues, such as plastics, plant wastes and others.
- The vast accumulation of plant waste in areas where there is a high density of greenhouses has forced them to be considered as a residue.
- Proper assessment of the environmental impact of greenhouse cultivation requires methods that integrate its different aspects from a global perspective. One of the most widespread methods of environmental impact assessment is 'life cycle assessment' (LCA), and this is of great interest for assessing the greenhouse cultivation system.
- The small number of existing studies on Mediterranean greenhouses, using the LCA methodology, highlight the interest in reusing the drainage waters in substrate crops and underline the low environmental impact of these greenhouses in terms of energy use, compared with the sophisticated greenhouses of Northern Europe.

15

Postharvest

15.1 Introduction

Postharvest losses of fresh horticultural products usually exceed 25% of the total production and are caused by inappropriate control of the physical, physiological and microbiological conditions during storage and commercialization (Lioutas, 1998).

The weight loss after harvest of fresh horticultural products is caused, mainly, by water loss through evaporation, which depends on the temperature and humidity of the surrounding environment and on the temperature of the product. The respiratory processes also contribute to the weight loss, but to a lesser extent, and are quite dependent on the temperature, increasing with it.

Vegetable water loss causes a quality decrease in the form of product wilt, discoloration and loss of firmness. This water loss in some fruit vegetables, such as tomato, originates in the peduncles mainly, because the skin is practically impermeable, being possible to compare a tomato fruit to a container filled with water, because the water content may be as high as 95% (Scheer, 1994). Other vegetables such as cucumber, whose skin is much more permeable, are more sensitive to dehydration. Therefore, cucumbers are usually packed inside a plastic film to limit the postharvest water losses, as it is also common with leafy vegetables (Photo 15.1).

15.2 Postharvest Respiratory Metabolism

The respiration process involves the combining of oxygen (O_2) from the air with organic molecules in the plant tissues (usually one type of sugar), downgrading them to form several intermediate compounds and, eventually, CO_2 and water. The energy produced during respiration is used for other metabolic processes.

Most of the postharvest technology is directed towards decreasing respiration and other metabolic reactions, to maintain the quality of the product by manipulating the external environment. In general, the shelf life of the storage of products varies inversely with their respiration rate, so products with a lower respiration rate have a longer shelf life and vice versa.

The most important factors affecting postharvest are: (i) *temperature*; (ii) *humidity*; (iii) *composition of the atmosphere*; and (iv) *physical stress* (Saltveit, 2003a).

Within the physiological range of most crops (between 0 and 30°C) an increase in temperature causes an exponential increase in respiration; that is, the respiratory rate is doubled, for every 10°C increase in temperature (Van't Hoff law).

High temperatures outside this interval (0–30°C) induce a decrease in respiration

Photo 15.1. Covering cucumbers in a plastic film prevents dehydration and subsequent loss of quality.

and the death of the tissue, if they are sustained beyond a certain minimum time period. Nevertheless, many tissues tolerate high temperatures for a small interval (minutes), and this is occasionally used as a means to control insects and diseases.

Generally, the application of low temperatures that limit transpiration is the most common technology to extend the life of vegetable products. The temperature decrease of the product, besides decreasing respiration, also reduces the evaporation of water from the product to the surrounding air.

Regarding the composition of the atmosphere, this is important because the maintenance of aerobic respiration requires adequate O_2 levels. Generally, low O_2 levels of the order of 2–3% induce a beneficial decrease in the respiration rate and other metabolic reactions, in the same way as an increase in the CO_2 levels (Salveit, 2003a).

Physical stress produced by an injury of the tissue generates a respiration increase, normally associated with a higher production of ethylene.

The respiration rate of tissue which has a vegetative or flower meristem (asparagus, broccoli) is very high, which means that many products that are harvested while they are actively growing have high respiration rates. After the harvest, the respiration rate decreases, slowly in non-climacteric fruits and quickly in vegetative tissue and unripe fruits. In climacteric fruits the respiration rate increases if they are harvested during the ripening process.

15.3 Ripening

Fruit ripening involves a number of physiological, biochemical and structural processes that occur during the last developmental stages which confer the proper organoleptic and nutritional characteristics for consumption.

The changes in colour and firmness, the decrease in acidity, the increase in sugars and the generation of volatiles are some of the most common changes during fruit ripening (Seymour *et al.*, 1993).

Climacteric fruits are characterized by the production of ethylene during the ripening process, associated with an increase in respiration, something that does not occur significantly in non-climacteric fruits (Biale and Young, 1981).

Ripening of the fruit follows a genetically determined pattern, during which ethylene has a direct and essential implication, as a ripening regulatory hormone, although there are ripening processes that are independent of ethylene.

In some tomato cultivars (long life) the ripening pattern has been altered, limiting the production of ethylene by means of breeding (Zacarías and Alferez, 2000). Among the vegetables grown in greenhouses, tomato, watermelon and melon are examples of climacteric fruit (Photo 15.2).

The respiration process after harvest may negatively affect the quality of the product, which is usually at its best at the time of harvest for non-climacteric fruit and in the majority of leafy vegetables. In climacteric fruits, the synthesis of pigments and volatiles (e.g. lycopene in tomato), chlorophyll loss and the conversion of starch into sugar, which develop after harvest, are necessary processes to achieve maximum quality (Salveit, 2003a).

The most relevant difference between climacteric and non-climacteric fruit is the ability to ripen once harvested. Climacteric fruit can ripen once separated from the plant, if harvested at a certain ripening stage. This is not the case for non-climacteric fruit, in which harvest stops the fruit-ripening process, with the exception of the degradation of the chlorophylls and the synthesis of carotenoids (Martínez-Madrid *et al.*, 2000). Therefore, non-climacteric fruits must be harvested when they reach their optimum ripening point. For these reasons, in the ripening process of climacteric fruits the physiological and consumption ripening points are different.

15.4 Ethylene

Ethylene (C_2H_4) is a gas, colourless at ambient temperature, which is normally produced under certain conditions in organic compounds.

The synthesis of ethylene is enhanced in mature reproductive tissues of climacteric vegetable species. For its synthesis ethylene requires O_2 and to become active it requires low CO_2 levels. The presence of low concentrations of ethylene (ppb) is enough to induce its effects.

The presence of ethylene has the following effects, among others: (i) it induces the ripening of climacteric fruits and of some non-climacteric fruits; (ii) it activates the synthesis of anthocyanins in fruits (modifying their colour); (iii) it contributes

Photo 15.2. Tomato is a climacteric fruit that can ripen after harvest.

to the destruction of chlorophyll with subsequent yellowing; and (iv) it promotes senescence (Salveit, 2003b).

A change induced by ethylene can be beneficial for one species (ripening of tomato fruit) or harmful for another species (yellowing in broccoli).

The application of ethylene must be done at a proper temperature and at the proper time of development. Field application of ethylene is usually done by applying products that release ethylene (such as ethephon), whereas in the store, pure ethylene or ethylene generated from alcohol, is used in sealed chambers.

In order to avoid the action of ethylene in greenhouses or in closed environments (store chambers, transport vehicles) proper ventilation is required to prevent the accumulation of biologically active levels of this gas (Photo 15.3). Other ways of avoiding the effects of ethylene are: (i) to prevent its contact with plant tissues; (ii) storage at low temperature; or (iii) using inhibitors of the perception of ethylene, such as CO_2, silver thiosulfate (used for flowers) and MCP (methylcyclopropane) (Salveit, 2003b). In the same way, it is possible to block the response of the plant tissues to ethylene using controlled atmospheres or modified atmosphere packaging (MAP; see section 15.5).

In the storage and transport of fruits and vegetables it is essential to properly combine the horticultural products that will be mixed in the same enclosure, to avoid undesired effects caused by ethylene generated by some of them (climacteric fruits) on the others and, if ethylene is produced, to prevent its accumulation by means of ventilation, absorption or blockage.

15.5 Postharvest Handling

The ideal management of vegetables starts with proper handling at harvest, an operation that should be done preferably in the morning, when the ambient temperature is lower. The harvested product must be protected from the sun, and whenever possible, to proceed immediately to their pre-cooling (fast cooling before processing) if such facilities are available.

The most popular pre-cooling procedure is by forced air, which circulates air at low temperature (Tompson, 2003).

Many vegetables are sensitive to cooling (chilling), that is they get damaged if

Photo 15.3. Handling and packaging buildings must be ventilated to avoid the action of ethylene.

exposed to low temperature (but above the freezing point) for a certain minimum period of time. Depending on their origin, tropical and subtropical fruits have their threshold for chilling at 10–15°C, whereas the threshold is lower for fruits that originate in temperate areas.

At 7°C chilling damage occurs in cucumber, aubergine, pepper, melon or ripened tomatoes, whereas for green tomatoes damage occurs at higher temperatures (Wang, 2003). Chilling damage can be very relevant if the low temperatures last a long time. If the duration of low temperatures is short, normal metabolic capacity in these plants is limited or cancelled, affecting their shelf life, although the damage is only evident when the product goes back to normal temperatures.

In general, tomatoes can be conserved well with a RH of 90%, but the optimal thermal regime varies depending on the ripening stage, the recommendation being for less than 15°C for green and early pink-colour stage tomatoes, and less than 10°C for late pink-colour stage and ripened fruits (Chaux and Foury, 1994a, b). The storage temperature allows for regulating the ripening speed; for instance, for pink-colour stage tomatoes, a temperature of 10°C allows them to ripen in 10–20 days, whereas at 20°C ripening is shortened to 8–10 days (Chaux and Foury, 1994b).

The optimal storage temperature of greenhouse cucumbers is from 12 to 13°C, because lower temperatures cause the fruits to wilt and higher temperatures accelerate their respiration and dehydration (Chaux and Foury, 1994b). Covering them in a plastic film extends their shelf life.

The optimal storage temperature of leafy vegetables is lower than for fruit vegetables. In general, while temperatures of 0–2°C, with RH of 90–98%, are optimal for some (artichokes, asparagus, broccoli, cabbage, cauliflower, Chinese cabbage, endives, lettuce, carrots and cantaloupe melons), the fruit vegetables are better conserved at 7–10°C, with RH of 85–95% (Tompson and Kader, 2003). The usual greenhouse vegetables are all very sensitive to freezing (Wang, 2003).

In fresh products, postharvest treatments with high temperatures can be of interest to control pests and insects and fungal diseases, before their storage or long-distance shipping. Washing the peppers with water at temperatures between 50 and 65°C, while simultaneously brushing them, has proved to be efficient for the control of postharvest diseases; the same is true for the treatment of tomatoes with hot water at 50°C for 2 min (Lurie and Klein, 2003).

A proper environmental humidity has a notable influence on maintaining the postharvest quality of fruits and vegetables, especially during cold storage. An inappropriate humidity can increase the incidence of fungal diseases, alter the organoleptic characteristics and induce the cracking or cork-like texture of the fruits and vegetables (Scharz, 1994).

Storage in a controlled atmosphere involves the modification of the normal composition of the air (78% N_2, 21% O_2 and 0.03% CO_2), in order to have less than 8% of O_2 and more than 1% of CO_2, while keeping low temperature and adequate humidity according to the product being stored, which decreases the respiration rate of the product and the production of ethylene (Kader, 2003).

An optimum atmospheric composition delays: (i) the loss of chlorophyll (green colour); (ii) the biosynthesis of carotenoids (yellow and orange colours) and anthocyanins (red and blue colours); and (iii) the biosynthesis and oxidation of phenolic compounds (brown colour) (Kader, 2003). Low levels of O_2 and/or high concentrations of CO_2 in the air affect the flavour, decreasing the loss of acidity, the conversion of starch into sugar, the interconversions of sugars and the biosynthesis of volatiles that affect the flavour and aromas, resulting in an improvement of the nutritional flavour, as the ascorbic acid and other vitamins remain in the fruits (Kader, 2003).

In addition to a delay in senescence, storage under a controlled atmosphere decreases the sensitivity to ethylene (if the O_2 level is below 8%, and if the CO_2 level is above 1%), and can be useful to control pests and diseases.

Disadvantages of storage in a controlled atmosphere include: (i) irregular ripening in some cases; (ii) the modification of the organoleptic characteristics (as a result of anaerobic respiration); and (iii) the increase, sometimes, of physiological disorders and chilling damage (Mir and Beaudry, 2003).

When the aim is to accelerate the ripening of a product, such as tomato, ripening chambers may be used, adding ethylene and keeping a proper temperature. These chambers are often used in ripening citrus (degreening).

MAP (modified atmosphere packaging) of fresh vegetables allows for isolating fresh products, with active respiration, in plastic film packages to modify the O_2 and CO_2 levels in the atmosphere inside the package. At the time of packaging, levels of 2–3% of O_2 and 5% of CO_2 are usual (Kader *et al.*, 1989).

In addition, MAP decreases the dehydration of the product and insulates it from the external environment, limiting its exposure to pathogens and contaminants, which contributes to maintaining its quality. It is often used in fresh cut products.

The modification of the atmosphere in MAP requires active respiration of the plant tissues, on one side, and the existence of a barrier that prevents the gas exchange, on the other side. The creation of such barriers is achieved by using plastic films of a characteristic permeability (that controls the entrance and exit of O_2 and CO_2 into/from the package) and by means of micro-perforated plastic films (Mir and Beaudry, 2003).

In MAP, the control of temperature is essential. The decrease of water losses prevents the product becoming desiccated but in some cases may increase how much it wilts. Not all products are suitable for MAP.

A variant of MAP is partial vacuum packaging, which keeps the normal composition of the atmosphere, but at a lower than normal pressure in impermeable packages, at low temperature (Gorris *et al.*, 1994). This system stabilizes the quality, decreasing the metabolic activity of the products avoiding the increase of undesired microorganisms.

15.6 Quality

The notion of a food product's quality is, simultaneously, complex and relative; it is complex because the quality of a product cannot be determined by a single property, but from the combination of all its physical, chemical and sensorial properties; and it is relative because this combination of factors that define it must be such that determines acceptance by the consumer (Martínez-Madrid *et al.*, 2000). Fresh vegetable products maintain, after their harvest, a metabolic activity that is essential for the preservation of their quality. In Chapter 16, other aspects of quality will be discussed. In this section sensorial quality aspects are described.

The sensorial quality is dictated by a number of external and internal factors. The external factors include the attributes related to the appearance, such as colour, form, size and firmness, and are subject to physical and visual properties, being appreciated by the consumer through the senses of sight and touch, whereas the attributes related to flavour, aroma and texture, which are sensed by the taste and smell, are included among the internal factors of sensorial quality (Martínez-Madrid *et al.*, 2000).

The essential criteria to evaluate the sensorial quality in fruits and vegetables are the colour, flavour, aroma and texture (Photo 15.4). The flavour and the aroma are the most subjective and difficult to evaluate qualitative aspects. The flavour can be evaluated by taste and smell, and is mainly composed of sweetness, acidity and aroma, that correspond to the sugars, acids and volatiles, respectively (Baldwin, 2003). Other components of the flavour are bitterness, salinity and astringency. The acidity and the aroma modify the perception of sweetness, one of the most important components of the flavour in fruits and vegetables. The perception of the non-volatile components of flavour (sweet, acid, salty and bitter) takes place on the tongue and the aromas

Photo 15.4. The essential criteria to evaluate the sensorial quality of horticultural products are the colour, flavour, aroma and texture.

are detected by the nose; both perceptions are integrated in the brain, being difficult to distinguish between them (Baldwin, 2003).

The genetic characteristics are the main determinants of flavour and aroma of fresh horticultural products, although they are influenced, but to a lesser degree, by cultural practices and the pre-harvest conditions, as well as by the ripening stage at harvest and any subsequent handling.

The organoleptic quality of non-climacteric fruits generally decreases after harvest, whereas climacteric fruits may reach their best quality after being harvested, if they are harvested after the beginning of the ripening process.

The sensorial evaluation of flavour and aroma of a product is usually done by taster panels. Consumer preferences vary depending on socio-economic, ethnic and geographical conditions.

The sugars that supply the sweet flavour are fructose (the sweetest), sucrose and glucose (the least sweet). Organic acids, such as citric acid in tomato, provide the acid taste.

In the majority of melon cultivars the main sugar is sucrose and the most common acids are citric and malic acids, whereas in watermelon sucrose predominates and, in some cultivars, there are high levels of fructose, and malic acid is the only relevant acid (Baldwin, 2003).

In tomato, the total content of soluble solids and the acidity determine its taste. The most abundant sugars are glucose and fructose, at approximately equal levels, with citric acid being present in greater quantities than malic acid. There is also the presence of a large number of volatile compounds (more than 400) from which 16 contribute more effectively to the taste and aroma (Baldwin, 2003).

The texture is a qualitative attribute that is critical for the acceptance of fruits and vegetables, that is, for the perception that the consumer has of the qualitative characteristics. The texture involves the structural and mechanical properties of an edible product and its sensorial perception in the hand or in the mouth (Abbot and Harker, 2003). The texture is related to a series of chemical compounds responsible for the perception of the structure, such as pectin, cellulose, hemicelluloses and proteins (Martínez-Madrid *et al.*, 2000).

Sometimes, the term texture includes some mechanical properties, which cannot be of interest to the consumer, such as resistance to mechanical damage or transport. The texture is altered throughout the shelf life of the product, so it can only be referred to at the time of evaluation.

Measurements of texture, nowadays, are considered critical indicators of the non-visual aspects of quality. The complexity of the texture allows for its complete measurement only by means of sensorial evaluation (valuation panels), although instrumental measurements are preferable, whenever possible. There are many measurements that relate to textural attributes, normally the more precise ones being those that use destructive methods (Abbot and Harker, 2003).

Obtaining a high quality product depends on the expression of the genetic characteristics of the chosen cultivar under the ecological conditions in which it is cultivated.

The study of the nutritional value and the beneficial effects of fruits and vegetables on human health has become increasingly relevant in recent years (Desjardins and Patil, 2007; Patil *et al.*, 2009).

The control of the pre-harvest conditions, of an environmental nature (temperature, humidity, radiation, soil, rain) and cultural nature (nutrition, irrigation, pruning), is not enough to achieve a good quality product, as the ripening stage of the fruits at the time of harvesting is the factor that plays an essential role in the sensorial qualitative characteristics. This is because the production of compounds such as the aromas that contribute to the flavour take place, mainly, in the advanced stages of the ripening process (Martínez-Madrid *et al.*, 2000).

An early harvest has advantages for distribution of the product, as the texture is maintained for a longer period extending the shelf life, but this is to the detriment of its sensorial quality, at least in non-climacteric fruits.

15.7 Food Safety: Traceability

Food safety is increasingly important for the consumer. The two main causes of lack of food safety are: (i) microbial toxins; and (ii) the contamination by faecal microorganisms (Sholberg and Conway, 2003). Another cause of lack of food safety is the presence of pesticide residues in the products, when the phytosanitary defence is inappropriate.

'Cold pasteurization', which is the use of non-thermal radiation, does not solve the majority of the microbiological problems, so prevention becomes the best method (Gorny and Zagory, 2003). Washing the fresh vegetables with chlorinated water before they are consumed is a recommended practice, because, although it does not sterilize, at the very least, microbial populations are drastically decreased.

The prevention of the product's contamination is the only way to minimize the risks and to achieve healthiness and food safety, so the systems that provide food safety are based in prevention programmes of good practices, during the production and processing of the food (Gorny and Zagory, 2003).

Among the good practices of prevention programmes traceability plays a key role. Traceability is a set of pre-established procedures that inform on the history, the location and trajectory of a product or a batch of products throughout the food chain, in all the production, processing and distribution phases (Photo 15.5). The establishment of traceability systems is a good solution to combat the mistrust of the consumers that has arisen from various different food crises, allowing for better control and warranty of quality and safety for the consumer (Belloso, 2003).

Traceability contributes to relaying to the consumer, through the commercial channel, detailed information of the history of the product. Despite its great interest to quality control and food safety, traceability

Photo 15.5. Traceability provides knowledge of the history of a product throughout the food chain, beginning at the greenhouse where it has been produced.

has notable complications of postharvest manipulation and logistics, which involves a high cost and affects the price to the consumer, who will have to pay more for a traced product. In some countries, where the consumer has great trust in its administration (official agencies) to take care of food safety, the need for the extra cost of traceability is questioned, for many products.

The new European regulation on food safety (Regulation 178/2002) that imposed that traceability of food should be ensured, became compulsory from 2005.

Nowadays, the importance of freshly cut horticultural products, a concept that involves fresh fruits and vegetables, cleaned, cut and packed, that maintain their natural properties and are ready to be consumed, has increased dramatically (Belloso, 2003). The term 'IV range products' is used in France and Spain, whereas in English-speaking countries they are known as 'fresh cut products', among other commercial terms.

In fresh cut products, in order to maintain the quality of fruit and vegetables that have been just freshly cut and packaged in modified atmosphere, it is essential to maintain a low temperature that decreases the enzymatic degradation of the product and limits the multiplication of microorganisms, with the exception of those that can grow and multiply at low temperatures such as *Listeria* (Gorny and Zagory, 2003). Therefore, prevention and implementation of good practice programmes are essential to maintain proper quality and to provide healthiness and food safety to the consumer.

15.8 Postharvest Pathologies

Several factors affect the postharvest pathology during cultivation, as they affect the crop condition, especially fertilization and the soil characteristics. High contents of nitrogen shorten the shelf life of the product. Calcium has a great relevance as well, due to its influence as a constituent of cell walls (Sholberg and Conway, 2003). The ripening stage at harvest also has an effect, as well as the handling and storage characteristics.

It is essential to maintain high sanitary conditions during the whole process of handling and postharvest storage. The complementary use of conventional or new antifungal treatments (radiation, UV light) may be of interest, but good management of temperature and humidity are essential to prevent the development of diseases.

Bacteria are more common pathogens in vegetables than in fruits, because vegetables are less acid, in general, than fruits (Sholberg and Conway, 2003).

In general, the alternatives to chemical control of postharvest pathogens are less efficient than many fungicides; so in the future it will be necessary to combine several alternative methods to develop an integrated strategy that can be efficient (Sholberg and Conway, 2003).

15.9 Summary

- Postharvest losses of fresh horticultural products are caused by an inappropriate control of the physical, physiological and microbiological deterioration during their storage and commercialization.
- Most postharvest technology is directed towards decreasing the respiration of the product in order to maintain its quality. The most important factors that affect the postharvest quality are temperature, humidity and composition of the atmosphere and the physical stress derived from wounds or scars in the plant tissues.
- The ripening process confers fruits with the proper organoleptic and nutritional characteristics for their consumption. The ripening of the fruit follows genetically determined patterns, in which the regulation of ethylene, the ripening hormone, plays an essential role.
- The most relevant difference between climacteric and non-climacteric fruits is their ability to ripen once harvested. Tomato, watermelon and melon are examples of climacteric fruits.
- Climacteric fruits generate ethylene during their ripening. Therefore the products that can be stored and transported together must be considered carefully, to avoid the undesired effects of ethylene.
- Good postharvest handling starts by harvesting the fruits preferably in the morning, when the temperature is lower, protecting the harvested products from the direct action of the sun and proceeding immediately to the pre-cooling facility, if such facilities are available, before processing.
- Low temperatures together with a proper environmental humidity are the simplest methods of postharvest conservation of horticultural products.
- The alteration of the normal composition of the atmosphere (decreasing the O_2 levels and increasing the CO_2) delays senescence and extends the shelf life of horticultural products. The technique of packaging in modified atmospheres is used increasingly for horticultural products for fresh consumption, especially in fresh cut products (fresh fruits and vegetables, cleaned, cut and packed, ready to consume).
- The essential criteria to evaluate the sensorial quality in fruits and vegetables are colour, flavour, aroma and texture.
- The organoleptic quality of non-climacteric fruits generally decreases after harvest, whereas climacteric fruits can reach their best quality after being harvested, if properly managed. Therefore, the ripening stage of the fruit at the moment of harvest is most important.
- Food safety is increasingly important for the consumer. The new legislation, which is compulsory in the European Union since 2005, ensures the traceability of food.
- The prevention and implementation of good practice programmes, both for production and for postharvest, are essential to achieve and maintain a proper quality and to provide food safety to the consumer.

16

Marketing

16.1 Introduction

The marketing of horticultural products, in a broad sense, is the process of getting the products from the farm to the consumer. It involves, therefore, the operations and transactions related to the movement, storage, processing and distribution of the products. The perishable nature of greenhouse horticultural products, preferably destined for short-term fresh consumption, gives a specific character to their marketing, as these products are not normally subject to long-term conservation processes (freezing, dehydration, canning).

Marketing adds value to the products for their positioning in space (placing them where the consumers are) and in time (offering them at the right moment), and in the form (adapting the product to the requirements and likes of the consumer) and the possession (making possible the transference of the product's ownership to the consumer).

Traditionally three major functions have been considered in the marketing of agricultural products: (i) stock; (ii) processing (preparation for consumption); and (iii) distribution. Stocking allows for grouping batches to reach a minimum volume, that allows for their preparation for the consumption (set of operations that provide utilities of form, time and space: standardization, packaging, storage, transport) and later distribution to the consumer.

Nowadays, it is frequent to mistake distribution for marketing, as the stock stage is normally non-existent as the growers, usually grouped into associative entities, do this. Equally, the preparation for consumption is usually assumed by the grower's associations or by the distribution companies.

The marketing process of horticultural products is usually considered as a set of elemental processes, or marketing services, among which we may highlight: transport, storage, industrialization, standardization and normalization, packaging, buying and selling, financing and risk assumption (Caldentey, 1972).

The marketing agents are the individuals or entities, which take part in the marketing processes, there being a distinction between direct agents, who become owners of the merchandise, and indirect agents, who never become owners.

During recent decades, and especially in high added value products such as greenhouse vegetables, the number of agents involved in their marketing has notably decreased, with an increasing participation of the growers in the process, and the subsequent sharing in the associated added value.

In relation to the way in which the product is sold, the consignment (commission selling) and the deposit sale modes are still relevant, although there is an increasing importance of the fixed price sale, with a previous supply contract in many cases, when the commercial relations between both parties and the quality of the product permit it.

16.2. Postharvest Alterations: Storage

The perishable nature of vegetables is a critical factor in the marketing process (Plate 27). Harvesting must be done at the proper time, keeping in mind that the consumption ripening point does not have to coincide with the physiological ripening (see Chapter 15). Once a certain ripening point is exceeded, the quality of these vegetables deteriorates very quickly.

In order to extend their postharvest life some products are harvested before ripening (for instance, green tomatoes), which may involve improper development and the consumer not enjoying the best of taste, aroma, texture and, therefore, the expected satisfaction. Depending on the length of the period between harvest and consumption, the best harvest time will have to be considered. So, postharvest management is critical to maximize the duration of the quality of the product.

The perishable nature of greenhouse horticultural products and their high value have allowed the profitable use of quick and expensive transport methods, such as refrigerated lorries or, even, planes, minimizing the storage time.

Fresh horticultural products maintain, after being harvested, a metabolic activity which is essential to preserving their quality. The changes during the ripening process of the fruits are very complex and contribute to maintain (and even enhance) the initial quality of the product. Characteristics that determine the quality of horticultural products include: (i) the colour, a result of modifications in the content of chlorophylls, carotenoids and anthocyanins; (ii) the firmness, derived from alterations in the cell walls; (iii) the taste, a consequence of the metabolism of the carbohydrates; and (iv) the characteristic aromas caused by the release of volatile compounds. In the initiation of these ripening processes several plant hormones are involved, mainly ethylene, apart from being regulated by specific ripening genes (see Chapter 15).

The intensity of the physiological processes associated with the ripening process is affected by external factors: mainly temperature, humidity and composition of the atmosphere (see Chapter 15). The most widespread use of technology to extend the storage period of plant products is the application of low temperatures, which limits respiration, the main postharvest physiological process.

Pre-cooling is the rapid cooling operation of just-harvested products, to decrease their ripening process and to limit their deterioration before storage or before sending them to the market.

Storage allows for more control of the ‘offer’ of horticultural products, within certain limits, and reduces their seasonality. Cold storage of perishable products aims to decrease their respiration to retard microbial activity and decrease the water losses, by means of regulating the temperature, oxygen and CO_2 levels.

The storage temperature must be constant, which must be permanently monitored because respiration (although reduced) generates heat that must be removed. An excessively low temperature may interfere with the ripening process. In tomatoes, for instance, maintaining temperatures above 13°C, and even up to 16–18°C is advisable, to induce the ripening of the tomatoes that were harvested green, for their transport to very distant markets (Janick, 1986).

The use of plastic films has contributed to the development of controlled or modified atmosphere packaging (MAP). These films, which are selectively permeable to oxygen and CO_2, maintain the desired gas composition, within certain limits, and extend the shelf life in combination with low temperatures. It is used with fresh cut products.

During the ripening process, fruits generate several volatile organic compounds (acids, alcohols, aldehydes), which have a marked influence in the organoleptic characteristics of the fruit. These volatiles usually have little influence on respiration, except for ethylene. Ethylene has an inducing effect on the ripening of immature fruits in the majority of species, speeding up their respiration and ripening, without influencing the already ripened fruits; for a long time it has been used commercially to speed up ripening (for instance, in the ripening of citrus and bananas).

Other preserving techniques, such as use of low or high pressure, or techniques based on the use of microwaves or radiant energy, for instance, are not widespread, and some are in the development stage (Dolado, 2000).

Chapter 15 discusses the technical aspects of vegetable postharvest management in more detail.

16.3 Standardization and Classification

Classification and calibration allows for describing the products without the need to see them, and they represent the basis of trade at a distance, without the physical presence of the commodities (Photo 16.1). Classification allows for the elimination of all the inappropriate products and for grouping them according to a pre-fixed quality pattern (size, colour, etc.).

The packaging of products, using a wide variety of materials, allows the products, which have already been classified, to get to the consumer in the right condition through the distribution chain. Pre-packaging in small units, of the size of buying unit used by the consumer, is widespread among fresh horticultural products; in this way its 'functionality' is improved by adapting its size to the ration demanded by the consumer.

Traditionally, quality improvement is based on product standardization. This is understood as the set of activities whose aim is to ensure that the agricultural products exceed a minimum level of quality and be presented to the market classified in

Photo 16.1. Classification and calibration are the basis of trade at a distance, without the physical presence of the commodities. The image shows a tomato classification machine.

different categories (according to characteristics of, e.g. size, etc.), in agreement with standards (uniform characteristics) established by previous evaluation of the product (Galdeano, 2002).

The establishment of commercial standards for horticultural products has consisted, essentially, of the application of quality regulations established at national and international levels.

16.4 Marketing Channels

The trade or marketing channels are the routes travelled by the product throughout the trade process, from the grower to the consumer.

Throughout the commercial channel, a value adding process occurs and the utility of the product increases. These increases in the value are much higher in greenhouse products than in any other agri-food product, such as, for instance, plant oils. Therefore, studying the marketing channels in protected cultivation is of maximum interest.

The main functions of the commercial channel are: (i) the exchange and transfer of the ownership of the commodity; (ii) the physical functions (processing, storage and transport of the product); and (iii) those functions that ease the marketing process (degree of product standardization, funding, risk assumption and market price formation).

There are a number of different kind of 'flows' that are generated through the trade channel, among which we can highlight, besides the goods and services, the monetary, financial, information and risk flows.

From the flows that pass through the trade channel, some, such as those related to the risk or the information are difficult to analyse, whereas others, such as the financial ones, linked to the changes in the price of the product are much more explicit and easy to analyse. Therefore, usually, the price transmission analysis is used between the different segments of the channel to measure *its performing efficiency*. From the price transmission, normally, it is interesting to study: (i) the level of *transmission of the price variations* (for instance, from grower to wholesaler, from grower to destination, from wholesaler to destination); (ii) the level of *symmetry or asymmetry* of the transmission; and (iii) the existence of *time lags*. In the same way, it is frequently of interest to complement these analyses with a '*causality*' analysis, for example whether the price at the origin determines the price at the destination (or at any intermediate level) or vice versa.

An acceptable level of price symmetric transmission, with a time lag that does not exceed the duration of the distribution process, indicates a good performance of the channel, which benefits all the involved trade agents, and, especially, the grower and the consumer. If the prices are well transmitted, we may assume that the same happens to other kind of flows.

Normally, it is considered that the larger the number of agents in the channel the more difficult it is to achieve a proper transmission, which in principle is logical. In the hypothetical extreme case of a single origin agent that would directly reach the destination, selling directly to the consumer, the price transmission would be, in principle, perfect. Nevertheless, a situation of monopoly would occur, which could involve a darkening of the channel in detriment of its transparency. In any case, there are few analyses of the channels in protected horticulture.

Trade channels are usually represented by means of graphs, which start at the grower and end at the destination market, the last step of the trade channel being represented by the consumer. The set of these possible routes is called 'marketing channels'. The charts can reflect volumetric flows of goods or other aspects (prices, margins).

Taking the case of Almeria as an example, the greenhouse vegetable grower, has two basic alternatives to market his or her products (Fig. 16.1): (i) trade them privately as an individual (mainly at auction) (Photo 16.2); or (ii) trade them through some sort of association (cooperative, an associated growers organization such as SAT),

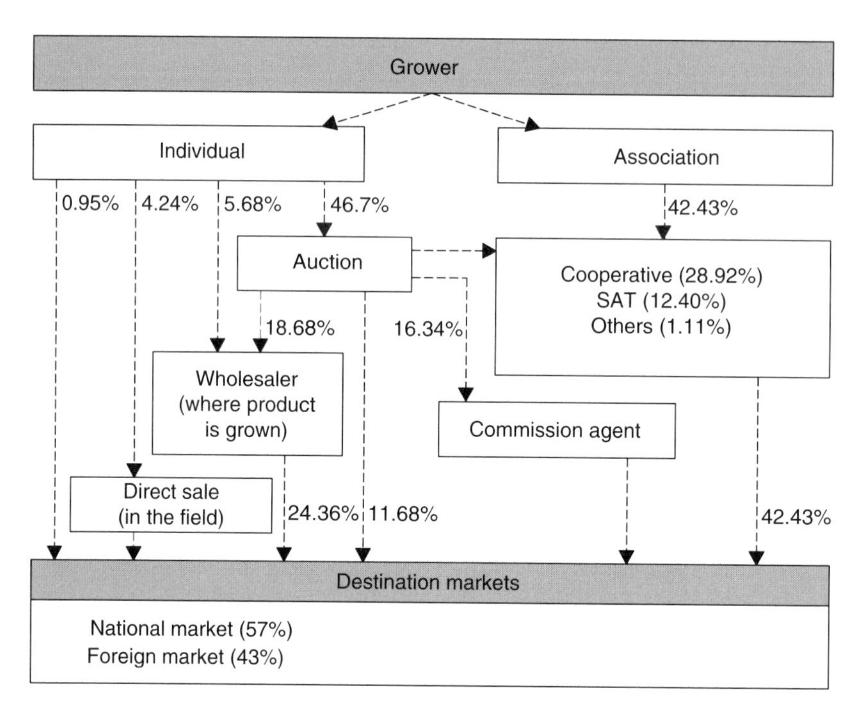

Fig. 16.1. Commercial channels for vegetables in Almeria (Martínez-Carrasco and Calatrava, 1996).

Photo 16.2. Years ago in the south of Spain, auctions took place only with unclassified products.

participating in the trade process. Most products are channelled through the auctions (46.7%), followed by the associations (42.4%), with a small relevant volume through wholesalers based where the product is grown and direct sale (Fig. 16.1). Direct sale in the field is only relevant for some products such as watermelon and melon (Martínez-Carrasco and Calatrava, 1996).

Export, mainly to European markets, involves around 43% of the production, allocating the rest (57%) to the national market, although part of it may be reissued to foreign markets (Martínez-Carrasco and Calatrava, 1996). In this text, we consider 'national' to refer only to the Spanish market, including the market of other countries belonging to the European Union within the 'foreign' market.

Generally, horticultural handling and trade companies in Spain nowadays are small, as is their corporate structure.

In Europe, the supermarket chains are the main vegetable distribution channels, and their importance is increasing. Today, Spain and The Netherlands are the main suppliers of the two main high-value vegetable markets in Europe: (i) the German market where price is essential; and (ii) the British market based on added value and quality more than the price (Wijnands, 2003). Non-European Mediterranean countries (Turkey, Morocco, Egypt) are becoming important competitors, especially for Spain, because their production calendars are closer to those of Spain than to those of The Netherlands.

16.5 Transport

In general, horticultural products are consumed close to the production area, given their perishable nature and the transport costs, expeditions to faraway countries being very limited. Around 95% of the European production of vegetables is consumed at no greater distances than 1000 km from the production regions (Wijnands, 2003), although some areas direct a larger part of their production to more distant markets, such as the south of Spain.

The energy required to produce 1 kg of greenhouse tomatoes in Ohio (north of the USA) is more than three times that of the energy used in Arizona (south of the USA), due to the better climate conditions in the latter region (Jensen and Malter, 1995). From an energy point of view, it is much cheaper to transport tomatoes cultivated in southern latitudes in the winter period in the USA, than to grow them near the consumption centres, at higher latitudes such as Ohio (Jensen and Malter, 1995). A similar situation happens in Europe with the production in climate control greenhouses in the north of Europe, in relation to the Mediterranean greenhouses, during the winter period. Obviously, the quality condition of the commodities play a key role in the consumer's perception, especially in products like tomato, and can contribute to increasing the acceptance of locally grown products, adapted to the consumer's demands for product quality.

Despite the great variability in the production costs, typical of Mediterranean greenhouses, Table 16.1 summarizes approximations of the production costs and the integral cost price in the market of tomato, pepper and cucumber for the best growers in The Netherlands and Spain.

The 'Dutch greenhouse agrosystem' has much higher levels of production than the 'Mediterranean greenhouse agrosystem', but its production costs are also much higher. In Spain, the marketing costs (difference between the integral cost price on the market and the production cost) are higher than in The Netherlands, mainly due to the higher transportation costs.

A similar situation occurs in Morocco with respect to Spain, for tomato costs (Table 16.2). Although the growing cycles in Morocco are restricted by the exportation calendars to the European Union, which limits their harvest nowadays, the production costs are lower than in Spain, but the higher transportation costs decrease the differences, limiting the competitiveness. The extension of the growing cycles in Morocco, as their exportation calendars are expanding due to recent commercial agreements (2010), means that they are increasing in their competitiveness in the European market, and this will continue to rise in the future if transport costs do not change in a serious way.

Table 16.1. Conventional greenhouse tomato, bell pepper (California type) and cucumber costs, for the best growers in The Netherlands (using climatized glasshouses) and Spain (using unheated plastic greenhouses) (1996) (adapted from Verhaegh, 1988; Verhaegh and de Groot, 2000; Castilla *et al.*, 2004).

			Costs (€ kg^{-1})	
Crop	Country	Yield (kg m^{-2})	Production	Market[a]
Tomato	Spain	18–20	0.26–0.27[b]	0.53–0.62
Tomato	The Netherlands	58–60	0.55[c]	0.73
Pepper (autumn)	Spain	6.2–6.6	0.52–0.57	0.94–0.98
Pepper (spring)	Spain	8.0–12.0	0.29–0.30	0.70–0.71
Pepper	The Netherlands	23.0–27.0	1.18	1.50
Cucumber (autumn)	Spain	9.5–12.5	0.20–0.23	0.49–0.52
Cucumber	The Netherlands	65.0	0.49	0.65

[a]The integral cost prices on the market refer to the German market (Frankfurt) and single-use packages.
[b]Unclassified.
[c]Classified.

Table 16.2. Estimated conventional greenhouse tomato costs in Morocco (Agadir) for the limited export cycles to the European market (year 1998). The production cycles are short, coupled to the exportation costs. (Source: J. Calatrava-Requena, unpublished data, project CICYT-SEC-944-0391.)

	Costs (€ kg^{-1})	
Yield (kg m^{-2})	Production	Market[a]
6.0	0.23–0.24	0.66–0.69
11.0	0.15–0.16	0.58–0.61

[a]The integral cost prices on the market refer to the Perpignan market (France).

Therefore, in vegetable protected cultivation, the location of the production areas is a key factor in their economic feasibility, both due to the affect on production of local climatic conditions and transport costs.

16.6 Distribution

Distribution is the last step in the marketing process, which allows the agri-food products to be available to the consumers (Plate 28).

In recent years, distribution in Europe has been characterized by a decrease in the number of wholesale points and an increase in their size, due to an increase in commercial integration (in chains, franchises, etc.) with subsequent larger sized marketing chains. In addition products are now distributed internationally across the globe (Planells and Mir, 2000; Langreo, 2002). All this has meant that the negotiating position of the distribution chains with respect to the growers has improved, highlighting the need for grower's organizations to join force against the distribution chains.

Nowadays, the way to reach the final consumer is through the large distribution chains, which are undergoing a process of change marked by distribution chains with a higher dimension, concentration, internationalization and an increase in their market shares (Planells and Mir, 2000, 2002).

The evolution of fresh fruits and vegetables in the large distribution companies is linked to the development of packaging in small units, which allows the distribution companies to offer several brands, providing food safety as a main reason to keep the consumers' confidence and retain them as permanent customers (Photo 16.3).

This development of the large distribution companies, together with the changes in the behaviour of the consumers, will continue determining the vegetable exportation sector in Spain in the near future. The demand for fresh vegetables, throughout the whole year and in large quantities, will be complemented with requirements for increasing quality, food safety and traceability,

Photo 16.3. Packaging in small units is increasingly used in fresh fruits and vegetables.

which will be key factors in maintaining the competitiveness in the increasingly global European markets.

Traditionally the export markets have been a destination for a small part of the Spanish vegetable production, due to the perishable nature of the product and the distance. However, the demand for fresh vegetables has expanded throughout the whole year which, together with the liberalization of the European markets, has increased the importance of exports.

Except for some specialty products (cucumber, pepper and melon, among others), it is usual for growers to produce commercial types that are accepted in both national and foreign markets so that there are two destinations for production.

The European vegetable markets are characterized by: (i) a saturation of the offer; (ii) an increase in the quality control requirements and of new values (codification, traceability); (iii) an adjustment of the growing seasons (requirement for year-round supply); (iv) a certain blocking of the demand; and (v) the globalization and liberalization of the markets (Planells and Mir, 2000).

In increasingly competitive markets, the differentiation of the products is one of the classical measures used by economic agents. The differentiation in fresh vegetables is based on covering the consumer's expectations regarding organoleptic quality and food safety, by means of the good policy of quality and traceability (see Chapter 15).

Other concerns regarding environmental, ethical and social aspects of the production of vegetables are receiving increasing attention from consumers.

Stabilization in the demand for fresh vegetables in Europe is due, mainly, to: (i) the abundance and variety in the offer of food in general and vegetables in particular (Photo 16.4); (ii) the reduction in the population increase rates; and (iii) a decrease in the food expense within the consumer's budget (Wijnands, 2003). Despite this, for some vegetables such as tomato and pepper, the consumption increases are due to: (i) less time being available to cook (as the housewife works out of the family home so she uses food such as these fresh vegetables that are easy to prepare); (ii) the good image of food safety of fresh vegetables in view of increasing consumer concerns about food safety; (iii) the availability of fresh vegetables in the markets throughout the whole year; (iv) the increasing habit of eating smaller quantities but several times a day (to adapt to the working hours); and (v) a demand for more varied and attractive products with diversity in presentation and flavours (Wijnands, 2003).

Although fruits and vegetables represent less than 10% of the sales of big distribution companies, they are a key factor in the frequency of visits to the supermarket from the consumers, which justifies their interest from the large distribution companies (García-Azcarate and Mastrostefano, 2002).

The diversification in the offer of fresh vegetables has been more based on new presentations of already known products, with several colours, shapes or sizes (tomato types, coloured peppers, mini-vegetables) or in the use of quality distinctions (such as integrated production or organic production), than in the introduction of new horticultural species.

The inclusion of the marketing costs in the final price of the product oscillates depending on the destination market (national or foreign) due to differences in the transportation costs. In Spain, the production costs of export vegetables involves 35–50% of the wholesaler's market prices (Caballero and Miguel, 2002), whereas, in The Netherlands, growers receive 25–30% of

Photo 16.4. The range and variety of fresh fruits and vegetables offered in the European markets are enormous.

the price paid by the consumers and in North America around 21% (Wijnands, 2003).

16.7 Quality

The quality is the set of properties and characteristics of a product, process or service that grant its aptitude to satisfy established or implicit needs.

The quality is a combination of attributes, properties or characteristics that provide the product value, depending on the destined use. The appearance, the texture, the firmness, the organoleptic characteristics and the nutritional value are components of the quality. The relative importance of each component of the quality depends on the product and how it is consumed, varying among growers, distributors and consumers. For the grower, a product must provide high yields and have a good appearance, be easy to grow and have a good resistance to transportation, whereas for the wholesaler the qualitative attributes of appearance, firmness and shelf life prevail. In addition, the consumer values the healthiness and the nutritional value. In general, consumers place a high value on fresh fruits and vegetables as a healthy and natural food (Photo 16.5).

Many of the qualitative attributes of fresh horticultural products are subjective, which makes their evaluation even more complex. The consumer usually judges the quality of a product by its external appearance and, if there is no other information (different production methods, differentiating labels), will deduce that a product with a good appearance will have a good internal quality. The colour, the size, the uniformity and the absence of defects are basic aspects of the appearance, together with a good presentation, in proper packages that contain a standardized product.

The food safety and hygiene in the production process demanded by the consumer have made necessary the establishment of rigorous production protocols (which specify growing methods, traceability, etc.) that guarantee the healthiness of the products.

Transport over long distances in some cases means that some vegetables are harvested before the commercial ripening point; so that when they reach the consumer they are in a suboptimal organoleptic condition.

Photo 16.5. Consumers place a high value on fresh fruits and vegetables as a natural and healthy food.

The greenhouse growing conditions (light, temperature, irrigation, nutrition, salinity) affect the quality (Welles, 1999), so their management must be optimized.

The qualitative attributes of vegetables vary with the species. In some products such as cucumber, whose shelf life (length of time a food is given before it is unsuitable for consumption) is the main quality attribute, a qualitative evaluation is simple, whereas for others such as tomato, it is more complicated, especially regarding its organoleptic characteristics.

In some cases, such as tomato, achieving a high quality may involve a decrease in the yield, so a compromise between quality and quantity in the production must be reached.

In general, the quality regulations of fruits and vegetables in Europe are based on the external quality (appearance and condition) and in the critical concentrations of pesticides and nitrates (Schnitzler and Gruda, 2003).

It is evident that the future relies on the quality, but is must be economically feasible quality. The fixing of integral quality systems is a clear priority in the production of greenhouse vegetables. The technical aspects of the vegetable quality are described in Chapter 15.

16.8 Quality Management

Quality must become a priority in the agri-food business, because, besides it being a social demand, nowadays in Europe the costs involved in the absence of quality (defective products, claims, lawsuits) may be estimated at between 5 and 20% of the sales value (Rivera, 2000).

Guaranteeing the quality to the customers requires, in practice, having a 'quality guarantee system', that is the 'organization structure, the procedures and required resources to start a quality guarantee system' (Regulation ISO-8402). Quality control in the agri-food system is the set of pre-established actions to ensure the consumer's trust in the demanded quality. The 'quality system' is a strategic and organizational model articulated to carry out the desired quality control.

The quality system of a company must be 'certified' by a competent body (i.e. Asociación Española de Normalización y Tipificación (AENOR) in Spain), so that the company can show their customers that certification, as a prestige label supporting the quality of their products. Regulation ISO-9001 integrates and updates the previous regulations on the subject.

The final objective of agri-food certification is to provide confidence and mutual appreciation between the retailer and the customer. Nowadays, the most widespread quality system under certification in Spanish horticultural processing plants is the Controlled Production system, described in the regulation series UNE-155001 of AENOR (the Spanish Association for Standardization and Certification) that includes regulations on non-processed fruits and vegetables, destined for consumption in their natural state (Sierra, 2003). In addition, there are Integrated Production Regulations, in different regions (Aparicio *et al.*, 2003).

The development of these quality certifications has been determined, to a large extent, by the quality regulations imposed by the so called Euro-Retailer Produce Working Group (EUREP), which represents the most important association of agri-food distribution chains in the European Union, who have established a 'good agricultural practices code' known as EUREP-GAP.

Nowadays, the globalization of the horticultural commodities export has promoted the GLOBAL-GAP concept, which is extending all over the world, as a necessary means of production in order to have access to the foreign markets.

The UNE-155001 regulations, developed by AENOR for horticultural crops responds to the requirement of the large distribution chains with respect to the production systems, with the aim of: (i) ensuring the protection or safety of the consumer, of the grower and the environment; (ii) regulating the use of pesticides; (iii) limiting the use of chemical inputs; (iv) forcing the elimination of the different wastes in an environmentally friendly way; and (v) favouring efficiency in the use of inputs (water, fertilizers) and other complementary measures. Nowadays, the UNE-155001 regulations are approved by EUREP-GAP.

The integrated production system of the Andalusia region (Spain) that has been introduced by the Andalusia grower's organizations (with the logo 'Integrated Production-Andalusia') follows the guidelines of the International Organization for Biological Control (IOBC) in relation to growing techniques and integrated management of pests, for greenhouse vegetable cultivation. For more details on this system, see Aparicio *et al.* (2003).

In addition to EUREPGAP certification, there are other systems in Europe among which we may highlight the 'Bio-Label', which is only used in organic growing.

Summarizing, the quality differentiation strategies in the marketing of vegetables have determined the introduction of quality certification systems that aim to inform the consumer and gain his/her trust so the consumer gives 'commercial value' to the product offered, resulting in an appropriate selling price.

16.9 Future Prospects

If they are to survive in the increasingly globalized markets, agricultural companies must make the necessary effort to adapt to the new scenario of increasing competition, the concentration of the distribution companies, the new consumer's requirements regarding quality, healthiness and food safety, on the one side, and the social and environmental requirements of the production process on the other. For agricultural companies this will involve making structural changes and adapting their corporate and commercial strategies (especially increasing their size through different ways, such as joining or integrating with other companies) (Planells and Mir, 2000; Torrente, 2000), always keeping an eye on the changing trends in the markets.

16.10 Summary

- Marketing is, in a broad sense, the process in which the products are taken from the farm to the consumer.
- Nowadays it is frequent to confuse distribution with marketing. Distribution is the last stage of the marketing process.
- The perishable nature of vegetables is a critical factor in the marketing process. The postharvest technology (cold storage, controlled atmosphere packaging, etc.) allows for maintaining the quality of the product and extending its shelf life.
- The standardization of the horticultural products becomes the basis of trade at a distance, as the physical presence of the goods is not necessary for the transaction.
- Commercial channels are the routes along the marketing process, covered by the product, from the grower to the consumer. All these routes constitute the distribution circuit.
- The use of expensive transport carriers is frequent in greenhouse horticultural products, due to their high added value.
- Transport costs are very significant in the total cost composition of marketing,

and are a determining factor in the competitiveness of the horticultural product in foreign markets.

- Nowadays, the distribution companies across Europe are undergoing a process of amalgamation, increasing in size, internationalization and their market shares.
- The fresh vegetable demand in Europe has extended to the whole year, with increasing quality, food safety and healthiness, as well as traceability requirements from the consumers.
- These demands have determined the implementation of 'quality systems' at the production and marketing levels, with the aim of ensuring protection or safety to the consumer, the grower and the environment.
- The horticultural production companies must make the necessary effort to adapt to the new scenario (competition, concentration of the distribution companies, consumer's requirements) if they are to survive in the increasingly globalized markets.

17

Greenhouse Production Strategies

17.1 Introduction

The increase in the greenhouse area experienced in several parts of the world during recent decades (in the Mediterranean area, China, Korea, Mexico) is based on the use of plastic greenhouses and mainly oriented towards vegetable production.

The selection of the production sites has been determined by production costs and the integral cost prices in the market, which depend mainly on the local climate conditions and the transport distance to the market (see Chapter 14) among other factors.

The conventional concept of the greenhouse as a collector of solar energy, based on the greenhouse effect, in which the inside temperature is increased with respect to that of the open field and in which crops are protected from inclement weather, is today considered obsolete. In many areas the 'windbreak effect' or 'shelter effect' are more important than the 'greenhouse effect' itself, at least during certain periods of the year (Castilla, 1994). In very arid regions, the insulation of the protected enclosure from the external environment, which is very hot and dry, involves generating an 'oasis effect' as the main function (Sirjacobs, 1988), as well as the 'shading effect'. In tropical areas protection from rain, the 'umbrella effect' prevails (Wittwer and Castilla, 1995).

From the point of view of greenhouse management, we may consider three management levels: (i) *strategic*; (ii) *tactical*; and (iii) *operational* (Van Straten and Challa, 1995). At the strategic level, decisions on the investment in structure and equipment determine the future possibilities for greenhouse climate control. At a tactical level, before the start of the growing season the grower chooses, depending on the different predictions (weather, commercial, etc.), the species and growing calendar, defining expected patterns of crop development and production. This tactical plan determines the conditions that need to be controlled at the operational level.

The control system is the instrument for the grower to follow his or her tactical plan and, if necessary, modify it to correct any deviations that may arise with respect to the original predictions (weather, commercial). The control system is an operational management tool. The time span for the strategic decisions is many years, because the investments in the structure and equipment will have an influence throughout the whole lifespan of the enterprise. The tactical plan spans the duration of the growing season and the operational decisions have a brief operational influence (from several minutes to a week).

17.2 Crop Productivity and Production Costs

In the Mediterranean area, the crop production strategy, in the past, was to adapt the plants to a suboptimal microclimate instead of optimizing the microclimate by means of more expensive structures and equipment. This trend has a limit, and nowadays it is considered necessary to improve the technological level of the greenhouses to improve production, especially in qualitative terms, to be able to compete in an increasingly globalized market (Plate 29).

The difficulty lies in finding a balance between higher investments in crop production infrastructure (greenhouse and equipment) and competitive production costs (see Chapter 14).

17.3 Destination of the Produce

When planning the development of a greenhouse sector, on a large scale, there are two important matters to consider (Jensen and Malter, 1995): (i) the targeted market (i.e. will the produce be delivered to national or foreign markets); and (ii) the type of production (i.e. will the products be edible horticulture or ornamentals) (Plate 30).

In general, whatever the destination of the produce, it should be assumed that the best climatic conditions and the lowest costs are the basis for this development, with special consideration to the transport costs when the destination markets are far away (see Chapter 14).

However, nowadays, consumer demand for high quality products year round, hinders the access of seasonal suppliers to the markets (Jensen and Malter, 1995), because the well-established suppliers enjoy a more privileged access to the markets, especially, to foreign markets. Therefore, in many cases, the national market must be considered a first choice, as it involves less risk and less transport and preparation costs. On the other hand, national markets are sometimes a destination for second quality products.

17.4 Greenhouse Production Options

Vegetables represent the most common greenhouse crop option in the Mediterranean area (90% of the greenhouse area in Spain, 84% in Italy, 95% in Turkey), with the exception of Israel (Castilla, 2002).

The climatic conditions, besides those of the market, influence the chosen species (Plate 31). In France, for instance, the most cultivated greenhouse species in the last few years has been lettuce, due to its lower thermal requirements, displacing tomato. Cultivation of several species in the family *Cucurbitaceae* is avoided during certain periods in colder areas of the Mediterranean Basin, as these species are more temperature demanding than members of the family *Solanaceae* (Castilla, 2002).

The cultivation of cut flower species or ornamentals is of much lower economic importance than vegetables and the cultivation of trees is focused mainly on bananas, while other species are quite infrequent (Castilla, 2002).

At the production level, vegetable growers may choose between different options (Wijnands, 2003): (i) to produce an acceptable quality at a low cost; (ii) to produce high quality at an acceptable cost; (iii) to produce on contract, and in accordance with the requirements of the purchaser, the cost being of lesser concern as long as the product fulfils the desired quality criteria; and lastly, although of much less importance, (iv) to produce innovative products, such as vegetables demanded by certain ethnic groups for niche markets (i.e. for specific segments of the market).

17.5 Production Strategies and Tactics in Mediterranean Climates

17.5.1 General aspects

In Mediterranean areas, the use of greenhouses changes depending on the prevailing local climatic conditions (basically radiation and temperature) and the local sociological and economic conditions. There is a clear need to define the optimal greenhouse for a specific environment and to outline the plant characteristics that would best fit the specific greenhouse/environment combination (De Pascale and Maggio, 2008).

Crop rotation is not frequently practised, and as the majority of cultivated species belong to the botanical families *Solanaceae* and *Cucurbitaceae* (with similar problems of soil-borne diseases), it would not be effective anyway. Besides, economic reasons dictate the practice of monoculture. The diversification of production is a desirable objective, but difficult to reach in practice (Castilla, 2002).

17.5.2 Biological aspects

In Mediterranean greenhouses, the production strategies are more focused on biological aspects of the crops than in highly technical greenhouses where the focus is on proper greenhouse climate control (Alpi and Tognoni, 1975; Tognoni and Serra, 1989). Breeding programmes have been focused on improving production, precocity and fruit set at low temperatures (Nuez, 1986). Cultivars more adapted to low temperatures, more resistant to parasites and more efficient in the use of inputs (water, light, etc.) are technological tools to save energy in protected cultivation (Tognoni and Serra, 1989). Recent developments such as the long shelf-life tomatoes denote the importance of better biological knowledge to improve production strategies. Other aspects with room for improvement are: (i) the photosynthetic efficiency; (ii) the optimization of plant architecture to improve the interception and use of light; and (iii) the intelligent management of the climate parameters inside the greenhouse (Tognoni and Serra, 1989).

17.5.3 Strategies and tactical management

The essential operation in the management of the Mediterranean greenhouse microclimate is ventilation, due to its influence on the temperature, humidity and air composition. Shading, heat accumulation and maintaining a low, even temperature are also important in some cases (Monteiro, 1990). The management of the microclimate through transpiration control during the summer can be significantly improved by means of a dynamic manipulation of ventilation, shading and water fogging (Boulard *et al.*, 1991).

Strategies for more efficient and economic heating include: (i) managing the night temperatures depending on the wind conditions in heated greenhouses (Bailey, 1985); (ii) the use of several minimum temperature thresholds during the night (Toki *et al.*, 1978); and (iii) the use of variable minimum night temperatures depending on the radiation levels of the previous day (Gary, 1989). Developing and implementing recommendations for the management of climate parameters (mainly temperature and humidity) at a local level, adapted to the existing greenhouse characteristics, are necessary.

By comparison with other regions in the north of Europe, in Mediterranean climates the selection of species better adapted to the greenhouse microclimate (light and temperature) has promoted the use of members of the family *Cucurbitaceae*, which are more temperature demanding than species in the *Solanaceae*. The selection of cultivars is based on their acceptance in foreign markets and local cultivars are, frequently, left behind; there is a risk that these genetic resources adapted to local conditions will disappear. The resistance and/or tolerance to diseases have induced the use of a greater number of hybrids. Soil-borne pathologic problems have enhanced the use of plants grafted on to tolerant rootstocks (mainly melon, watermelon and tomato).

The adaptation to low radiation and temperature conditions has involved a great diversity of cycles and transplant dates (Castilla *et al.*, 1992). In a similar way, the periods of excessively high temperatures are avoided. Sometimes the strategy consists of storing the fruit during the autumn on the plants (pepper) allowing for their ripening later in the winter (Monteiro and

Portas, 1986). The appearance of long shelf-life cultivars has changed the growing and marketing strategies.

Other cultivation techniques, such as pruning and leaf removal, must be adapted to the conditions of the chosen growing cycle. The plant density, usually lower in winter, must be increased in the short growing cycles of spring and summer to optimize the interception of radiation. Chemical treatments for fruit set (tomato, aubergine, squash) have been replaced by the use of bumblebees and bees to facilitate pollination, without affecting the fruit's quality.

17.5.4 Future perspectives

At the start of a new century, the main problems of greenhouse cultivation in mild climate areas, such as the Mediterranean, are: (i) the production costs; (ii) consumer questions on product quality; and (iii) society's concerns on the environmental impact of the entire operation (La Malfa and Leonardi, 2001).

The globalization of the markets has increased the competitiveness between productive regions, highlighting the need to increase the quality of the products, by means of better climate control in the greenhouses. It is necessary to find a compromise solution between higher production costs in better-equipped greenhouses and the agronomic performance, to produce the required quality at competitive prices (Castilla *et al.*, 2004; Giacomelli *et al.*, 2008). It is necessary, therefore, not only to generate information and deepen our knowledge of these systems (Valls, 2002) but also to transfer the proven new information and technology to the growers and train them on how to properly use the new technologies.

The location of the greenhouses is a key issue, both for its influence on production costs, depending on the climatic conditions, and for its effect on the transport costs to the markets (Plate 32).

17.6 Summary

- The selection of greenhouse production areas has been determined by the production and marketing costs.
- The conventional greenhouse concept, based on the 'greenhouse effect', is considered obsolete nowadays. In many areas the 'windbreak effect' or the 'shelter effect' are more important than the 'greenhouse effect' itself; and depending on the local climate conditions, the 'oasis effect' or the 'umbrella effect' are also considered very important, at least during certain periods of the year.
- Two important aspects to consider when planning the development of a greenhouse sector are: (i) the targeted market (national or foreign markets); and (ii) the type of production (edible products or ornamentals).
- In mild climate areas, such as the Mediterranean Basin, the greenhouse production strategies have been related more to the biological aspects of the crops rather than to proper greenhouse climate control.
- The globalization of the markets and the increasing competitiveness between productive regions highlight the need to increase the quality of the products. For this, a compromise solution must be found between the higher production costs in better-equipped greenhouses and the associated agronomic performance, to provide the required product quality to the markets at competitive prices.

Appendix 1

A.1 Chapter 2

A.1.1 Calculation of the zenith angle (Jones, 1983; Villalobos *et al.*, 2002)

$$\cos\theta = \sin(LAT)\sin(DEC) + \cos(LAT)\cos(DEC)\cos(h) \qquad (2.5)$$

where:
θ = Zenith angle (Fig. 2.8)
LAT = Latitude
DEC = Solar declination: angular position of the Sun at noon with respect to the plane of the Equator (ranges between −23.5° y + 23.5° approximately)
h = Hourly angle (ranges from 0 to 360°) being 0° at noon
Referred to the northern hemisphere

$$DEC = 23.5 \cdot \cos\left[\frac{360[DJ-172]}{365}\right] \qquad (2.6)$$

DJ = Julian day
For instance, for a latitude of 37°N:
In the summer solstice: $DJ = 172$
$DEC = 23.5$
$\theta = 13.49°$
In the winter solstice: $DJ = 355$
$DEC = -23.5$
$\theta = 60.5°$
In the equinoxes: $\theta = 37°$

A.1.2 Calculation of global radiation as a function of insolation

Global solar radiation can be estimated if insolation data are available (Rosenberg *et al.*, 1983) and for Spain it is calculated using the following equation (Villalobos *et al.*, 2002):

$$R_s = \left(0.25 + 0.50\frac{n}{N}\right)R_a \qquad (2.7)$$

where:
R_s = Solar global radiation
R_a = Extraterrestrial radiation = solar constant
n = Real number of sunshine hours per day
N = Maximum number of sunshine hours (see next section)

A.1.3 Day length (Jones, 1983; Villalobos et al., 2002)

The half of a day length (h_s) is:

$$h_s = \frac{\pi}{180} \times arc.cos\left[-tan(LAT) \times tan(DEC)\right] \qquad (2.8)$$

where:
h_s = Half of a day length, in radians.

$$N = 24 \times \frac{h_s}{\pi} \qquad (2.9)$$

The N values would be (for latitude 37°N):

21 December, 9.4 h
21 March, 12.0 h
21 June, 14.5 h

A.1.4 Wien's law

The product of the temperature of a radiant surface (in K) and the wavelength of the emitted radiation expressed in micrometres (microns) is constant, and equal to 2897 K·µm (Rosenberg *et al.*, 1983).

As the Sun's temperature is of the order of 5800 K, the dominant wavelength will be $\frac{2897}{5800}$, this is, around 0.5 µm, equivalent to 500 nm.

The temperature of the greenhouse is of the order of 300 K (or 27°C). So, the dominant wavelength in the radiation emitted by the greenhouse would be $\frac{2897}{300}$, this is, around 9.5 µm, equivalent to 9500 nm.

A.1.5 Wavelength and frequency

The wavelength L, expressed in metres, and the frequency f, expressed in Hertz, are related by:

$$L \times f = constant \quad (2.10)$$

This being constant for the speed of light (2.9979×10^8 m s^{-1}).

Table 2.3. The electromagnetic spectrum (Bot and Van de Braak, 1995).

Type of radiation	Wavelength (nm)	Frequency (Hertz)
Gamma rays	<0.01	>3 × 10^{19}
X-rays	0.01–10	3 × 10^{19}–3 × 10^{16}
Ultraviolet	10–390	3 × 10^{16}–7.7 × 10^{14}
Visible light	380–760	7.9 × 10^{14}–3.9 × 10^{14}
Solar radiation	300–2,500	10^{15}–1.2 × 10^{14}
Infrared	760–3 × 10^5	3.9 × 10^{14}–10^{11}
Thermal infrared (300 K)	2,500–25,000	1.2 × 10^{14}–1.2 × 10^{13}
Microwaves	3 × 10^5–3 × 10^8	10^{12}–3 × 10^9
Radio waves	>10^8	<3 × 10^9

A.1.6 Hellman's equation

To calculate the wind velocity (U_H) at a certain height (H), as a function of the wind velocity at a height of 10 m (U_{10}), Hellman's equation may be used (Wacquant, 2000):

$$U_H = U_{10}\left[0.233 + 0.656\log\left(H + 4.75\right)\right] \quad (2.11)$$

A.1.7 Saturation vapour pressure

Water vapour pressure in the atmosphere at saturation (e_s):

$$e_s = 0.61078 \times \exp\left[\frac{17.27 \times T}{T + 237.3}\right] \quad (2.12)$$

where:
T = Temperature (°C)
e_s = Water vapour pressure in the atmosphere at saturation in kPa

As the temperature rises, the air can contain more water vapour, as the water vapour saturation pressure e_s increases (see Fig. 2.17).

A.2 Chapter 3

A.2.1 Thermal integral

The thermal integral, thermal time or physiological time is the summation of the temperatures above a certain threshold or base temperature (Villalobos *et al.*, 2002):

$$IT = \Sigma\left(T_i - T_1\right) \quad (3.1)$$

where:
IT = Thermal integral
T_i = Temperature of the process (daily average)
T_1 = Threshold or base temperature

The thermal integral is expressed in degree days (°C·day). A degree day is equal to 1°C above the threshold temperature, through a 24 h period.

In order to accurately use the concept of the thermal integral, the response of the development velocity versus the temperature must be linear, and the temperatures considered must be above the base temperature and below the optimal temperature. Above this, the growth rate decreases, so the thermal time must be corrected (Villalobos *et al.*, 2002).

For instance, López (2003) quantified the base temperature for Mediterranean greenhouse bean crops at 6°C, establishing a thermal integral from sowing until the beginning of harvest of 757 degree days. The value of the base temperature may vary, depending on the cultivar used (Mauromicale *et al.*, 1988).

A.3 Chapter 4

A.3.1 Diffuse solar radiation inside a greenhouse

Plastic films that have a diffusing (haze) effect on direct solar radiation change the proportions of direct and diffuse radiation inside the greenhouse, in relation to that existing in the open field. The higher proportion of diffuse light is of interest for two reasons: (i) to increase the uniformity of the spatial distribution of radiation inside the greenhouse; and (ii) to improve the light-use efficiency, because diffuse light penetrates the inside of the canopy better due to its non-directional nature.

The diffuse radiation inside a greenhouse can be estimated knowing the external radiation conditions and two parameters: (i) the diffuse radiation enrichment coefficient (DIF_i/DIF_o); and (ii) the conversion factor (β) from direct to diffuse radiation (Baille *et al.*, 2003), which depends mainly on the covering material and the greenhouse characteristics.

$$\beta = \frac{DIF^*}{IDIR_o} = \frac{\left(DIF_i - \tau_{dif} \times DIF_o\right)}{IDIR_o} \qquad (4.1)$$

where:
DIF_i = Diffuse solar radiation inside the greenhouse
DIF_o = External diffuse solar radiation
τ_{dif} = Transmissivity to diffuse radiation (equals the global transmissivity in completely cloudy conditions)
$IDIR_o$ = External direct solar radiation
DIF^* = Amount of direct solar radiation converted into diffuse radiation inside the greenhouse

A.4 Chapter 5

A.4.1 Conduction

The heat transmitted by conduction (Fig. 5.1) between two parallel plane isothermal surfaces (1 and 2), in a perpendicular direction to both surfaces (Urban, 1997a) is:

$$q_c = \lambda \times S\,(T_2 - T_1)/d \qquad (5.3)$$

where:
q_c = Heat transmitted by conduction through the surface S per time unit (W)
T_1 = Lower temperature (°C)
T_2 = Higher temperature (°C)
d = Distance between the two plane surfaces 1 and 2 (m)
λ = Thermal conductivity coefficient (W m^{-1} °C^{-1})
S = Area (m^2)

The heat flux increases with the thermal conductivity and with the temperature differences and decreases with the width of the material.

Table 5.3. Thermal conductivity coefficient (λ) of some common materials (Wacquant, 2000).

Material	λ (W m^{-1} °C^{-1})
Steel	45–70
Aluminium	200
Polyethylene	0.3
Glass	0.9
Wood	0.2
Humid clay soil	1.2

A.4.2 Convection without phase change

The heat transmitted from a surface to a moving fluid can be estimated as (Montero *et al.*, 1998):

$$q_{conv} = h_c \times S\,(T_S - T_F) \qquad (5.4)$$

where:
h_c = Heat transfer convection coefficient (W m^{-2} °C^{-1})
S = Surface crossed by the heat flux (m^2)

$T_S - T_F$ = Temperature difference between the surface and the fluid outside the contact zone (°C)
q_{conv} = Heat transmitted by convection per unit time (W) from a surface S to the fluid.

A.4.3 Evaporation and condensation

The heat transmitted when water evaporates or condenses is given by (Montero *et al.*, 1998):

$$Q = \Delta H \times \frac{h_c}{C_P} \times S(X_S - X_A) \quad (5.5)$$

where:
Q = Heat transmitted per time unit (W)
ΔH = Latent heat of vapourization (J kg^{-1})
h_c = Heat transmission convection coefficient (W m^{-2} °C^{-1})
C_p = Dry air specific heat (J kg^{-1} °C^{-1})
S = Evaporation or condensation surface (m^2)
X_S = Absolute humidity on the surface (kg kg^{-1})
X_A = Absolute humidity in the greenhouse air (kg kg^{-1})

A.4.4 Radiation

The radiation emitted by a body depends on its surface temperature and its nature, according to the Stefan–Boltzman law (Monteith and Unsworth, 1990).

$$\Phi_{rad} = \varepsilon \times \sigma \times T^4 \quad (5.6)$$

Φ_{rad} = Energy flux (W m^{-2})
ε = Emissivity of the surface (dimensionless)
σ = Stefan–Boltzman constant (5.67 10^{-8} W m^{-2} K^{-4})
T = Surface temperature, in Kelvin (K)

At low temperatures the energy emitted will be low, increasing as the temperature increases.

It is admitted that the radiation emitted by a body has the same intensity in all directions. If two bodies are at different temperatures, they will exchange radiation (Fig. 5.2). The net transmission is expressed (Urban, 1997a):

$$Q_{1-2} = S_1 \times F_{1-2} \times \sigma(T_1^4 - T_2^4) \times \varepsilon_1 \times \varepsilon_2 \quad (5.7)$$

Q_{1-2} = Heat exchanged by radiation from body 1 that reaches body 2 (W)
S_1 = Body surface (m^2)
F_{1-2} = Energy fraction emitted from body 1 that reaches body 2 (dimensionless)
T_1 and T_2 = Surface temperatures of bodies 1 and 2 (K)
σ = Stefan–Boltzman constant (5.67 10^{-8} W m^{-2} K^{-4})
ε = Emissivity of bodies 1 and 2

To calculate the night radiation heat losses through a greenhouse cover, if we consider the whole greenhouse as one of the bodies (body 1) and the sky as the other (body 2), the equation would be (Montero *et al.*, 1998):

$$\Phi_{1-2} = \varepsilon_1 \times \sigma(T_2^4 - T_1^4) \quad (5.8)$$

where T_2 is the sky temperature (a supposed black body), as $\varepsilon_2 = 1$ and $F_{1-2} = 1$.

The sky temperature may be estimated (T_2) (under clear sky conditions and low ambient humidity):

$$T_2 = 0.055 \times T_a^{1.5} \quad (5.9)$$

where T_a is the air temperature (K).

A.4.5 Air renewal

The heat losses by air renewal in a greenhouse (Q_{ren}) can be approximately calculated (Montero *et al.*, 1998):

$$Q_{ren} = m \times C_p(T_i - T_e) \quad (5.10)$$

Q_{ren} is in watts where:
m = Air mass renewed (kg s^{-1})
C_p = Air specific heat (J kg^{-1} °C^{-1})
T_i = Internal air temperature (°C)
T_e = External air temperature (°C)

The air mass can be calculated knowing the air exchange rates (volumes of greenhouse air renewed per hour), the greenhouse volume and the air density.

The most difficult thing to establish is the air exchange rate. In greenhouses that are closed tight it is less than 1, in greenhouses that are closed but not so tightly it reaches several units, and in open greenhouses, 10 or more. The external wind has a large influence, as well as the temperature differences between inside and outside and, therefore, pressure differences.

In a first approach, the leakage losses can be estimated with the equation (Bordes, 1992):

$$Q_{ren} = 0.35 \times R \times V(T_e - T_i) \quad (5.11)$$

R = Air exchange rate (h^{-1})
V = Greenhouse volume (m^3)
Q_{ren} = in watts (W)

A.4.6 Energy balance

Assuming, in a simplified way, that the solar energy that enters the greenhouse is used only to heat the greenhouse and for evapotranspiration (neglecting the energy used for photosynthesis, among other simplifications) the instantaneous energy balance would be, approximately (Montero *et al.*, 1998):

Solar radiation – Evapotranspiration
+ Heating = Overall losses
+ Air renewals

$$S_s \times \tau \times R_s(1 - t) + Q_c = K(T_i - T_e) \times S_c + m \times C_p(T_i - T_e) \quad (5.12)$$

S_s = Soil surface (m^2)
τ = Greenhouse transmissivity to solar radiation (expressed as a decimal per unit)
R_s = Instantaneous external solar radiation ($W\ m^{-2}$), or irradiance
t = Proportion of radiation used for transpiration (expressed as a decimal per unit)
Q_c = Heat supplied by the heating system per time unit (W)
K = Global heat transfer coefficient ($W\ m^{-2}\ °C^{-1}$)
T_i = Greenhouse (internal) air temperature (°C)
T_e = External air temperature (°C)
S_c = Greenhouse cover area (roof and sidewalls; m^2)
m = Air mass renewed by ventilation or infiltration ($kg\ s^{-1}$)
C_p = Air specific heat ($J\ kg^{-1}\ °C^{-1}$)

This equation is only approximate, for standard conditions (in the measurement of K) and with the mentioned exceptions.

For an approximate calculation of the heating, a simplified energy equation may be applied, for the night when the requirements are higher ($R_s = 0$), resulting in:

Heating = Overall losses + Air renewal

$$Q_c = K(T_i - T_e) \times S_c + m \times C_p(T_i - T_e) \quad (5.13)$$

A.4.7 Specific heat of a body

The specific heat of a body is the amount of energy that must be supplied to 1 kg of this body to increase its temperature by 1°C.

The specific heat of water (at 15°C) is 4.186 $kJ\ kg^{-1}\ °C^{-1}$ and the specific heat of dry air (at 20°C) is 1 $kJ\ kg^{-1}\ °C^{-1}$.

A.4.8 Latent heat of vaporization

The latent heat of vaporization for a liquid is the amount of energy that must be supplied to 1 kg of this liquid to convert it from liquid to gas.

The latent heat of vaporization of water (at 20°C) is 2445 $kJ\ kg^{-1}$.

A.4.9 Global heat transfer coefficient

The global heat transfer coefficient of a greenhouse covering material (K) is (Papadakis *et al.*, 2000):

$$K = \frac{1}{\frac{1}{h_i} + \frac{1}{h_e} + \sum_j \frac{d_j}{k_j}} \quad (5.14)$$

K = Global heat transfer coefficient of a greenhouse covering material ($W\ m^{-2}\ °C^{-1}$)
h_i = Convection heat transfer coefficient between the inner side of the greenhouse cover and the internal air ($W\ m^{-2}_{cover}\ °C^{-1}$)
h_e = Convection heat transfer coefficient between the outer side of the greenhouse cover and the external air ($W\ m^{-2}_{cover}\ °C^{-1}$)
d_j = Thickness of the element j of the greenhouse covering material (m)

k_j = Thermal conductivity of the element j of the greenhouse covering material (expressed in W m^{-1} °C^{-1})

A.5 Chapter 6

A.5.1 Interception of radiation by the canopy: extinction coefficient

The radiation intensity intercepted by a certain layer of the canopy is proportionally inverse to the LAI accumulated above such a layer (Montsi and Saeki, 1953). The attenuation of the radiation follows the Lambert–Beer law (Giménez, 1992):

$$I_e = I \times e^{-K(\mathrm{LAI})} \quad (6.3)$$

where:
I_e = Radiation intensity that passes through the considered layer
I = Radiation incident on the top of the canopy
e = Mathematical constant (approximately equal to 2.71828)
LAI = Leaf area index accumulated in the considered layer
K = Radiation extinction coefficient

The extinction coefficient (K) is not a constant value, because it depends on the age of the crop, the average inclination of the crop's leaves, the type of solar radiation (direct or diffuse) and the solar elevation. Fully developed crops with horizontal leaves have higher K values, close to 1.0 (Fig. 6.11) compared with crops with erect leaves, which are more vertical, where K may decrease to values around 0.3 (e.g. grasses, gladiolus, garlic). The extinction coefficient for diffuse light for cucumber and tomato is of the order of 0.85 (Gijzen, 1995a).

When cultivating in rows, especially in pair or double rows, the effect of the passage on radiation interception is important when the height of the plant is smaller than the width of the passage, notably affecting photosynthesis (Gijzen and Goudriaan, 1989). Under these conditions, it is beneficial to use white mulch that reflects the light that reaches the soil, directing it towards the crop again (Gijzen, 1995a), but this technique must only be used if the soil or substrate temperature is suitable, such as in heated greenhouses, because it may involve a yield decrease in the case of undesirable cooling of the soil or of the substrate in unheated greenhouses (Lorenzo *et al.*, 2001, 2005).

A.5.2 Radiation absorbed by the crop

The PAR absorbed by the crop (Fig. 6.12) is (Baille, 1999):

$$PAR_{abs} = (PAR_i - PAR_t) + (PAR_{t,r} - PAR_r) \quad (6.4)$$

where:
PAR_{abs} = PAR absorbed by the canopy
PAR_i = PAR incident over the canopy
PAR_t = PAR transmitted to the soil level
$PAR_{t,r}$ = PAR reflected from the soil
PAR_r = PAR reflected by the canopy or albedo

Some authors neglect $PAR_{t,r}$, considering that it is not relevant, except in the case of very low density crops. The absorption efficiency of the radiation by the plant is very influenced by the epidermal characteristics of the leaf and its chlorophyll content.

The use of reflecting materials over the greenhouse soil, such as white plastic mulch, aims to increase $PAR_{t,r}$ in order to raise PAR_{abs} (Fig. 6.12).

A.5.3 Growth parameters

Table 6.1. Main parameters used in the analysis of growth (adapted from Berninger, 1989).

Name	Accepted abbreviation	Units
Relative growth rate	RGR	g g^{-1}
Leaf weight ratio	LWR	g (leaf) g^{-1} (total)
Leaf area ratio	LAR	m^2 (leaf) g^{-1} (total)
Specific leaf area	SLA	m^2 (leaf) g^{-1} (leaf)
Specific leaf weight	SLW	g (leaf) m^{-2} (leaf)
Leaf area index	LAI	m^2 (leaf) m^{-2} (soil)
Net assimilation rate	NAR	g m^{-2} h^{-1}

The relative growth rate (RGR) is expressed in g g^{-1} (or similar) for the considered period.

The leaf area ratio (LAR) is the ratio between the leaf area and the total weight of the plant, whereas the specific leaf area (SLA) is the ratio between the leaf area and the weight of the leaves. The specific leaf weight (SLW) is the inverse of the specific leaf area (SLA). The leaf area index (LAI) is one of the most common parameters. The leaf area duration (LAD) is the integral of the LAI through time and is another parameter used in longer term growth studies.

The net photosynthesis assimilation rate (NAR) is the weight of CO_2 fixed per unit leaf area in a certain time.

The NAR can also be expressed as weight of dry matter produced (instead of the CO_2 fixed).

The SLA is an indicator of quality in leafy vegetables, in which a thinner leaf is more appreciated, and it is sensitive to climate parameters, such as radiation (Giménez *et al.*, 2002) and CO_2 (Enoch, 1990).

A.5.4 Fruit harvest and biomass indexes

Table 6.2. Estimated values of the harvest index (HI, expressed in decimals per unit), dry matter content of the harvestable product (DMC, expressed in decimals per unit) and EFF (ratio of the amount of fresh product formed to total dry matter produced by the plant: HI/DMC), for vegetables cultivated in a heated greenhouse. (Source: Challa and Bakker, 1999.)

Crop	HI	DMC	EFF
Tomato	0.7	0.06	12.0
Cucumber	0.7	0.035	20.0
Sweet pepper (red)	0.8	0.085	9.4
Aubergine	0.8	0.07	11.4
Radish	0.9	0.05	18.0
Lettuce (butterhead)	0.85	0.05	17.0

A.5.5 Use of radiation in a typical greenhouse ecosystem

Table 6.3. Use of solar radiation in a typical greenhouse ecosystem (adapted from Baille, 1999, according to Warren-Wilson, 1972).

Hypothesis	Proportion of energy available
External global radiation	100%
External PAR (50%)	50% External PAR
Greenhouse transmissivity (70%)	35% PAR greenhouse
Absorbed PAR (50%)	17.5% PAR absorbed
Dry matter fixed by photosynthesis (14.3%)	2.5% Gross photosynthesis
Minus photorespiration and maintenance (40%)	1.5% Net photosynthesis
Minus growth respiration (33.3%)	1.0% Total dry matter energy
Harvest index (50%)	0.5% Dry matter energy in fruit

With these hypotheses, the efficiency values in the use of radiation are of the order of 0.5% of the harvested dry matter with respect to the external global radiation, or 1% of the total dry matter with respect to the external global radiation (Baille, 1999). The energetic equivalence of dry matter is 16 MJ kg^{-1} dry matter.

A.6 Chapter 8

A.6.1 Wind effect in natural ventilation

In a multi-tunnel greenhouse with a single vent or with several vents with the same ventilation characteristics, half of the ventilation area is supposed to act as air inlet and the other half as air outlet (Papadakis *et al.*, 1996). According to this simplification the volumetric flux can be linearly expressed as a function of the outside wind (Muñoz *et al.*, 1999; Pérez-Parra *et al.*, 2003a):

$$\Phi_W = \frac{S}{2} \times C_d \times C_W^{\frac{1}{2}} \times U \qquad (8.4)$$

where:

Φ_w = Air volumetric flux caused by the wind effect

S = Total ventilation area of the greenhouse
C_d = Discharge coefficient of the vent
C_w = Global wind effect coefficient
U = External mean wind velocity

A.6.2 Thermal effect in natural ventilation

With the same previous simplification (half of the ventilation area acts as an air inlet and the other half as an outlet), the volumetric flux of air caused by the thermal effect is (Muñoz *et al.*, 1999):

$$\Phi_T = \frac{S}{2} \times C_d \left[2g \frac{\Delta T}{T} \times \frac{H}{4} \right]^{\frac{1}{2}} \quad (8.5)$$

where:
Φ_T = Air volumetric flux caused by the thermal effect
S = Total ventilation area of the greenhouse
C_d = Discharge coefficient of the vent
g = Gravity acceleration
ΔT = External – internal air temperature difference
T = External air temperature
H = Vertical height of the vent

The thermal effect in natural ventilation usually has little relevance in coastal areas, where the wind effect clearly predominates, due to the frequent presence of wind.

A.6.3 Wind loads

The wind loads over the greenhouse structure are calculated using the equation (Zabeltitz, 1999):

$$F_w = A \times C \times P_w \quad (8.6)$$

where:
F_w = Wind force
A = Greenhouse area under wind pressure or suction
C = Wind pressure coefficient
P_w = Wind dynamic pressure

The wind dynamic pressure depends on the effective height of the greenhouse (Zabeltitz, 1999). See Fig. 8.3.

A.6.4 Air flow reduction when a screen is placed on a greenhouse vent

As a first approach, the reduction of the air flow through a screen can be quantified as follows (J.I. Montero, personal communication):

$$\frac{\Phi}{\Phi_m} = \varepsilon \left(2 - \varepsilon\right) \quad (8.7)$$

where:
Φ = Volumetric air flow through a vent without screen
Φ_m = Volumetric air flow through a vent with screen
ε = Screen porosity (screen holes area per unit area)

A.7 Chapter 9

A.7.1 CO_2 units

The most commonly used units to quantify the air CO_2 content are (Nederhoff, 1995):

Dimension	Unit
Air CO_2 index, in volume:	1 vpm = 0.0001% (volume)
Air CO_2 index, in moles:	1 ppm = 1 µmol mol^{-1}
Partial pressure	1 Pa = 10 mbar

At normal greenhouse temperatures at sea level, it is not necessary to make corrections to convert CO_2 units. The equivalences (at 20°C and 101.3 kPa) between units are:

1 vpm = 1 ppm = 0.101 Pa

A.8 Chapter 11

A.8.1 Crop water stress index

The crop water stress index (CWSI) is based on the measurement of the crop canopy temperature and allows for knowing whether the ET rate of a crop is below its maximum level (Idso *et al.*, 1981):

$$\text{CWSI} = \frac{(T_c - T_a) - (T_c - T_a)_{\min}}{(T_c - T_a)_{\max} - (T_c - T_a)_{\min}} \quad (11.11)$$

where:
$(T_c - T_a)$ is the difference between the crop temperature (T_c) and the air temperature (T_a) in the moment of the measurement.
$(T_c - T_a)_{min}$ corresponds to a crop without water stress and $(T_c - T_a)_{max}$ to the maximum value of the difference, when the stress is maximum.

The critical value of CWSI at which irrigation should be started is 0.25 (Villalobos *et al.*, 2002).

A.8.2 Irrigation water quality

The electrical conductivity of the water (EC_w) and the sodium adsorption ratio (SAR) are the most common indices used to evaluate the salinity and alkalinity of water.

The conductivity increases with the dissolved salt content; by convention, it is measured at 25°C.

The SAR is:

$$\text{SAR} = [Na] \times \left[\frac{Ca + Mg}{2}\right]^{-1/2} \quad (11.12)$$

where $[Na]$ is the sodium concentration and $[Ca + Mg]$ is the sum of the concentrations of calcium and magnesium, expressed in milliequivalents per litre (meq l^{-1}).

There are several ways to evaluate the quality of irrigation water. Table 11.9 (in Chapter 11) summarizes the guidelines proposed by the FAO (Ayers and Westcot, 1987).

A.8.3 Estimation of the evapotranspiration (ET_0) in a greenhouse

The radiation method estimates the ET_0 as a function of the global solar radiation R_s (Doorenbos and Pruitt, 1976):

$$ET_0 = a + b \times R_s \quad (11.13)$$

where a and b are coefficients that depend on the climatic conditions.

For greenhouses, an adaptation of the Penman–Monteith method, proposed by Baille *et al.* (1994) and used in soilless crops, is based on the solar radiation, the VPD (vapour pressure deficit) and the leaf development to calculate the evapotranspiration of the crop:

$$ET_C = A\left(1 - e^{-K \cdot (\text{LAI})}\right) \times R_s + B \times f(\text{LAI}) \times VPD \quad (11.14)$$

where A and B are two coefficients, K is the radiation extinction coefficient and f(LAI) a function dependent of the LAI (leaf area index).

A.9 Chapter 12

A.9.1 Transmissivity models

Solar radiation transmissivity models used in greenhouse, in chronological order up until 1998. (Source: Soriano, 2002.)

Year	Author	Model
1967	Manbeck	Study of the transmissivity in simple modules
1970	Bowman	Transmissivity to solar radiation
1970	Deltour	Transmissivity in simple modules
1971	Takakura	Transmissivity in simple modules
1971	Stoffers	Transmissivity for different roof slopes and orientations
1973	Basiaux	Transmissivity in simple modules
1975	Kozai	Transmissivity for different roof slopes and orientations
1977	Palmer	Light interception in canopies comparing orientations
1978	Critten	Transmissivity in Venlo glasshouses
1983	Bot	Transmissivity in Venlo glasshouses
1983	Kurata and Critten	Improvement achieved with Fresnel prisms and reflecting ribbons in Venlo
1984	Nederhoff	Light interception
1989	Gueymard	PAR atmospheric transmissivity, direct, diffuse and global
1990	Baille and Baille	Simple transmissivity model without polarization or reflection
1990	Kurata	Transmissivity in Venlo
1993	Tachsmitchian	Light interception by a crop considering the effect of the sidewalls
1993	Zwart	Transmissivity to radiation in a multi-span greenhouse
1995	Li	Transmissivity in 'lean to' greenhouses
1995	Heuvelink	Transmissivity in multi-span Venlo glasshouse
1996	Saeki	Light interception by the canopy
1997	Jeroen	Effect of artificial light supply on assimilation
1997	Feuerman	Liquid radiation filters for different wavelengths
1997	Koning	Supplementary light supply
1998	Stoffers	Zig-zag multi-span greenhouses
1998	Pieters	Effect of the type of condensation on the transmissivity of short-wave radiation
1998	Sun-Zhung Fu	Radiation in a 'lean to' greenhouse

Appendix 2
Symbols and Abbreviated Forms

A	Angle; also designates coefficient
A	Area subject to wind pressure or suction; also designates coefficient
AC	Capillary ascension
ADC	Analogical digital converter
AENOR	Asociación Española de Normalización y Tipificación (Spanish Association for Standardization and Certification)
AH	Absolute humidity (of air)
ASAE	American Society of Agricultural Engineers
ASHRAE	American Society of Heating, Refrigeration and Air Conditioning Engineers
ATP	Adenosine triphosphate (energetic organic molecule)
b	Coefficient
B	Coefficient
BCMV	Bean common mosaic virus
BER	Blossom end rot
BnYDV	Bean yellow disorder virus
BPYV	Beet pseudo-yellows virus
BYMV	Bean yellow mosaic virus
C	Wind pressure coefficient
C/N	Carbon/nitrogen ratio
C_2H_4	Ethylene
C3	Type of plant metabolism
C4	Type of plant metabolism
$C_6H_{12}O_6$	Carbon hydrate (glucose, fructose)
$Ca(NO_3)_2$	Calcium nitrate
CAM	Type of plant metabolism
C_d	Vent discharge coefficient
CEC	Cation exchange capacity
CFD	Computational Fluid Dynamics
CGR	Crop growth rate
CMV	Cucumber mosaic virus
CO	Carbon monoxide

CO_2	Carbon dioxide
C_p	Specific heat
CVYV	Cucumber vein yellowing virus
C_w	Wind effect global coefficient
CWSI	Crop water stress index
CYSDV	Cucurbit yellow stunting disorder virus
d	Distance between two plane surfaces; also designates day
DEC	Solar declination (angle)
DIF	Day/night temperature difference
*DIF**	Direct solar radiation converted into diffuse radiation inside the greenhouse
DIF_i	Diffuse solar radiation inside the greenhouse
DIF_o	Open field diffuse solar radiation
DJ	Julian day
DMC	Dry matter of the harvestable product
DPI	Derived proportional integral (type of regulator)
E	Evaporation; also designates East (cardinal point)
E_0	Pan evaporation
E_a	Water application efficiency
e_a	Water vapour partial pressure
EC	Electrical conductivity
EC_e	Electric conductivity of the saturated soil extract
EC_w	Water electrical conductivity
EN	European norm (regulation)
e_s	Water vapour saturation pressure
ESP	Exchangeable sodium percentage
ET	Evapotranspiration
ET_0	Reference evapotranspiration
ET_c	Crop evapotranspiration
E_u	Emission uniformity (of water in drip irrigation)
EUREP	Euro-Retailer Produce Working Group
EVA	Ethylene vinyl acetate
f	Frequency
FAO	Food and Agriculture Organization of the United Nations
FC	Field capacity
FDR	Frequency domain reflectometry
FIAPA	Fundación para la Investigación Agraria en la Provincia de Almería
F_w	Wind force
g	Gravity
GAP	Good agricultural practices
H	Average height of the greenhouse
h	Solar elevation angle; also designates hour and hourly angle
H_2O	Water
HALS	Hindered amine light stabilizers (photostabilizers used as plastic additives)
h_c	Convective heat transfer coefficient
HDPE	High density polyethylene
HFLI	High frequency localized irrigation
HI	Harvest index (dimensionless, expressed as a decimal or as a percentage)
h_S	Duration of half a day (radians)
i	Direct radiation angle of incidence
I	Solar radiation intensity (irradiance)
$IDIR_i$	Direct solar radiation inside the greenhouse

$IDIR_o$	Open field direct solar radiation
I_e	Radiation intensity that passes through the considered canopy layer
I_o	Solar radiation intensity (irradiance) on a perpendicular plane to the radiation direction
IOBC	International Organization for Biological Control
IP	Integral proportional (type of regulator)
IPM	Integrated pest management
IPP	Integrated production and protection
IR	Infrared
IR-PE	Infrared or thermal polyethylene
IT	Thermal integral
K	Global heat transfer coefficient of a greenhouse cover; also designates the radiation extinction coefficient of a canopy
K	Kelvin
K_2SO_4	Potassium sulfate
K_c	Crop coefficient
KCl	Potassium chloride
KNO_3	Potassium nitrate
K_p	Pan coefficient
K_s	Soil water storage efficiency coefficient
L	Wavelength; 'l' also designates litre
LAI	Leaf area index
LAR	Leaf area ratio
LAT	Latitude
LCA	Life cycle analysis or life cycle assessment
LD-PE	Long duration polyethylene
LDPE	Low density polyethylene
LED	Light emitting diode
LF	Leaching fraction
LPG	Liquefied petroleum gas
LVDT	Lineal variable displacement transducer
LWR	Leaf weight ratio
Ly	Langley
m	Air mass renewed when ventilating a greenhouse, per unit time
MAP	Modified atmosphere packaging
meq	Milliequivalent
min	Minute
MIR	Medium infrared
MNSV	Melon necrotic spot virus
N	Maximum number of sunny hours per day; N also designates the north (cardinal point)
n	Real number of sunny hours per day
NADPH	Energetic organic molecule (nicotine adenine dinucleotide phosphate)
NAR	Net assimilation rate
NFT	Nutrient film technique
NGS	New growing system (type of NFT)
$(NH_4)_2HPO_4$	Biammonium phosphate
$(NH_4)H_2PO_4$	Monoamonium dihydrogen phosphate
$(NH_4)_2SO_4$	Ammonium sulfate
NIR	Near infrared
NO_x	Nitrogen oxides

NRAW	Not readily available water
O_2	Oxygen
P	Phytochrome
P	Proportional (type of regulator)
PAR	Photosynthetically active radiation
PAR_{abs}	PAR absorbed by the crop
PAR_i	PAR incident on the crop
PAR_r	PAR reflected by the crop or albedo
PAR_t	PAR transmitted at soil level
$PAR_{t,r}$	PAR reflected from the soil level
PC	Polycarbonate
PE	Polyethylene
P_e	Effective rain
PET	Polyethylene terephthalate
P_{FR}	Phytochrome form
pH	Measure of the acidity of a solution
PMMA	Polymethyl methacrylate
PMMoV	Pepper mild mottle virus
Pn	Net productivity or net biomass gain or net photosynthesis
P_R	Phytochrome form
P_{TOTAL}	Total phytochrome
PVC	Polyvinyl chloride
PVF	Polyvinyl fluoride
PVY	Potato virus Y
P_w	Wind dynamic pressure
PWP	Permanent wilting point
Q	Heat exchanged per unit time
Q_c	Greenhouse heating requirements
q_c	Heat transmitted by conduction per unit time
q_{conv}	Heat transmitted by convection per unit time
Q_{ren}	Heat transmitted by air renewal per unit time
Q_s	Monthly heating consumption
R	Hourly greenhouse renewal rate (volumes per hour); also designates respiration
R/FR	Red/far red ratio of radiation
R_a	Extraterrestrial radiation (or solar constant)
RAW	Readily available water
R_b	Gross irrigation water requirements
RGR	Relative growth rate
RH	Relative humidity (of air)
R_n	Net irrigation
R_s	Global solar radiation
RW	Reserve water
s	Second
S	Surface (m^2); also designates south (cardinal point)
SAR	Sodium adsorption ratio
SAT	Associated growers organization
SBMV	Southern bean mosaic virus
S_c	Greenhouse cladding area
SLA	Specific leaf area
SLW	Specific leaf weight

SM	Solid material
SO_2	Sulfur dioxide
SqMV	Squash mosaic virus
S_s	Soil area
t	Proportion of the radiation used for transpiration
T	Transpiration; Temperature; Tangent to the drop in the contact point
T (°C)	Temperature, Celsius degrees
T (°F)	Temperature, Fahrenheit degrees
T (K)	Temperature, Kelvin degrees
T_1	Threshold temperature
T_a	Air temperature
TBSV	Tomato bushy stunt virus
T_c	Set point temperature
TDR	Time domain reflectometry
T_e	External air temperature
T_F	Fluid temperature
T_i	Air temperature inside the greenhouse; also designates temperature of a process
TMV	Tobacco mosaic virus
ToCV	Tomato chlorosis virus
ToMV	Tomato mosaic virus
T_s	Surface temperature
TSWV	Tomato spotted wilt virus
TYLCV	Tomato yellow leaf curl virus
U	Average external wind velocity
U	Wind velocity
U_{10}	Wind velocity at 10 m height
U_h	Wind velocity at a height h
UNE	Spanish norm (regulation)
UV	Ultraviolet
UV-PE	Ultraviolet polyethylene
V	Greenhouse ventilation rate per soil area unit; also designates the volume of the greenhouse
VPD	Vapour pressure deficit
W	West (cardinal point)
WMV-2	Watermelon mosaic virus-2
WTO	World Trade Organization
WUE	Water use efficiency
X	Air absolute humidity; also designated as AH
X_A	Greenhouse ambient absolute humidity
X_S	Surface absolute humidity
ZYMV	Zucchini yellow mosaic virus
α	Absorptivity or absorption coefficient
β	Conversion factor for direct solar radiation into diffuse radiation (in the greenhouse)
γ	Azimuth (angle)
Δ	Difference
ΔH	Vaporization latent heat
ΔT	Temperature difference
ε	Emissivity
ε_b	Solar radiation into biomass conversion efficiency

ε_i	Plant solar radiation interception efficiency
$(\varepsilon_i \times \varepsilon_b)$	Crop solar radiation use efficiency
θ	Zenith angle; also designates soil moisture content
λ	Thermal conductivity coefficient
ρ	Reflectivity or reflection coefficient
σ	Stefan–Boltzman constant (5.67 10^8 W m^{-2} K^{-4})
Σ	Summation
τ	Transmissivity or transmission coefficient
τ_{dif}	Transmissivity to diffuse radiation
Φ	Energy flux
Φ_{rad}	Radiation energy flux
Φ_T	Air volumetric flux by thermal effect
Φ_w	Air volumetric flux by wind effect
Ψ_g	Gravitational potential
Ψ_m	Matrix potential
Ψ_s	Osmotic potential
Ψ_{soil}	Soil water potential
′	Minute
″	Second
°	Degree
°C	Celsius degree
°F	Fahrenheit degree
K	Kelvin

Appendix 3
Units and Equivalences

Length

1 nanometre (nm) = 10^{-9} metres (m)
1 micron or micrometre (μm) = 10^{-6} metres (m)
1 millimetre (mm) = 10^{-3} metres (m)
1 centimetre (cm) = 10^{-2} metres (m)
1 centimetre (cm) = 0.394 inches
1 metre (m) = 1.094 yards
1 metre (m) = 3.281 feet
1 metre (m) = 39.37 inches
1 kilometre (km) = 1000 metres (m)
1 kilometre (km) = 0.621 miles
1 inch = 2.54 centimetres (cm)
1 foot = 30.48 centimetres (cm)
1 yard = 0.914 metres (m)
1 mile = 1.609 kilometres (km)

Area

1 square metre (m^2) = 1.196 square yards
1 square metre (m^2) = 10.764 square feet
1 hectare (ha) = 10,000 square metres (m^2)
1 hectare (ha) = 2.47 acres
1 acre = 0.4047 hectares (ha)

Volume

1 cubic metre (m^3) = 1000 litres (l)
1 cubic metre (m^3) = 1.308 cubic yards

1 cubic metre (m^3) = 35.3198 cubic feet
1 cubic yard = 0.765 cubic metres (m^3)
1 cubic foot = 0.0283 cubic metres (m^3)
1 litre (l) = 0.2210 gallons (imperial measurement)
1 litre = 0.2642 gallons (USA)
vpm = volume per million = 0.0001% (volume)

Mass

1 metric ton (t) = 1000 kilograms (kg)
1 kilogram (kg) = 1000 grams (g)
1 kilogram (kg) = 2.205 pounds
1 ounce = 0.454 kilograms (kg)
1 ounce = 28.35 grams (g)
ppm = parts per million = 1 μmol mol^{-1}
ppb = parts per billion

Thickness of Plastic Films

Inches	Millimetres	Microns	Gauges
0.001	0.0254	25.4	100
0.002	0.0508	50.8	200
0.004	0.1016	101.6	400
0.006	0.1524	152.4	600
0.008	0.2032	203.2	800
0.010	0.2540	254.0	1000

Temperature

T (°C) Temperature in centigrade or Celsius degrees
T (°F) Temperature in Fahrenheit degrees
T (K) Temperature in Kelvin

$$T\,(°C) = T\,(K) + 273°C$$

$$T\,(°F) = \frac{9}{5}T\,(°C) + 32$$

$$T\,(°C) = \frac{5}{9}\left[T(°F) - 32\right]$$

Pressure

1 kilopascal (kPa) = 1000 Pa
1 pascal (Pa) = 1 newton per square metre ($N\ m^{-2}$)
1 milibar (mb) = 100 Pa
1 centibar (cb) = 1000 Pa = 1 kilopascal (kPa)
1 bar (b) = 0.987 atmospheres = 0.1 megapascales (MPa)

Energy and Power

1 joule (J) = 0.239 calories (cal)
1 calorie (cal) = 4.18 joules (J)
1 watt (W) = 1 joule per second ($J\ s^{-1}$)
1 kilowatt-hour (kWh) = 3600 kilojoules (kJ)
1 kilowatt-hour (kWh) = 3.6 megajoules(MJ)
1 kilowatt-hour (kWh) = 860 kilocalories (kcal)

Radiation

Solar constant:
1353 $W\ m^{-2}$ = 4871 $kJ\ m^{-2}\ h^{-1}$ =
1940 $cal\ cm^{-2}\ min^{-1}$ = 1940 $Ly\ min^{-1}$
Radiation intensity (irradiance):
1 $W\ m^{-2}$ = 1.434 10^{-3} $cal\ cm^{-2}\ min^{-1}$
1 $cal\ cm^{-2}\ min^{-1}$ = 697.3 $W\ m^{-2}$
Cumulated radiation:
1 langley (Ly) = 1 $cal\ cm^{-2}$ = 41.84 $kJ\ m^{-2}$
1 kilolangley (kLy) = 1000 Ly
1 $kWh\ m^{-2}$ = 3.6 $MJ\ m^{-2}$
1 $kWh\ m^{-2}$ = 860 $kcal\ m^{-2}$ = 86 Ly

Water Lamina

1 litre per square metre ($l\ m^{-2}$) = 1 mm = 10 $m^3\ ha^{-1}$

Prefixes

n (nano) = 10^{-9}
μ (micro) = 10^{-6}
m (milli) = 10^{-3}
k (kilo) = 10^{3}
M (mega) = 10^{6}
G (giga) = 10^{9}

List of Tables

Appendix 1

List of Figures

List of Photos

List of Plates

Plate 10. Transmissivity of different greenhouse roof geometries of single-span greenhouses at the winter solstice (21 December), at the latitude of Belgium (51°N), depending on the orientation (east–west or north–south) (adapted from Nisen and Deltour, 1986).

Plate 11. Spectral activities of different photo-biological processes (adapted from Whatley and Whatley, 1984).

Plate 12. Net photosynthesis of an isolated leaf, of a leaf within a canopy and a whole canopy (adapted from Urban, 1997a).

Plate 13. (a) Monthly energy consumption (Q, in MJ m^{-2}) for different temperature set points (T_c) in a low-cost greenhouse in Almeria. (b) Yearly accumulated energy consumption (Q_s, in MJ m^{-2}) for different temperature set points (T_c). Data refer to an average year using hot air heating (adapted from López, 2003).

Plate 14. Increases in the air temperature of a greenhouse with respect to the outside air, depending on the incident global radiation for different ventilation conditions. Light green, 11% of ventilation area, only sidewall; green, 13% of roof ventilation area; red, 22% of sidewall ventilation area; blue, 11% and 13% of sidewall and roof ventilation areas, respectively. (Source: J.I. Montero.)

Plate 15. Air renewal rate as a function of velocity and wind direction in a multi-tunnel greenhouse with hinged-type vents (with flap) without obstacles (a) and vents implemented with an anti-thrips screen (b) (from Muñoz, 1998).

Plate 16. Ventilation studies of scale models of greenhouses, by means of a flow visualization technique, using liquids of different colours and densities. The scale model is mounted downwards. (Source: J.I. Montero.)

Plate 17. Scheme of a scalar temperature field of a low-cost greenhouse, of five spans, with only roof vent (top) or combined roof and side ventilation (bottom), which allows for the visualization of the best ventilated zones (blue colours) and the worst ventilated zones (red colours). Windward wind, 3 m s^{-1}. CFD technique, simulation of fluid dynamics (data provided by 'Las Palmerillas' Experimental Station, Cajamar Foundation, Almeria, and by J.I. Montero).

Plate 18. The greenhouse that uses a screen as cladding material is known as a screenhouse.

Plate 19. High greenhouse air temperatures (with respect to the exterior temperature) as a function of the ventilation rate with a well-developed crop, under different conditions. Green, With shading; blue, with shading and air humidification; red, without shading or fogging. (Source: J.I. Montero.)

Plate 20. Set point values for carbon enrichment (CO_2) depending on the demand of heat (A) and the solar radiation intensity level (A and B) depending on the vent opening conditions (adapted from Nederhoff, 1995). x1 represents the lower degree of vent opening that determines changes in the CO_2 enrichment set point. x2 and x3 represent the vent opening that determine the minimum CO_2 enrichment set point when radiation is below a preset level (x2) or when radiation exceeds a preset level (x3). See explanation in the text (Chapter 9).

Plate 21. Substrate cultivation has spread widely.

Plate 22. The volumetric relations of the water and air content of a P-2 type perlite (grain size between 0 and 5 mm), depending on the height and width of the container holding it, represented in cross-section (adapted from Caldevilla and Lozano, 1993). RAW, Readily available water; NRAW, not readily available water; RW, reserve water.

References

Abad, M. and Noguera, P. (1998) Sustratos para el cultivo sin suelo y fertirrigación. In: Cadahía, C. (ed.) *Fertirrigación: Cultivos Hortícolas y Ornamentales.* Ediciones Mundi-Prensa, Madrid, pp. 287–342.

Abbot, J.A. and Harker, F.R. (2003) Texture. In: Gross, K.C., Yang, C.Y. and Saltveit, M. (eds) *The Commercial Storage of Fruits, Vegetables and Florist and Nursery Stocks.* USDA Agriculture Handbook no. 60. USDA, Washington, DC, pp. 81–99.

Abdel-Ghany, A.M. and Al-Helal, I.M. (2010) Characterization of solar radiation transmission through plastic shading nets. *Solar Energy Materials and Solar Cells* 94, 1371–1378.

Aberkani, K., Hao, X., Halleux, D., Papadopoulos, A.P., Dorais, M., Vineberg, S. and Gosselin, A. (2011) Energy saving achieved by retractable liquid foam between double polyethylene films covering greenhouses. *Transactions of the ASABE* 54(1), 275–284.

Acock, B., Acock, M.C. and Pasternak, D. (1990) Interactions of CO_2 enrichment and temperature on carbohydrate production and accumulation in muskmelon leaves. *Journal of the American Society for Horticultural Science* 115, 525–529.

Adams, P. and Ho, L.C. (1995) Uptake and distribution of nutrients in relation to tomato fruit quality. *Acta Horticulturae* 412, 374–387.

Aldrich, R.A. and Bartok, J.W. (1994) *Greenhouse Engineering.* Cooperative Extension, NRAES-33 (Natural Resouce, Agriculture and Engineering Service), Ithaca, New York.

Allen, L.J. (1990) Plant responses to rising carbon dioxide and potential interactions with air pollutants. *Journal of Environmental Quality* 19(1), 15–34.

Alpi, A. and Tognoni, F. (1975) *Cultivo en Invernadero.* Ediciones Mundi-Prensa, Madrid.

American Society of Agricultural Engineers (ASAE) (1984) *Heating, Ventilating and Cooling Greenhouses. ASAE Standards.* ASAE, St Joseph Charter Township, Michigan.

American Society of Agricultural Engineers (ASAE) (1988) *Heating, Ventilating and Cooling Greenhouses. ASAE Standards.* ASAE, St Joseph Charter Township, Michigan.

American Society of Agricultural Engineers (ASAE) (2002) *Heating, Ventilating and Cooling Greenhouses. ASAE Standards.* ASAE, St Joseph Charter Township, Michigan.

American Society of Heating, Refrigerating and Air Conditioning Engineers (ASHRAE) (1989) *Fundamentals Handbook.* ASHRAE, Atlanta, Georgia.

Ammerlaan, J.C.J., Ruijs, M.N.A., Van Uffelen, R.L.M., Van der Maas, A.A., Jonkman, B. and Ogier, J.P. (2000) Glasshouse of the future. *Acta Horticulturae* 536, 215–222.

Antón, A. (2004) *Utilización del Análisis del Ciclo de Vida en la Evaluación del Impacto ambiental del cultivo bajo invernadero Mediterráneo.* Doctoral thesis, Universitat Politècnica de Catalunya, Barcelona.

Antón, A. and Montero, J.I. (2003) Life cycle assessement: a tool to evaluate and improve the environmental impact of Mediterranean greenhouses. *Acta Horticulturae* 614, 35–40.

Antón, A., Montero, J.I., Muñoz, P. and Castells, F. (2003) Cuantificación del impacto ambiental mediante análisis del ciclo de vida en relación con el uso del agua en invernadero. In: Fernández, M., Lorenzo, P. and Cuadrado, I.M. (eds) *Mejora de la Eficiencia del Uso del Agua en Cultivos Protegidos*. Dirección General de Investigación y Formación Agraria, Hortimed, FIAPA, Cajamar, Almería, Spain, pp. 495–510.

Aparicio, V., Rodríguez, M.P., Gómez, V., Sáez, E., Belda, J.E., Casado, E., Lastres, J. and Torres, M. (1995) *Plagas y Enfermedades de los Principales Cultivos Hortícolas de la Provincia de Almería: Control Racional*. Dirección General de Investigación Agraria, Junta de Andalucía, Seville.

Aparicio, V., Belda, J.E., Casado, E., García, M.M., Gómez, V., Lastres, J., Mirasol, E., Roldán, E., Sáez, E., Sánchez, A. and Torres, M. (1998) *Plagas y Enfermedades en Cultivos Hortícolas de la Provincia de Almería: Control Racional*. Dirección General de Investigación Agraria, Junta de Andalucía, Seville.

Aparicio, V., Rodríguez, M.P. and Manzanares, C. (2003) Producción integrada en cultivos hortícolas bajo abrigo: Andalucía. In: Camacho. F. (ed.) *Técnicas de Producción en Cultivos Protegidos*. Cajamar, Almería, Spain, pp. 225–243.

Aranda, E. (1994) Maquinaria de aplicación. In: Moreno, R. (ed.) *Sanidad Vegetal en la Horticultura Protegida*. Dirección General de Investigación Agraria, Junta de Andalucía, Seville, pp. 67–80.

Ayers, R.S. and Westcot D.W. (1976) *Calidad del Agua para la Agricultura*. Estudio FAO: Riego y drenaje no. 29. FAO, Rome.

Ayers, R.S. and Westcot, D.W. (1987) *Calidad del Agua para la Agricultura*. Estudio FAO: Riego y Drenaje no. 29 1a Revisión. FAO, Rome.

Aznar, J.A. (2000) Industria y servicios auxiliares a la agricultura intensiva. In: Fernández, M. and Cuadrado, I.M. (eds) *Comercialización de Productos Hortofrutícolas II*. Dirección General de Investigación y Formación Agraria de la Junta de Andalucía, FIAPA, Caja Rural, Almería, Spain, pp. 537–553.

Baeza, E.J. (2007) Optimización del diseño de los sistemas de ventilación en invernaderos tipo parral. Doctoral thesis, Escuela Politécnica Superior, Departamento de Ingeniería Rural, Universidad de Almería, Almería, Spain.

Bailey, B.J. (1985) Wind dependent control of greenhouse temperature. *Acta Horticulturae* 174, 381–386.

Bailey, B.J. (1998) Modelización de la gestión del clima en invernadero. In: Pérez, J. and Cuadrado, I. (eds) *Tecnología de Invernaderos II*. Dirección General de Investigación Agraria, FIAPA, Caja Rural, Almería, Spain, pp. 417–456.

Bailey, B.J. (1999) The use of models in greenhouse environmental control. *Acta Horticulturae* 491, 93–99.

Bailey, B.J. and Richardson, G.M. (1990) A rational approach to greenhouse design. *Acta Horticulturae* 281, 111–117.

Baille, A. (1993) Artificial light sources for crop production. In: Varlet-Granchet, C., Bonhomme, R. and Sinoquet, H. (eds) *Crop Structure and Light Microclimate*. Institut National de la Recherche Agronomique (INRA) Editions, Paris, pp. 107–119.

Baille, A. (1995) Serres plastiques, climat et production. *PHM Revue Horticole* 357, 15–29.

Baille, A. (1999) Energy cycle. In: Stanhill, G. and Enoch, H.Z. (eds) *Greenhouse Ecosystems*. Elsevier, Amsterdam, pp. 265–286.

Baille, A. and Tchamitchian, M. (1993) Solar radiation in greenhouses. In: Varlet-Granchet, C., Bonhomme, R. and Sinoquet, H. (eds) *Crop Structure and Light Microclimate*. Institut National de la Recherche Agronomique (INRA) Editions, Paris, pp. 93–105.

Baille, M., Baille, A. and Laury, J.C. (1994) A simplified model for predicting evapotranspiration of nine ornamental species vs. climate factors and leaf area. *Scientia Horticulturae* 59, 217–232.

Baille, A., López, J.C., Cabrera, J., Gonzalez-Real, M.M. and Pérez-Parra, J. (2003) Characterization of the solar diffuse component under 'parral' type greenhouses. *Acta Horticulturae* 614, 341–346.

Bakker, J.C. (1991) Analysis of humidity effects on growth and production of glasshouse fruit vegetables. PhD dissertation, Wageningen Agricultural University, Wageningen, The Netherlands.

Bakker, J.C. (1995) Greenhouse climate control: constraints and limitations. *Acta Horticulturae* 399, 25–35.

Bakker, J.C. and Van Holsteijn, G.P.A. (1995) Screens. In: Bakker, J.C., Bot, G.P.A., Challa, H. and Van de Braak, N.J. (eds) *Greenhouse Climate Control: an Integrated Approach*. Wageningen Pers, Wageningen, The Netherlands, pp. 185–195.

Bakker, J.C., Bot, G.P.A., Challa, H. and Van de Braak, N.J. (eds) (1995) *Greenhouse Climate Control: an Integrated Approach.* Wageningen Pers, Wageningen, The Netherlands.

Baldwin, E.A. (2003) Fruit and vegetable flavor. In: Gross, K.C., Yang, C.Y. and Saltveit, M. (eds) *The Commercial Storage of Fruits, Vegetables and Florist and Nursery Stocks.* USDA Agriculture Handbook no. 60. USDA, Washington, DC, pp. 114–132.

Barahona, F.J. and Gómez-Vázquez, J. (1985) Influence of pesticides on the degradation of polyethylene film greenhouse cladding on the Andalusian coast. *Plasticulture* 63, 3–10.

Bar-Yosef, B., Sagiv, B. and Eliah, E. (1980) *Fertilization and Irrigation of Winter Tomatoes Grown in a Glasshouse in the Besor Area. Preliminary Report 775.* Volcani Center, Bet Dagan, Israel.

Belda, J.E. and Cabello, T. (1994) Afidos plaga (Homóptera: Aphididae). In: Moreno, R. (ed.) *Sanidad Vegetal en la Horticultura Protegida.* Dirección General de Investigación Agraria, Junta de Andalucía, Seville, pp. 155–178.

Belloso, O.M. (2003) Trazabilidad y legislación de los productos de la IV gama. In: Lobo, M.G. and González, M. (eds) *Productos Hortofrutícolas Mínimamente Procesados.* Instituto Canario de Investigaciones Agrarias, La Laguna, Tenerife, pp. 175–184.

Benoit, F. (1990) *Economic Aspects of Ecologically Sound Soilless Growing Methods.* European Vegetable R&D Centre Report, Sint-Katelijne-Waver, Belgium.

Berenguer, J.J., Escobar, I. and Cuartero, J. (2003) Opciones de producción de tomate cereza en invernadero. *Horticultura* 172, 34–37.

Berninger, E. (1989) *Cultures Florales de Serre en Zone Mediterraneenne Francaise: Elements Climatiques et Physiologiques.* Institut National de la Recherche Agronomique (INRA), Pepinieristes, Horticulteurs, Maraichers (PHM), Paris.

Biale, J.B. and Young, R.E. (1981) Respiration and ripening in fruits: retrospect and prospect. In: Fried, F. and Rhodes, M.J.C. (eds) *Recent Advances in the Biochemistry of Fruits and Vegetables.* Academic Press, Orlando, Florida, pp. 1–39.

Bielza, P. (2000) Fundamentos del control integrado de plagas y enfermedades. In: Alarcón, A.A. (ed.) *Tecnología Para Cultivos de Alto Rendimiento.* Novedades Agrícolas, Torre Pacheco, Murcia, Spain, pp. 453–459.

Biro, R.L. and Jaffe, M.J. (1984) Thigmomorphogenesis: ethylene evolution and its role in the changes observed in mechanically perturbed bean plants. *Physiologia Plantarum* 62, 289–296.

Blain, J., Gosselin, A. and Trudel, M.J. (1987) Influence of HPS supplemental lighting on growth and yield of greenhouse cucumbers. *HortScience* 22, 36–38.

Blom, J.V.D. (2002) La introducción artificial de la fauna auxiliar en cultivos agrícolas. *Boletín de Sanidad y Plagas* 28, 109–120.

Boodley, J.N. (1996) *The Commercial Greenhouse.* Delmar Publishers, Albany, New York.

Bordes, P. (1992) Les plastiques et la maitrise du climat en productions vegetales: bases generales. In: *Les Plastiques en Agriculture.* Comité des Plastiques en Agriculture (CPA), Paris, pp. 163–355.

Bot, G.P.A. (1983) Greenhouse climate: from physical processes to a dynamic model. PhD thesis, Wageningen Agricultural University, Wageningen, The Netherlands.

Bot, G.P.A. and Van de Braak, N.J. (1995) Energy balance. In: Bakker, J.C., Bot, G.P.A., Challa, H. and Van de Braak, N.J. (eds) *Greenhouse Climate Control: an Integrated Approach.* Wageningen Pers, Wageningen, The Netherlands, pp. 135–141.

Boulard, T., Baille, A. and Gall, F.L. (1991) Etude de differentes methodes de refroidissement sur le climat et la transpiration de tomates de serre. *Agronomie* 11, 543–553.

Boyer, J.S. (1982) Plant productivity and the environment. *Science* 218, 443–448.

Bressler, E. (1977) Trickle-drip irrigation principles and application to soil-water management. *Advances in Agronomy* 29, 343–393.

Breuer, J.J.G. and Knies, P. (1995) Ventilation and cooling. In: Bakker, J.C., Bot, G.P.A., Challa, H. and Van de Braak, N.J. (eds) *Greenhouse Climate Control: an Integrated Approach.* Wageningen Pers, Wageningen, The Netherlands, pp. 179–185.

Briassoulis, O., Waaijenberg, D., Grataud, J. and Von Elsner, B. (1997a) Mechanical properties of covering materials for greenhouses. Part 1: General overview. *Journal of Agricultural Engineering Research* 67, 81–96.

Briassoulis, O., Waaijenberg, D., Grataud, J. and Von Elsner, B. (1997b) Mechanical properties of covering materials for greenhouses. Part 2: Quality assessment. *Journal of Agricultural Engineering Research* 67, 171–217.

Brougham, R.W. (1956) Effects of intensity of defoliation on regrowth of pasture. *Australian Journal of Agricultural Research* 7, 377–387.

Brouwer, R. (1981) Effects of environmentals conditions on root functioning. *Acta Horticulturae* 119, 91–101.

Bunt, A.C. (1988) *Media and Mixes for Container-grown Plants.* Unwin Hyman Ltd, London.

Caballero, P. and Miguel, M.D. (2002) Costes e investigación en la hortofruticultura Mediterránea. In: Alvarez-Coque, J.M.G. (ed.) *La Agricultura Mediterránea en el Siglo XXI.* Instituto Cajamar, Almería, Spain, pp. 222–244.

Cabello, T. and Belda, J.E. (1994) Noctuidos plaga (Lepidoptera: Noctuidae). In: Moreno, R. (ed.) *Sanidad Vegetal en la Horticultura Protegida.* Dirección General de Investigación Agraria, Junta de Andalucía, Seville, pp. 179–211.

Cabello, T. and Benítez, E. (1994) *Frankliniella occidentalis* (Pergande). In: Moreno, R. (ed.) *Sanidad Vegetal en la Horticultura Protegida.* Dirección General de Investigación Agraria, Junta de Andalucía, Seville, pp. 241–259.

Cabrera, F.J., Baille, A., Lopez, J.C., Gonzalez-Real, M.M. and Perez-Parra, J. (2009) Effects of cover diffuse properties on the components of greenhouse solar radiation. *Biosystems Engineering* 103, 344–356.

Cadahía, C. (1998) *Fertirrigación: Cultivos Hortícolas y Ornamentales.* Ediciones Mundi-Prensa, Madrid.

Calatrava-Requena, J., Cañero, R. and Ortega, J. (2001) Productivity and cultivation cost analysis in plastic greenhouses in the Níjar (Almería) area. *Acta Horticulturae* 559, 737–744.

Caldentey, P. (1972) *Comercialización de Productos Agrarios.* Agricola Española, Madrid.

Caldevilla, E.M. and Lozano, M.G. (1993) *Cultivos sin Suelo: Hortalizas en Clima Mediterráneo.* Ediciones de Horticultura, Reus, Spain.

Calvert, A. and Slack, G. (1974) Light-dependent control of day temperature for early tomato crops. *Acta Horticulturae* 51, 163–168.

Cañero, R. and Calatrava-Leyva, J. (2001) Production functions for plastic covered pepper and tomato in the coastline of Almería: an analysis of productive efficiency. *Acta Horticulturae* 559, 725–730.

Cañero, R., Calatrava, J., Cabello, T. and Castilla, N. (1994) Análisis de costes variables en cultivos en invernadero. *Hortofruticultura* V(2), 27–33.

Castellano, S., Russo, G. and Scarascia-Mugnozza, G.S. (2006) The influence of construction parameters on radiometric performances of agricultural nets. *Acta Horticulturae* 718, 283–290.

Castellano, S., Hemming, S. and Russo, G. (2008a) The influence of colour on radiometric performances of agricultural nets. *Acta Horticulturae* 801, 227–236.

Castellano, S., Scarascia-Mugnozza, G.S., Briassoulis, D., Mistriotis, A., Hemming, S. and Waaijenberg, D. (2008b) Plastic nets in agriculture: a general review of types and applications. *Applied Engineering in Agriculture* 24(6), 799–808.

Castilla, N. (1985) Contribución al estudio de los cultivos enerenados en Almería: necesidades hídricas y extracción de nutrientes del cultivo de tomate de crecimiento indeterminado en abrigo de polietileno. Doctoral thesis, Universidad Politécnica de Madrid, Madrid, Spain.

Castilla, N. (1987) Manejo del riego por goteo en invernadero. In: *Nuevas Tecnologías en la Producción Hortícola de Invernadero.* Universidad de Córdoba, Córdoba, pp. 77-97.

Castilla, N. (1989) Programación del riego por goteo en invernadero plástico sin calefacción. *Plasticulture* 82, 59–63.

Castilla, N. (1991) El cultivo forzado en España. S.E.C.H. In: Rallo, L. and Nuñez, F. (eds) *La Horticultura Española en la C.E.E.* Ediciones de Horticultura, Reus, España, pp. 284–291.

Castilla, N. (1994) Greenhouses in the Mediterranean areas: technological level and strategic management. *Acta Horticulturae* 361, 44–56.

Castilla, N. (1995) Manejo del tomate en cultivo intensivo con suelo. In: Nuez, F. (ed.) *El Cultivo del Tomate.* Ediciones Mundi-Prensa, Madrid, pp. 189–225.

Castilla, N. (2000) *Improved Irrigation Management of Greenhouse Vegetables.* FAO, Rome.

Castilla, N. (2001) La radiación solar en invernadero en la costa Mediterránea española. In: *Incorporación de Tecnología al Invernadero Mediterráneo.* Cajamar, Almería, Spain, pp. 35–47.

Castilla, N. (2002) Current situation and future prospects of protected crops in the Mediterranean region. *Acta Horticulturae* 582, 135–147.

Castilla, N. and Fereres, E. (1990) Tomato growth and yield in unheated plastic greenhouse under Mediterranean climate. *Agricoltura Mediterranea* 120(I), 31–40.

Castilla, N. and Hernández, J. (1995) Protected cultivation in the Mediterranean area. *Plasticulture* 107, 13–20.

Castilla, N. and Hernández, J. (2005) The plastic greenhouse industry of Spain. *Chronica Horticulturae* 45(3), 15–20.

Castilla, N. and Hernández, J. (2007) Greenhouse technological packages for high-quality crop production. *Acta Horticulturae* 761, 285–297.

Castilla, N. and Lopez-Galvez, J. (1994) Vegetable crops response to the improvement of low-cost plastic greenhouses. *The Journal of Horticultural Science* 69(5), 915–921.

Castilla, N. and Montalvo, T. (1998) Programación del riego. In: Cadahía, C. (ed.) *Fertirrigación: Cultivos Hortícolas y Ornamentales.* Ediciones Mundi-Prensa, Madrid, pp. 266–286.

Castilla, N. and Montero, J.I. (2008) Environmental control and crop production in Mediterranean greenhouses. *Acta Horticulturae* 797, 25–36.

Castilla, N., Bretones, F. and Jorge, G. (1977) Nuevos materiales plásticos para invernaderos. *Agricultura* 554, 451–462.

Castilla, N., Jiménez, C. and Ferreres, E. (1986) Tomato root development on sand mulch, plastic greenhouse in Almería. *Acta Horticulturae* 191, 113–121.

Castilla, N., Elías, F. and Fereres, E. (1990a) Caracterización de condiciones climáticas y de relaciones suelo-agua-raíz en el cultivo enarenado del tomate en invernadero en Almería. *Investigación Agraria: Producción y Protección Vegetal* 5(2), 259–271.

Castilla, N., Elías, F. and Fereres, E. (1990b) Evapotranspiración de cultivos hortícolas en invernadero en Almería. *Investigación Agraria: Producción y Protección Vegetal* 5(1), 117–125.

Castilla, N., Bretones, F. and Lopez-Galvez, J. (1991) Cucumber growth and yield in unheated plastic greenhouse. *Agricoltura Mediterránea* 121, 166–172.

Castilla, N., Tognoni, F. and Olympos, C. (1992) *Vegetable Production Under Simple Structures in Southern Europe.* Food and Fertilizer Technology Centre Bulletin no. 348. Food and Fertilizer Technology Centre, Taipei, Republic of China.

Castilla, N., Gallego, A. and Cruz-Romero, G. (1996) Greenhouse melon response to plastic mulch and drip irrigation. *Acta Horticulturae* 458, 263–267.

Castilla, N., Hernández, J., Quesada, F.M., Morales, M.I., Guillén, A., Escobar, I. and Montero, J.I. (1999) Alternative asymmetrical greenhouses for the Mediterranean area of Spain. *Acta Horticulturae* 491, 83–86.

Castilla, N., Hernández, J., Quesada, F.M., Morales, M.I., Guillén, A., Soriano, M.T., Escobar, I., Antón, A. and Montero, J.I. (2001) Comparison of asymmetrical greenhouse types in the Mediterranean area of Spain. *Acta Horticulturae* 559, 183–186.

Castilla, N., Hernández, J. and Abou-Hadid, A.F. (2004) Strategic crop and greenhouse management in mild winter climate areas. *Acta Horticulturae* 633, 183–196.

Challa, H. (2001) Modellig for present production problems in greenhouse horticulture in mild winter climates. *Acta Horticulturae* 559, 431–440.

Challa, H. and Bakker, J.C. (1995) Crop growth. In: Bakker, J.C., Bot, G.P.A., Challa, H. and Van de Braak, N.J. (eds) *Greenhouse Climate Control: an Integrated Approach.* Wageningen Pers, Wageningen, The Netherlands, pp. 15–123.

Challa, H. and Bakker, M. (1999) Potential production within the greenhouse environment. In: Stanhill, G. and Enoch, H.Z. (eds) *Greenhouse Ecosystems.* Elsevier, Amsterdam, pp. 333–348.

Challa, H. and Schapendok, A.H.C.M. (1984) Quantification of the light reduction in greenhouses on yield. *Acta Horticulturae* 148, 501–509.

Challa, H., Heuvelink, E. and Van Meeteren, U. (1995) Crop growth and development. In: Bakker, J.C., Bot, G.P.A., Challa, H. and Van de Braak, N.J. (eds) *Greenhouse Climate Control: an Integrated Approach.* Wageningen Pers, Wageningen, The Netherlands, pp. 62–84.

Chapman, H.D. (1973) *Diagnosis Criteria for Plants and Soils.* Chamman, Riverside, California.

Chaux, C. and Foury, C. (1994a) *Productions Legumieres: Generalités* (Tome 1). Tec-Doc, Paris.

Chaux, C. and Foury, C. (1994b) *Productions Legumieres* (Tome 3). Tec-Doc, Paris.

Cockshull, K.E. (1992) Crop environments. *Acta Horticulturae* 312, 7–85.

Cockshull, K.E., Graves, C.J. and Cave, C.R.J. (1992) The influence of shading on yield of glasshouse tomatoes. *The Journal of Horticultural Science* 67, 11–24.

Cohen, Y. (2003) Respuestas fisiológicas a la utilización de aguas salinas. In: Fernández, M., Lorenzo, P. and Cuadrado, I.M. (eds) *Mejora de la Eficiencia del Uso del Agua en Cultivos Protegidos.* Dirección General de Investigación y Formación Agraria, Hortimed, FIAPA, Cajamar, Almería, Spain, pp. 131–147.

Colino, J. and Martinez-Paz, J.M. (2002) El agua en la agricultura del Sureste español: productividad, precio y demanda. In: Alvarez-Coque, J.M.G. (ed.) *La Agricultura Mediterránea en el Siglo XXI.* Instituto Cajamar, Almería, Spain, pp. 199–221.
Comité des Plastiques en Agriculture (CPA) (1992) *Les Plastiques en Agriculture.* CPA, Paris, 581 pp.
Conellan, G.J. (2002) Selection of greenhouse design and technology options for high temperature regions. *Acta Horticulturae* 578, 113–117.
Coombs, J., Hall, D.O., Long, S.P. and Scurlock, J.M.O. (1985) *Techniques in Bioproductivity and Photosynthesis.* Pergamon Press, Oxford.
Cooper, A. (1979) *The ABC of NFT.* Grower Books, London.
Cornillon, P. (1977) Effect de la temperature des racines sur l'absorption des elements mineraux par la tomate. *Annales Agronomiques* 28, 409–423.
Cuadrado, M.I. (1994) Virosis. In: Moreno, R. (ed.) *Sanidad Vegetal en la Horticultura Protegida.* Dirección General de Investigación Agraria, Junta de Andalucía, Seville, pp. 369–396.
Cuadrado, I.M., Janssen, D., Velasco, L., Ruiz, L. and Segundo, E. (2001) First report of cucumber vein yellowing virus in Spain. *Plant Disease* 3, 336.
D'Amico, M., La Via, G. and Scuderi, A. (2003) An economic analysis of greenhouse vegetable production with inert soil in Sicily. *Acta Horticulturae* 614, 849–855.
Dalrymple, D.G. (1973) *Controlled Environment Agriculture: a Global Review of Greenhouse Food Production.* Foreign Agricultural Economics Report no. 89. USDA, Washington, DC.
Dasberg, S. (1999a) The root medium. In: Stanhill, G. and Enoch, H.Z. (eds) *Greenhouse Ecosystems.* Elsevier, Amsterdam, pp. 101–109.
Dasberg, S. (1999b) The nutrient cycle. In: Stanhill, G. and Enoch, H.Z. (eds) *Greenhouse Ecosystems.* Elsevier, Amsterdam, pp. 327–332.
Day, W. and Bailey, B.J. (1999) Physical principles of microclimate modification. In: Stanhill, G. and Enoch, H.Z. (eds) *Greenhouse Ecosystems.* Elsevier, Amsterdam, pp. 71–99.
De Liñán, C. (ed.) (1998) *Entomología Agroforestal.* Agrotécnicas, Madrid.
De Pascale, S. and Maggio, A. (2008) Plant stress management in semiarid greenhouse. *Acta Horticulturae* 797, 205–215.
De Pascale, S., Scarascia-Mugnozza, G., Maggio, A. and Schettini, E. (eds) (2008) *Proceedings of the International Symposium on High Technology for Greenhouse System Management: Greensys-2007. Acta Horticulturae* 801. International Society for Horticultural Science, Leuven, Belgium.
De Visser, A. and Vesseur, W.P. (1982) Daglicht, een van de vele factoren die de produktie bepalen [Daylight, one of the production influencing factors]. *Tuinderij* 62(9), 38–39.
Deltour, J. and Nissen, A. (1970) Les verres diffusants en couverture de serres. *Bulletin de la Recherche Agronomique* (Gembloux, Belgium) V1(2), 232–255.
Desjardins, Y. and Patil, B. (eds) (2007) *Proceedings of the I International Symposium on Human Health Effects of Fruits and Vegetables. Acta Horticulturae* 744. International Society for Horticultural Science, Leuven, Belgium
Díaz, T., Espí, E., Fontecha, A., Jiménez, J.C., López, J. and Salmerón, A. (2001) *Los Filmes Plásticos en la Producción Agrícola.* Ediciones Mundi-Prensa, Madrid.
Dolado, P.C. (2000) Nuevas tecnologías de procesado de alimentos vegetales. In: Fernández, M. and Cuadrado, I.M. (eds) *Comercialización de Productos Hortofrutícolas II.* Dirección General de Investigación y Formación Agraria de la Junta de Andalucía, FIAPA, Caja Rural, Almería, Spain, pp. 525–536.
Doorenbos, J. and Pruitt, W.O. (1976) *Las Necesidades de Agua de los Cultivos.* Estudio FAO: Riego y Drenaje no. 24. FAO, Rome.
Dorais, M. (ed.) (2002) *Proceedings of the Fourth International Symposium on Artificial Lighting. Acta Horticulturae* 589. International Society for Horticultural Science, Leuven, Belgium.
Dorais, M. (ed.) (2011) *Proceedings of the International Symposium on High Technology for Greenhouse Systems: GreenSys-2009. Acta Horticulturae* 893. International Society for Horticultural Science, Leuven, Belgium.
Dorais, M., Papadopoulos, A.P. and Gosselin, A. (2001a) Greenhouse tomato fruit quality. *Horticultural Reviews* 26, 239–319.
Dorais, M., Papadopoulos, A.P. and Gosselin, A. (2001b) Influence of EC management on greenhouse tomato yield and fruit quality: a review. *Agronomie* 21, 367–383.
Dorais, M., Ehret, D.L. and Papadopoulos, A.P. (2008) Tomato (*Lycopersicon esculentum*) health components: from the seed to the consumer. *Phytochemical Review* 7, 231–250.

Dufie, J.A. and Beckman, W.A. (1980) *Solar Engineering of Thermal Processes.* John Wiley & Sons, New York.
Ehleringer, J. and Pearcy, R.W. (1983) Variation in quantum yield for CO_2 uptake among C_3 and C_4 plants. *Plant Physiology* 73, 555–559.
Ehret, D.L. and Ho, L.C. (1986) The effects of salinity on dry matter partitioning and fruit growth in tomatoes grown in nutrient film culture. *The Journal of Horticultural Science* 61, 361–367.
Elad, Y. (1999) Plant diseases in greenhouses. In: Stanhill, G. and Enoch, H.Z. (eds) *Greenhouse Ecosystems.* Elsevier, Amsterdam, pp. 191–211.
Ellis, R.G. (1990) Low temperature heating of greenhouses. In: *Proceedings of an International Seminar and British–Israeli Workshop on Greenhouse Technology.* Bet-Dagan, Israel, pp. 43–51.
Engindeniz, S. and Tuzel, Y. (2003) Comparative economic analysis of organic tomato and cucumber production in greenhouse: the case of Turkey. *Acta Horticulturae* 614, 843–848.
Enoch, H.Z. (1986) Climate and protected cultivation. *Acta Horticulturae* 176, 11–20.
Enoch, H.Z. (1990) Crop responses to aerial carbon dioxide. *Acta Horticulturae* 268, 17–32.
Enoch, H.Z. and Enoch, Y. (1999) The history and geography of the greenhouse. In: Stanhill, G. and Enoch, H.Z. (eds) *Greenhouse Ecosystems.* Elsevier, Amsterdam, pp. 1–15.
Erwin, J.E. and Heins, R.D. (1995) Thermomorphogenic responses in stem and leaf development. *Hortscience* 30, 940–949.
Escobar, A. (1998) Aspectos ambientales de la agricultura en invernaderos. In: Pérez-Parra, J. and Cuadrado, I. (eds) *Tecnología de Invernaderos.* Consejería de Agricultura y Pesca, FIAPA, Caja Rural de Almería, Almería, Spain, pp. 487–512.
Espi, E., Salmerón, A., Fontecha, A., Garcia, Y. and Real, A.I. (2006) Plastic films for agricultural applications. *Journal of Plastic Film and Sheeting* 22(2), 85–102.
Fahnrich, I., Meller, J. and Zabeltitz, C.V. (1989) Influence de la condensation sur le transmission lumineuse et thermique des materiaux de coverture des serres. *Plasticulture* 84(4), 13–18.
Fereres, E. (1981) *Drip Irrigation Management.* University of California, Berkeley, California.
Fereres, E. and Orgaz, F. (2000) 'Efficiency of water use in crop production: contrasts between open and protected cultivation and future prospects'. Invited speech in International Society for Horticultural Science Symposium on Protected Cultivation in Mild Winter Climate, Cartagena, Spain (unpublished).
Fernández, M.D. (2003) Programación del riego mediante parámetros climáticos en cultivos hortícolas bajo invernadero en el Sudeste Español. In: Fernández, M., Lorenzo, P. and Cuadrado, I.M. (eds) *Mejora de la Eficiencia del Uso del Agua en Cultivos Protegidos.* Dirección General de Investigación y Formación Agraria, Hortimed, FIAPA, Cajamar, Almería, Spain, pp. 343–352.
Fernández, M.D., Orgaz, F., Fereres, F., López, J.C., Céspedes, A., Pérez. J., Bonachela, S. and Gallardo, M. (2001) *Programación de Riego de Cultivos Hortícolas Bajo Invernadero en el Sudeste Español.* Cajamar, Almería, Spain.
Fletcher, J.T. (1984) *Diseases of Greenhouse Plants.* Longman, New York.
Fuchs, M. (1990) Effect of transpiration on greenhouse cooling. In: *Proceedings of an International Seminar and British–Israeli Workshop on Greenhouse Technology.* Bet-Dagan, Israel, pp. 147–173.
Fuentes, J.L. (1999) *Iniciación a la Astronomía.* Ediciones Mundi-Prensa, Madrid.
Galdeano, E. (2002) *Impacto del Control de Calidad y de las Practicas Respetuosas con el Medio Ambiente en las Cooperativas Hortofrutícolas Andaluzas.* Universidad de Almería, Almería, Spain.
Gallardo, M. and Thompson, R. (2003a) Relaciones hídricas en suelo y planta. In: Fernández, M., Lorenzo, P. and Cuadrado, I.M. (eds) *Mejora de la Eficiencia del Uso del Agua en Cultivos Protegidos.* Dirección General de Investigación y Formación Agraria, Hortimed, FIAPA, Cajamar, Almería, Spain, pp. 71–93.
Gallardo, M. and Thompson, R. (2003b) Uso de sensores de planta para la programación del riego. In: Fernández, M., Lorenzo, P. and Cuadrado, I.M. (eds) *Mejora de la Eficiencia del Uso del Agua en Cultivos Protegidos.* Dirección General de Investigación y Formación Agraria, Hortimed, FIAPA, Cajamar, Almería, Spain, pp. 353–373.
García, E. (1994) Enemigos naturales. In: Moreno, R. (ed.) *Sanidad Vegetal en la Horticultura Protegida.* Dirección General de Investigación Agraria, Junta de Andalucía, Seville, pp. 57–65.
García, M. and Daverede, P. (1994) Le residu des fibres de coco: noveau substrat pour le culture hors sol. *PHM Revue Horticole* 348, 7–12.

García-Azcarate, T. and Mastrostefano, M. (2002) Algunas reflexiones sobre los retos del sector europeo de las frutas y hortalizas. In: Alvarez-Coque, J.M.G. (ed.) *La Agricultura Mediterránea en el Siglo XXI*. Instituto Cajamar, Almería, Spain, pp. 83–99.

García-Marí, F., Costa-Comellas, Y. and Fenaquet-Pérez, F. (1994) *Plagas Agrícolas*. Phytoma, Valencia.

Garijo, C. and Frapolli, E. (1994) Medidas culturales, mecánicas y medios físicos. In: Moreno, R. (ed.) *Sanidad Vegetal en la Horticultura Protegida*. Dirección General de Investigación Agraria, Junta de Andalucía, Seville, pp. 47–56.

Garnaud, J.C. (1987) A survey of the development of plasticulture: questions to be answered. *Plasticulture* 74, 5–14.

Gary, C. (1989) Interest of a carbon balance model for on-line growth control: the example of a daylight dependent night-temperature control. *Acta Horticulturae* 248, 265–268.

Gary, C. (1999) Modelling greenhouse crops: state of the art and perspectives. *Acta Horticulturae* 495, 317–321.

Geisenberg, C. and Stewart, K. (1986) Field crop management. In: Atherthon, J.G. and Rudich, J. (eds) *The Tomato Crop*. Chapman and Hall, London, pp. 511–557.

Giacomelli, G. and Ting, K.C. (1999) Horticultural and engineering considerations for the design of integrated greenhouse production systems. *Acta Horticulturae* 481, 475–481.

Giacomelli, G., Castilla, N., Van Henten, E., Meras, D.R. and Sase, S. (2008) Innovation in greenhouse engineering. *Acta Horticulturae* 801, 75–88.

Gijzen, H. (1995a) CO_2 uptake by the crop. In: Bakker, J.C., Bot, G.P.A., Challa, H. and Van de Braak, N.J. (eds) *Greenhouse Climate Control: an Integrated Approach*. Wageningen Pers, Wageningen, The Netherlands, pp. 16–35.

Gijzen, H. (1995b) Interaction between CO_2 uptake by the crop and water loss. In: Bakker, J.C., Bot, G.P.A., Challa, H. and Van de Braak, N.J. (eds) *Greenhouse Climate Control: an Integrated Approach*. Wageningen Pers, Wageningen, The Netherlands, pp. 51–62.

Gijzen, H. and Goudriaan, J. (1989) A flexible and explanatory model of light distribution and photosynthesis in row crops. *Agricultural and Forest Meteorology* 48, 1–20.

Giménez, C. (1992) Bases fisiológicas de la producción hortícola. In: Ramos, E. and Rallo, L. (eds) *Nueva Horticultura*. Ediciones Mundi-Prensa, Madrid, pp. 57–74.

Giménez, C., Otto, R.F. and Castilla, N. (2002) Productivity of leaf and root vegetable crops under direct cover. *Scientia Horticulturae* 94, 1–11.

Gislerød, H.R., Eidsten, I.M. and Mortensen, L.M. (1989) The interaction of daily lighting period and light intensity on growth of some greenhouse plants. *Scientia Horticulturae* 38, 295–304.

Goedhart, M., Nederhoff, E.M., Udink ten Cate, A.J. and Bot, G.P.A. (1984) Methods and instruments for ventilation rate measurements. *Acta Horticulturae* 148, 393–400.

Gómez, J. (1994) Enfermedades causadas por hongos de suelo. In: Moreno, R. (ed.) *Sanidad Vegetal en la Horticultura Protegida*. Dirección General de Investigación Agraria, Junta de Andalucía, Seville, pp. 277–292.

Gómez, M.D., Baille, A., Gonzalez-Real, M.M. and Mercader, J.M. (2003) Comparative analysis of water and nutrient uptake of glasshouse cucumber grown in NFT and perlite. *Acta Horticulturae* 614, 175–180.

Gonzalez, A., Rodríguez, R., Bañon, S., Franco, J.A., Fernandez, J.A, Salmerón, A. and Espi, E. (2003) Strawberry and cucumber cultivation under fluorescent photoselective plastic films cover. *Acta Horticulturae* 614, 407–414.

Gonzalez-Real, M.M., López, J.C., Cabrera, J., Baille, A. and Pérez-Parra, J. (2003) Variabilidad espacial de la rdiación solar en invernadero parral: medidas y modelo. *Actas II Congreso Nacional de Agro-Ingeniería*, Córdoba, pp. 239–240.

Gorny, J.R. and Zagory, D. (2003) Food safety. In: Gross, K.C., Yang, C.Y. and Saltveit, M. (eds) *The Commercial Storage of Fruits, Vegetables and Florist and Nursery Stocks*. USDA Agriculture Handbook no. 60. USDA, Washington, DC, pp. 133–149.

Gorris, L.G.M., Witte, Y.D. and Smid, E.J. (1994) Storage under moderate vacuum to prolong the keepability of fresh vegetables and fruits. *Acta Horticulturae* 368, 479–486.

Grange, R.I. and Hand, D.W. (1987) A review of the effects of atmospheric humidity on the growth of horticultural crops. *The Journal of Horticultural Science* 62, 125–134.

Gugumus, F.L. (2000) Greenhouse film stabilisation. In: Halim Hamid, S. (ed.) *Handbook of Polymer Degradation*. Marcel Dekker, New York, pp. 39–80.

Hall, D.O. and Rao, K.K. (1977) *Fotosíntesis.* Ediciones Omega, Barcelona.

Hanafi, A. (2003) Integrated production and protection today and in the future in greenhouse crops in the Mediterranean region. *Acta Horticulturae* 614, 755–765.

Hanan, J.J. (1998) *Greenhouses: Advanced Technology for Protected Cultivation.* CRC Press, Boca Raton, Florida.

Hand, D.W. (1990) CO_2 enrichment in greenhouses: problems of CO_2 acclimation and gaseosus air pollutants. *Acta Horticulturae* 268, 81–102.

Hart, J.W. (1988) Light and plant growth. In: Black, I.M. and Chapman, J. (eds) *Topics in Plant Physiology.* Unwin Hyman, London.

Hemming, S., Dueck, T., Janse, J. and Van Noort, F. (2008) The effect of diffuse light on crops. *Acta Horticulturae* 801, 1293–1300.

Hernández, J. (1996) El semiforzado del cultivo de col china (*Brassica pakinensis* (Lour) Rupr) mediante cubiertas flotantes. Caracterización microclimática y evaluación agronómica. Doctoral thesis, Universidad de Granada, Granada, Spain.

Hernández, J. (2002) Regulación del clima en invernaderos de multiplicación y producción. In: Ballester-Olmos, J.F. (ed.) *Nuevas Tecnologías en la Viverística de Plantas Ornamentales.* Universidad Politécnica de Valencia, Valencia, pp. 89–110.

Hernández, J. and Castilla, N. (2000) Los invernaderos mediterráneos en España: III. Paquetes tecnológicos disponibles. *Horticultura* 142, 37–39.

Hernández, J., Morales, M.I., Soriano, T., Escobar, I. and Castilla, N. (2001) Bean response to mulching in unheated plastic greenhouse. *Acta Horticulturae* 559, 79–84.

Heuvelink, E. and González-Real, M.M. (2008) Innovation in plant–greenhouse interactions and crop management. *Acta Horticulturae* 801, 63–74.

Hicklenton, P.R. (1988) *CO_2 Enrichment in the Greenhouse.* Timber Press, Portland, Oregon.

Hobson, G.E. (1988) How the tomato lost its taste. *New Scientist* 119, 46–50.

Howell, J.A. (1990) Relationships between crop production and transpiration, evapotranspiration and irrigation. In: Stewart, B.A. and Nielsen, D.R. (eds) *Irrigation of Agricultural Crops.* Agronomy monograph no. 30. American Society of Agronomy, Crop Science Society of America, Soil Science Society of America, Madison, Wisconsin, pp. 391–434.

Hsiao, T.C. (1973) Plant responses to water stress. *Annual Review of Plant Physiology* 24, 519–570.

Hsiao, T.C. and Xu, L.K. (2007) Predicting water use efficiency of crops. *Acta Horticulturae* 537, 199–206.

Huguet, J.G., Li, S.H., Lorendau, J.Y. and Pellous, G. (1992) Specific micromorphometric reactions of fruit trees to water stress and irrigation scheduling automotion. *The Journal of Horticultural Science* 67, 631–640.

Huijs, J.P.G. (1995) Supplementary lighting. In: Bakker, J.C., Bot, G.P.A., Challa, H. and Van de Braak, N.J. (eds) *Greenhouse Climate Control: an Integrated Approach.* Wageningen Pers, Wageningen, The Netherlands, pp. 202–205.

Idso, S.B., Jackson, R.D., Pinter, P.J. and Hatfield, J.L. (1981) Normalizing the stress-degree-day parameter for environmental variability. *Agricultural Meteorology* 24, 45–55.

Ito, T. (1999) The greenhouse and hydroponic industries in Japan. *Acta Horticulturae* 481, 761–764.

Jackson, R.D. (1982) Canopy temperature and crop water stress. *Advances in Irrigation* 1, 43–85.

Jaffe, M.J. (1976) Thigmomorphogenesis: a detailed characterization of the response of beans (*Phaseolus vulgaris*, L.) to mechanical stimulation. *Zeitschrift fur Pflanzenphysiologie* 77, 437–453.

Jaffrin, A. and Urban, L. (1990) Optimization of light transmission in modern greenhouse. *Acta Horticulturae* 281, 25–33.

Janick, J. (1986) *Horticultural Science.* W.H. Freeman and Co., New York.

Janse, J. (1984) Invloed van licht op de kwaliteit van tomaat en komkommer [Effects of light on quality of tomato and cucumber]. *Groenten en Fruit* 40(18), 28–31.

Jarvis, W.R. (1997) *Control de Enfermedades en Cultivos de Invernadero.* Ediciones Mundi-Prensa, Madrid.

Jensen, M.H. and Malter, A.J. (1995) *Protected Horticulture: a Global Review.* World Bank, Washington, DC.

Jolliet, O. (1999) The water cycle. In: Stanhill, G. and Enoch, H.Z. (eds) *Greenhouse Ecosystems.* Elsevier, Amsterdam, pp. 303–326.

Jones, H.G. (1983) *Plants and Microclimate.* Cambridge University Press, Cambridge, UK.

Jouet, J.P. (2004) La situación de la plasticultura en el mundo. *Plasticulture* II-5(123), 48–57.

Kader, A.A. (1996) Maturity, ripening and quality relationships of fruit-vegetables. *Acta Horticulturae* 434, 249–255.
Kader, A.A. (2000) Quality of horticultural products. *Acta Horticulturae* 518, 15–16.
Kader, A.A. (2003) Controlled atmosphere storage. In: Gross, K.C., Yang, C.Y. and Saltveit, M. (eds) *The Commercial Storage of Fruits, Vegetables and Florist and Nursery Stocks.* USDA Agriculture Handbook no. 60. USDA, Washington, DC, pp. 32–34.
Kader, A.A., Zagory, D. and Kerbel, E.D. (1989) Modified atmosphere packaging of fruits and vegetables. *Critical Reviews in Food Science and Nutrition* 28, 1–30.
Kamp, P.G.H. and Timmerman, G.J. (1996) *Computerized Environmental Control in Greenhouses.* IPC-Plant, Ede, The Netherlands.
Kan, Y., Chang, Y.C.A., Choi, H.S. and Gu, M. (2012) Current and future status of protected cultivation techniques in Asia. *International Society for Horticultural Science (ISHS) International Symposium on High Tunnel Horticultural Crop Production*, Pennsylvania, USA, October 2011. *Acta Horticulturae* (in press). ISHA, Leuven, Belgium.
Katan, J. (1981) Solar heating (solarization) of soil for control of soilborne pests. *Annual Review of Phytopathology* 19, 211–236.
Kempes, F. (2003) Sistemas cerrados en cultivos sin suelo: elementos y técnicas. In: Fernández, M., Lorenzo, P. and Cuadrado, I.M. (eds) *Mejora de la Eficiencia del Uso del Agua en Cultivos Protegidos.* Dirección General de Investigación y Formación Agraria, Hortimed, FIAPA, Cajamar, Almería, Spain, pp. 55–69.
Kimball, B.A. (1986) Influence of elevated CO_2 on crop yield. In: Enoch, H.Z. and Kimball, B.A. (eds) *Carbon Dioxide Enrichment of Greenhouse Crops*, Volume II. CRC Press, Boca Raton, Florida, pp. 105–115.
Kittas, C. and Baille, A. (1998) Determination of the spectral properties of several greenhouse cover materials and evaluation of specific parameters related to plant response. *Journal of Agricultural Engineering Research* 71, 193–202.
Kittas, C., Baille, A. and Giaclaras, P. (1999) Influence of covering material and shading on the spectral distribution of light in greenhouses. *Journal of Agricultural Engineering Research* 73, 341–351.
Kozai, T. and Sase, S. (1978) A simulation of natural ventilation for a multi-span greenhouse. *Acta Horticulturae* 87, 39–40.
Kramer, P.J. (1983) *Water Relations of Plants.* Academic Press, San Diego, California.
Kurata, K. (1992) Two-dimensional analysis of irradiance distribution at canopy foliage in relation to the diffusivity of films of plastic houses. *Acta Horticulturae* 303, 113–120.
La Malfa, G. and Leonardi, C. (2001) Crop practices and techniques: trends and needs. *Acta Horticulturae* 559, 31–42.
Lambers, H., Chapin, F.S. and Pons, T.L. (1998) *Plant Physiological Ecology.* Springer-Verlag, New York.
Langhams, R.W. (1990) *Greenhouse Management.* Alcion Press, Ithaca, New York.
Langhams, R.W. and Tibbitts, T.W. (1997) *Plant Growth Chamber Handbook.* Iowa State University Press, Ames, Iowa.
Langreo, A. (2002) Nuevas formas de distribución de alimentos. In: Alvarez-Coque, J.M.G. (ed.) *La Agricultura Mediterránea en el Siglo XXI.* Instituto Cajamar, Almería, Spain, pp. 103–123.
Levanon, D., Motro, B. and Marchain, U. (1986) Organic materials degradation for CO_2 enrichment of greenhouse crops. In: Enoch, H.Z. and Kimball, B.A. (eds) *Carbon Dioxide Enrichment of Greenhouse Crops*, Volume II. CRC Press, Boca Raton, Florida, pp. 123–145.
Li, Y.L., Stanghellini, C. and Challa, H. (2001) Effect of electrical conductivity and transpiration on production of greenhouse tomato (*Licopersicum ascullentum* L.). *Scientia Horticulturae* 88, 11–29.
Lioutas, T.S. (1988) Challenges of controlled and modified atmosphere pockaging: a food company's perspective. *Food Technology* 42(9), 78–86.
López, J.C. (2003) Sistemas de calefacción en invernaderos cultivados de judía en el litoral Mediterráneo. Doctoral thesis, Universidad de Almería, Almería, Spain.
López, J.C., Lorenzo, P., Medrano, E, Sánchez-Guerrero, M.C., Pérez, J., Puerto, H.M. and Arco, M. (2000) *Calefacción de Invernaderos en el Sudeste Español.* Caja Rural de Almería, Almería, Spain.
López, J.C., Pérez, J., Montero, J.I. and Antón, A. (2001) Air infiltration rate of Almería type greenhouse. *Acta Horticulturae* 559, 229–231.

López, J.C., Baille, A., Bonachela, S. and Pérez-Parra, J. (2003a) Effects of heating strategies on earliness and yield of snap beans (*Phaseolus vulgaris* L.) grown under 'parral' plastic greenhouses. *Acta Horticulturae* 614, 439–444.

López, J.C., Pérez, C., Pérez-Parra, J. and Cabrera, F.J. (2003b) Evaluación de dos sistemas de ahorro de energía para un cultivo de pepino en invernadero parral. In: *Actas del X Congreso Nacional de la Sociedad Española Ciencias Hortícolas.* Nacional de la Sociedad Española Ciencias Hortícolas, Pontevedra, Spain, pp. 392–394.

Lorenzo, P. and Castilla, N. (1995) Pepper yield response to plant density and radiation in unheated plastic greenhouse. *Acta Horticulturae* 412, 330–334.

Lorenzo, P., Sánchez-Guerrero, M.C., Medrano, E., Escobar, I. and García, M. (1997a) Evaluación de la incorporación de sistemas de calefacción en la horticultura intensiva baja cubierta de plástico en el sur mediterráneo. *Actas de Horticultura* 17, 371–378.

Lorenzo, P., Sánchez-Guerrero, M.C., Medrano, E., Escobar, I. and García, M. (1997b) Horticultura intensiva en el sur mediterráneo: gestión del clima. *Horticultura* 119, 80–83.

Lorenzo, P., Sánchez-Guerrero, M.C., Medrano, E., Pérez, J. and Maroto, C. (1997c) El enriquecimiento carbónico en invernadero del sur mediterráneo. *Horticultura* 118, 66–67.

Lorenzo, P., Sánchez-Guerrero, M.C., Medrano, E., Aguilar, F.J., Pérez, J. and Castilla, N. (1999) Soilless cucumber response to mulching in unheated Mediterranean greenhouse. *Acta Horticulturae* 491, 401–403.

Lorenzo, P., Medrano, E., Sánchez-Guerrero, M.C., Castilla, N. and Pérez, J. (2001) Cucumber growth and yield as affected by mulching in soilless culture in unheated greenhouse. *Acta Horticulturae* 559, 107–112.

Lorenzo, P., Sánchez-Guerrero, M.C., Medrano, E., García, M.L., Caparrós, I. and Jiménez, M. (2003) El sombreado móvil exterior: efecto sobre el clima del invernadero, la producción y l a eficiencia del uso del agua y la radiación. In: Fernández, M., Lorenzo, P. and Cuadrado, I.M. (eds) *Mejora de la Eficiencia del Uso del Agua en Cultivos Protegidos.* Dirección General de Investigación y Formación Agraria, Hortimed, FIAPA, Cajamar, Almería, Spain, pp. 207–230.

Lorenzo, P., Sánchez-Guerrero, M.C., Medrano, E., Soriano, T. and Castilla, N. (2005) Responses of cucumber to mulching in an unheated plastic greenhouse. *The Journal of Horticultural Science and Biotechnology* 80(1), 11–17.

Louvet, J. (1984) Effets des facteurs climatiques sur les maladies en culture legumiere. In: *Agrometeorologie et Produtions Legumieres.* Institut National de la Recherche Agronomique (INRA), Paris, pp. 183–197.

Loveless, A.R. (1983) *Principles of Plant Biology for the Tropics.* Longman Group, Harlow, UK.

Lurie, S. and Klein, J.D. (2003) Temperature preconditioning. In: Gross, K.C., Yang, C.Y. and Saltveit, M. (eds) *The Commercial Storage of Fruits, Vegetables and Florist and Nursery Stocks.* USDA Agriculture Handbook no. 60. USDA, Washington, DC, pp. 35–44.

Magán, J.J. (2003) Efectos de la salinidad sobre el tomate en cultivo en sustrato en las condiciones del Sureste peninsular: resultados experimentales. In: Fernández, M., Lorenzo, P. and Cuadrado, I.M. (eds) *Mejora de la Eficiencia del Uso del Agua en Cultivos Protegidos.* Dirección General de Investigación y Formación Agraria, Hortimed, FIAPA, Cajamar, Almería, Spain, pp. 169–187.

Magán, J.J., López, J.C., Granados, T., Pérez-Parra, J., Soriano, T., Romero-Gámez, M. and Castilla, N. (2011) Global radiation differences under a glasshouse and a plastic greenhouse in Almeria (Spain): preliminary report. *Acta Horticulturae* 907, 125–130.

Maher, M.J. (1976) Growth and nutrient content of a glasshouse tomato crop grown in peat. *Scientia Horticulturae* 4, 23–26.

Marcelis, L.F.M. (1989) Simulation of plant–water relations and photosynthesis of greenhouse crops. *Scientia Horticulturae* 41, 9–18.

Marcelis, L.F.M. and De Koning, A.N.M. (1995) Biomass partitioning in plants. In: Bakker, J.C., Bot, G.P.A., Challa, H. and Van de Braak, N.J. (eds) *Greenhouse Climate Control: an Integrated Approach.* Wageningen Pers, Wageningen, The Netherlands, pp. 84–92.

Marco, I. (2001) Los plásticos como cubiertas de invernaderos y túneles. *Plasticulture* 119, 14–25.

Marfá, O. (2000) *Recirculación en Cultivos Sin Suelo.* Ediciones de Horticultura, Reus, Spain.

Martínez-Carrasco, F. (2001) El sistema de comercialización de la horticultura intensiva almeriense: un análisis del comportamiento de precios y márgenes. Doctoral thesis, Universidad de Almería, Almería, Spain.

Martínez-Carrasco, F. and Calatrava, J. (1996) La structure de commercislisation en amont des produits horticoles d'Almería: analyse d'une enquete aux horticulteurs. *Mediterranean Colloquium on Protected Cultivation*, Agadir, Morocco.
Martínez-Madrid, M.C., Flores, F. and Romojara, F. (2000) Factores que determinan la calidad de la producción hortofrutícola. In: Marrero, A. and Lobo, H.G. (eds) *Actas V Simposio Nacional y II Ibérico de Post-Recolección de Frutos y Hortalizas*. Instituto Canario de Investigaciones Agrarias, La Laguna, Tenerife, pp. 219–224.
Martínez-Paz, J. and Calatrava-Requena, J. (2001) Economic evaluation of water in protected horticulture in the Almería area. *Acta Horticulturae* 559, 731–735.
Martínez-Raya, A. and Castilla, N. (1993) Pepper growth and yield in an unheated plastic greenhouse. *Agricoltura Mediterranea* 123, 43–46.
Mastalerz, J.W. (1977) *The Greenhouse Environment*. John Wiley & Sons, New York.
Matallana, A. and Montero, J.I. (1989) *Invernaderos: Diseño, Construcción y Ambientación*. Ediciones Mundi-Prensa, Madrid.
Mauromicale, G., Cosentino, S. and Copan, V. (1988) Validity oj termal unit summations for purposes of prediction in *Phaseolus vulgaris* L. cropped in Mediterranean environments. *Acta Horticulturae* 229, 321–331.
McCree, K.J. (1972) Test of current definitions of photosynthetically radiation ageinst leaf photosynthesis data. *Agricultural Meteorology* 10, 443–453.
Meca, D., López, J.C., Gázquez, J.C. and Pérez –Parra, J. (2003) Evaluación de dos pantallas de ahorro de energía para un cultivo de pepino en invernadero. *Actas II Congreso Nacional de Agro-Ingeniería*, Córdoba, pp. 251–252.
Medrano, E. (1999) Gestión del riego en cultivo de pepino '*Cucumis sativus* L.' en sustrato: evaluación de la transpiración durante la ontogenia. Doctoral thesis, Universidad Politécnica de Madrid, Madrid, Spain.
Medrano, E., Lorenzo, P. and Sánchez-Guerrero, M.C. (2003) Gestión de riego en sustratos. In: Fernández, M., Lorenzo, P. and Cuadrado, I.M. (eds) *Mejora de la Eficiencia del Uso del Agua en Cultivos Protegidos*. Dirección General de Investigación y Formación Agraria, Hortimed, FIAPA, Cajamar, Almería, Spain, pp. 321–342.
Meneses, J.F. and Castilla, N. (2009) Protected cultivation in Iberian horticulture. *Chronica Horticulturae* 49(4), 37–39.
Messiaen, C.M., Blancard, D., Rouxel, F. and Lafon, R. (1995) *Enfermedades de las Hortalizas*. Institut National de la Recherche Agronomique (INRA) Ediciones Mundi-Prensa, Madrid.
Mir, N. and Beaudry, R.M. (2003) Modified atmosphere packaging. In: Gross, K.C., Yang, C.Y. and Saltveit, M. (eds) *The Commercial Storage of Fruits, Vegetables and Florist and Nursery Stocks*. USDA Agriculture Handbook no. 60. USDA, Washington, DC, pp. 45–54.
Moe, R. (ed.) (2006) *Proceedings of the Fifth International Symposium on Artificial Lighting. Acta Horticulturae* 711. International Society for Horticultural Science, Leuven, Belgium.
Moe, R., Fjeld, T. and Mortensen, L.M. (1992) Stem elongation and keeping quality in poinsettia as affected by temperature and supplementary lighting. *Scientia Horticulturae* 50, 127–136.
Mohr, H. (1984) Criteria for photoreceptor involvement. In. Smith, H. and Holmes, M.G. (eds) *Techniques in Photomorphogenesis*. Academic Press, London, pp. 13–14.
Moller, M., Cohen, S., Pirkner, M., Israeli, Y. and Tanny, J. (2010) Trnasmission of short-wave radiation by agricultural screens. *Biosystems Engineering* 107, 317–327.
Monci, F., Garcia-Andres, S., Sánchez, F., Morcones, E., Espi, E. and Salmerón, A. (2004) Tomato yellow leaf curl disease control with UV-blocking plastic covers in comercial plastichouses of Southern Spain. *Acta Horticulturae* 633, 537–542.
Montalvo, T. (1998) Cabezal de riego. In: Cadahía, C. (ed.) *Fertirrigación: Cultivos Hortícolas y Ornamentales*. Ediciones Mundi-Prensa, Madrid, pp. 247–263.
Monteiro, A. (1990) Greenhouses for mild-winter climates: goals and restraints. *Acta Horticulturae* 263, 21–32.
Monteiro, A. and Portas, C.M. (1986) Mild winter concept and cropping systems in solanacea protected cultivation. *Acta Horticulturae* 191, 21–34.
Monteith, J.L. and Unsworth, M.H. (1990) *Principles of Environmental Physics*. Edward Arnold, London.
Montero, J.I. and Antón, A. (2000a) Buoyancy driven ventilation in tropical greenhouses. *Acta Horticulturae* 534, 41–55.

Montero, J.I. and Antón, A. (2000b) Control climático del invernadero. *Horticultura* Special issue, 50–56.

Montero, J.I., Castilla, N., Gutiérrez, E. and Bretones, F. (1985) Climate under plastic in the Almería area. *Acta Horticulturae* 170, 227–234.

Montero, J.I., Antón, A. and Muñoz, P. (1998) Fundamentos. In: Pérez-Parra, J. and Cuadrado, I. (eds) *Tecnología de Invernaderos*. Consejería de Agricultura y Pesca, FIAPA, Caja Rural de Almería, Almería, Spain, pp. 253–266.

Montero, J.I., Antón, A., Hernández, J. and Castilla, N. (2001) Direct and diffuse light transmission of insect proof screens and plastic films for cladding greenhouses. *Acta Horticulturae* 559, 203–209.

Montero, J.I., Antón, A. and Muñoz, P. (2003) Nebulización: efectos sobre el microclima, producción y eficiencia en el uso del agua. In: Fernández, M., Lorenzo, P. and Cuadrado, I.M. (eds) *Mejora de la Eficiencia del Uso del Agua en Cultivos Protegidos*. Dirección General de Investigación y Formación Agraria, Hortimed, FIAPA, Cajamar, Almería, Spain, pp. 231–243.

Montero, J.I., Stanghellini, C. and Castilla, N. (2009) Greenhouse technology for sustainable production in mild winter climate: trends and needs. *Acta Horticulturae* 807, 33–44.

Montero, J.I., Son, J.E., Van Henten, E.J. and Castilla, N. (2010) Greenhouse engineering: new technologies and approaches. *Acta Horticulturae* 893, 51–63.

Montsi, M. and Saeki, T. (1953) Uber den lichfaktor in den pflanzengesellshaften und seine bedeutung für die stoffproduktion. *Japanese Journal of Botany* 14, 22–52.

Morales, M.I., Hernández, J., Castilla, N., Escobar, I. and Berenguer, J.J. (1998) Transmisividad de radiación solar en invernaderos de la costa mediterránea española. *Actas de Horticultura* 21, 33–36.

Morales, M.I., Hernández, J., Soriano, M.T., Martinez, F.M., Escobar, I., Berenguer, J.J. and Castilla, N. (2000) Optimización de la radiación en invernaderos mediterráneos para aumentar la calidad de las producciones. VII Jornadas del grupo de Horticultura de la SECH. *Actas de Horticultura* 21, 123–126.

Morard, P. (1984) Cultures diverses; tomate, concombre. In: Martín-Prevel, P., Gagnard, J. and Gauthier, P. (eds) *L'Analyse Vegetal dans le Controle de l'Alimentation des Plantes Temperées et Tropicales*. Lavoisier, Paris, pp. 760–771.

Morard, P. (1995) *Les Cultures Vegetales Hors Sol*. Publications Agricoles d'Agen, Paris, France.

Morard, P., Roncoble, A. and Merelle, F. (1991) L'analyse reguliere de tissus conducteurs: une nouvelle methode au service des producteurs. *PHM Revue Horticole* 314, 36–38.

Moreno, R. (ed.) (1994) *Sanidad Vegetal en la Horticultura Protegida*. Dirección General de Investigación Agraria, Junta de Andalucía, Seville.

Mortensen, L.M. and Roe, R. (1992) Effects of selective screening of the day light spectrum and of twilight on plant growth in greenhouses. *Acta Horticulturae* 305, 103–108.

Muñoz, P. (1998) Ventilación natural en invernaderos multitúnel. Doctoral thesis, Universidad de Lleida, Lleida, Catalonia, Spain.

Muñoz, P., Antón, A. and Montero, J.I. (1998) Estructuras de invernaderos: tipología y materiales. In: *Tecnología de Invernaderos*. Junta de Andalucía, Almería, España, pp. 65–99.

Muñoz, P., Montero, J.I., Antón, A. and Giuffida, F. (1999) Effects of insect-proof screens and roof openings on greenhouse ventilation. *Journal of Agricultural Engineering Research* 73, 171–178.

Muñoz, P., Antón, A., Nuñez, M., Paranjpe, A., Ariño, J., Castell, X., Montero, J.I. and Rieradevall, J. (2008) Comparing the environmental impacts of greenhouse versus open-field tomato production in the Mediterranean region. *Acta Horticulturae* 801, 1591–1596.

Nederhoff, E.M. (1984) Effects of CO_2 concentrations on photosynthesis, transpiration and production of greenhouse vegetable crops. PhD dissertation, Wageningen Agricultural University, Wageningen, The Netherlands.

Nederhoff, E.M. (1995) Carbon dioxide balance. In: Bakker, J.C., Bot, G.P.A., Challa, H. and Van de Braak, N.J. (eds) *Greenhouse Climate Control: an Integrated Approach*. Wageningen Pers, Wageningen, The Netherlands, pp. 151–155.

Nelson, P.V. (1985) *Greenhouse Operation and Management*. Prentice Hall, Upper Saddle River, New Jersey.

Nisen, A. and Deltour, J. (1986) *Considerations Pratiques sur la Transmisión du Rayonment Solaire et de la Chaleur pour le Materiaux Utilices en Serre comme Couverture, Ombrage et Ecran Thermique*. Institute pour la Recherche Scientifique dans l'Industrie et l'Agriculture (IRSIA), Brussels.

Nisen, A., Grafiadellis, M., Jiménez, R., La Malfa, G., Martínez-García, P.F., Monteiro, A., Verlodt, H., Villele, O., Zabeltitz, C.H., Denis, J.C., Baudoin, W. and Garnaud, J.C. (1988) *Cultures Protegees en Climat Mediterraneen*. FAO, Rome.

Nobel, P.S. (1974a) Boundary layers of air adjacent to cilindres: estimation of effective thickness and measurements on plant material. *Plant Physiology* 54, 177–181.

Nobel, P.S. (1974b) Water. In: Kennedy, D. and Park, R.B. (eds) *Introduction to Biophysical Plant Physiology*. W.H. Freeman and Company, San Francisco, California, pp. 44–91.

Nuez, F. (1986) Solanaceae breeding for protected cultivation. *Acta Horticulturae* 191, 317–330.

Oren-Shamir, M., Gussakovsky, E.E., Shpiegel, E., Nissim-Levi, A., Ratner, K., Ovadia, R., Giller, Y.U. and Shahak, Y. (2001) Coloured shade nets can improve the yield and quality of green decorative branches of *Pittosporum variegatum*. *The Journal of Horticultural Science and Biotechnology* 76(3), 353–361.

Papadakis, G., Mermier, M., Meneses, J.F. and Boulard, T. (1996) Measurements and analysis of air exchange rates in a greenhouse with continuous roof and side openings. *Journal of Agricultural Engineering Research* 63, 219–328.

Papadakis, G., Briassoulis, D., Mugnozza, G.S., Vox, G., Feuilloley, P. and Stoffers, J.A. (2000) Radiometric and thermal properties of, and testing methods for, greenhouse covering materials. *Journal of Agricultural Engineering Research* 77(1), 7–38.

Papadopoulos, A.T. and Hao, X. (1997a) Effects of greenhouse covers on seedless cucumber growth, productivity and energy use. *Scientia Horticulturae* 68, 113–123.

Papadopoulos, A.T. and Hao, X. (1997b) Effects of three greenhouse cover materials on tomato growth, productivity and energy use. *Scientia Horticulturae* 70, 165–178.

Papadopoulos, A.T. and Ormrod, D.P. (1988) Plant spacing effects on photosynthesis and transpiration of the greenhouse tomato. *Canadian Journal of Plant Science* 68, 1209–1218.

Papadopoulos, A.T. and Ormrod. D.P. (1991) Plant spacing effects on growth and development of the greenhouse tomato. *Canadian Journal of Plant Science* 71, 297–304.

Papadopoulos, A.P. and Pararajasinghma, S. (1997) The influence of plant spacing on light interception and use in greenhouse tomato (*Lycopersicon esculentum* Mill.): a review. *Scientia Horticulturae* 69, 1–29.

Pardossi, A. (2003) El manejo de la nutrición mineral en los cultivos sin suelo. In: Fernández, M., Lorenzo, P. and Cuadrado, I.M. (eds) *Mejora de la Eficiencia del Uso del Agua en Cultivos Protegidos*. Dirección General de Investigación y Formación Agraria, Hortimed, FIAPA, Cajamar, Almería, Spain, pp. 109–129.

Parra, S., Pérez, J.J. and Calatrava, J. (2001) Vegetal waste from protected horticulture in south-eastern Spain: characterisation of environmental externalities. *Acta Horticulturae* 559, 787–792.

Parrella, M.P. (1999) Arthropod fauna. In: Stanhill, G. and Enoch, H.Z. (eds) *Greenhouse Ecosystems*. Elsevier, Amsterdam, pp. 213–250.

Patil, B., Murano, P. and Amiot-Carlin, M.J. (eds) (2009) *Proceedings of the II International Symposium on Human Healath Effects of Fruits and Vegetables: FAVHEALTH-2007. Acta Horticulturae* 841. International Society for Horticultural Science, Leuven, Belgium.

Pearson, S., Wheldon, A.E. and Hadley, P. (1995) Radiation transmission and fluorescence of nine greenhouse cladding materials. *Journal Agricultural Engineering Research* 62, 61–70.

Penningsfeld, F. and Kurzmann, P. (1983) *Cultivos Hidropónicos y en Turba*. Ediciones Mundi-Prensa, Madrid.

Pérez, J., López, J.C. and Fernández, M.D. (2002) La agricultura del Sureste: situación actual y tendencias de las estructuras de producción en la horticultura almeriense. In: Alvarez-Coque, J.M.G. (ed.) *La Agricultura Mediterránea en el Siglo XXI*. Instituto Cajamar, Almería, Spain, pp. 262–282.

Pérez-Parra, J. (2002) Ventilación natural en invernadero tipo parral. Doctoral thesis, Universidad de Córdoba, Córdoba, Andalusia, Spain.

Pérez-Parra, J., Baeza, E.J., Montero, J.I. and López, J.C. (2003a) Determinación de modelos simples para la estimación del coeficiente de descarga de ventanas de invernadero con y sin malla anti-insectos. *Actas II Congreso Nacional de Agro-Ingeniería*, Córdoba, pp. 257–258.

Pérez-Parra, J., Baeza, E., Pérez, C., López, J.C. and Montero, J.I. (2003b) Influencia de las ventanas laterales sobre la ventilación natural en invernadero tipo parral. *Actas X Congreso Nacional de Ciencias Hortícolas*. Pontevedra, pp. 419–421.

Pérez-Parra, J., Gázquez, J.C., Sánchez, A. and López, J.C. (2003c) Evaluación de dos sistemas de sombreo de invernadero (malla móvil y blanqueo tradicional) en cultivo de pepino en ciclo de primavera. *Actas II Congreso Nacional de Agro-Ingeniería*, Córdoba, pp. 259–260.

Planells, J.M. and Mir, J. (2000) Comercialización: orientación de la producción al mercado. Situación actual de la distribución comercial europea. In: Fernández, M. and Cuadrado, I.M. (eds) *Comercialización de Productos Hortofrutícolas II*. Dirección General de Investigación y Formación Agraria de la Junta de Andalucía, FIAPA, Caja Rural, Almería, Spain, pp. 377–392.

Planells, J.M. and Mir, J. (2002) La agroexportación ante la nueva distribución alimentaria. In: Alvarez-Coque, J.M.G. (ed.) *La Agricultura Mediterránea en el Siglo XXI*. Instituto Cajamar, Almería, Spain, pp. 125–139.

Pollet, I.V., Pieters, J.G., Deltour, J. and Verschoore, R. (2005) Diffusion of radiation transmitted through dry and condensate covered transmitting materials. *Solar Energy Material and Solar Cells* 86, 177–196.

Raviv, M. (1988) The use of photoselective cladding materials as modifiers of morphogenesis of plants and pathogens. *Acta Horticulturae* 246, 275–284.

Raviv, M. and Lieth, J.H. (eds) (2008) *Soilless Culture: Theory and Practice*. Elsevier, San Diego, California.

Raviv, M., Medina, S., Shamir, Y. and Ner, B.Z. (1993) Very low medium moisture tension: a feasible criterion for irrigation control of container-grown plants. *Acta Horticulturae* 342, 111–119.

Rivera, L.M. (2000) La gestión de la calidad en el sector agroalimentarío. In: Fernández, M. and Cuadrado, I.M. (eds) *Comercialización de Productos Hortofrutícolas II*. Dirección General de Investigación y Formación Agraria de la Junta de Andalucía, FIAPA, Caja Rural, Almería, Spain, pp. 169–182.

Robledo, F. and Martín, L. (1981) *Aplicación de los Plásticos en la Agricultura*. Ediciones Mundi-Prensa, Madrid.

Rodríguez, M.D. (1994) Aleuródidos. In: Moreno, R. (ed.) *Sanidad Vegetal en la Horticultura Protegida*. Dirección General de Investigación Agraria, Junta de Andalucía, Seville, pp. 123–153.

Rose, C.N. (1979) *Agricultural Physics*. Pergamon Press, Oxford, UK.

Rosenberg, N.J., Blad, B.L. and Verna, S.B. (1983) *Microclimate: the Biological Environment*. John Wiley & Sons, New York.

Russel, R.S. (1977) *Plant Root Systems: their Function and Interaction with the Soil*. McGraw Hill, London.

Russell, G., Jarvis, P.G. and Montheith, J.L. (1989) Absorption of radiation by canopies and stand growth. In: Russell, G., Marshall, B. and Jarvis, P.G. (eds) *Plant Canopies: their Growth, Form and Function*. Cambridge University Press, Cambridge, UK, pp. 21–39.

Salisbury, F.B. (1985) Photoperiodism. *Horticultural Reviews* 4, 66–105.

Salmerón, A., Espi, E., Fontecha, A. and Garcia-Alonso, Y. (2001) Filmes agrícolas avanzados: un campo abierto. *Actas I Simposio Internacional de Plasticultura*. Valencia, pp. 58–65.

Saltveit, M.E. (2003a) Respiratory metabolism. In: Gross, K.C., Yang, C.Y. and Saltveit, M. (eds) *The Commercial Storage of Fruits, Vegetables and Florist and Nursery Stocks*. USDA Agriculture Handbook no. 60. USDA, Washington, DC, pp. 66–73.

Saltveit, M.E. (2003b) Ethylene effects. In: Gross, K.C., Yang, C.Y. and Saltveit, M. (eds) *The Commercial Storage of Fruits, Vegetables and Florist and Nursery Stocks*. USDA Agriculture Handbook no. 60. USDA, Washington, DC, pp. 74–80.

Sánchez-Guerrero, M.C., Portero, F., Medrano, E. and Lorenzo, P. (1998) Efecto del enriquecimiento carbónico sobre la producción y eficiencia hídrica en cultivo de pepino. *Actas de Horticultura* 21, 83–90.

Sánchez-Guerrero, M.C., Lorenzo, P., Medrano, E., García, M. and Escobar, I. (2001) Heating and CO_2 enrichment in inproved low-cost greenhouses. *Acta Horticulturae* 559, 257–262.

Sánchez-Guerrero, M.C., Lorenzo, P. and Medrano, F. (2003) Enriquecimiento carbónico: efecto sobre la producción y la eficiencia en el uso del agua. In: Fernández, M., Lorenzo, P. and Cuadrado, I.M. (eds) *Mejora de la Eficiencia del Uso del Agua en Cultivos Protegidos*. Dirección General de Investigación y Formación Agraria, Hortimed, FIAPA, Cajamar, Almería, Spain, pp. 245–258.

Sánchez-Guerrero, M.C., Lorenzo, P., Medrano, E., Castilla, N., Soriano, T. and Baille, A. (2005) Effect of variable CO_2 enrichment on greenhouse production in mild winter climates. *Agriculture and Forest Meteorology* 132, 244–252.

Sánchez-Guerrero, M.C., Lorenzo, P., Medrano, E., Baille, A. and Castilla, N. (2008) Effects of EC-based irrigation scheduling and CO_2 enrichment on water use efficiency of a greenhouse tomato crop. *Agricultural Water Management* 96(3), 429–436.

Santos, B., Rios, D. and Nazco, R. (2006) Climatic conditions in tomato screenhouses in Tenerife (Canary Islands). *Acta Horticulturae* 719, 215–221.

Savvas, D. and Passam, H. (2002) *Hydroponic Production of Vegetables and Ornamentals.* Embryo Publications, Athens, Greece.

Scharz, A. (1994) Relative humidity in cool stores: measurement, control and influence of discreet factors. *Acta Horticulturae* 368, 687–692.

Scheer, A. (1994) Reducing the water loss of horticultural and arable products during long term storage. *Acta Horticulturae* 368, 511–522.

Schnitzler, W.H. (2003) Impact assessment of horticulture on the environment. FAO Seminar of the Regional Working Group on Greenhouse Production in the Mediterranean Region. Nicosia, Cyprus.

Schnitzler, W.H. and Gruda, N. (2003) Quality issues of greenhouse production. *Acta Horticulturae* 614: 663–674.

Schnitzler, W.H., Woittke, M., Leonardi, C., Giuffrida, F., Tüzel, Y. and Tüzel, H.I. (2007) Ecoponics: efficient water use through environmentally sound hydrponic production of high quality vegetables for domestic and export markets in Mediterranean countries. Paper presented at the International Society for Horticultural Science (ISHS) International Symposium on Advances in Soil and Soilless Cultivation under Protected Environment. Agadir, Morocco (unpublished).

Scoullos, M.J. (2003) La gestión del agua dulce en el Mediterráneo. In: García-Orcoyen, C. (ed.) *Mediterráneo y Medio Ambiente.* Instituto Cajamar, Almería, Spain, pp. 157–178.

Seeman, U. (1974) *Climate Under Glass.* Technical Note no. 131. World Meteorological Organization, Geneva.

Segundo, E., Martín, G., Cuadrado, M.I. and Janssen, D. (2004) A new yellowing disease in *Phaseolus vulgaris* associated with a whitefly-transmitted virus. *Plant Pathology* 53, 517.

Segura, M.L., Parra, J.F., Lorenzo, P., Sánchez-Guerrero, M.C. and Medrano, E. (2001) The effects of CO_2 enrichment on cucumber growth under greenhouse conditions. *Acta Horticulturae* 559, 217–222.

Seymour, G.B., Taylor, J.A. and Tucker, G.A. (1993) *Biochemistry of Fruit Ripening.* Chapman and Hall, London.

Shahak, Y., Ratner, K., Zur, N., Offir, Y., Matan, E., Yehezkel, H., Messika, Y., Posalski, I. and Ben-Yakir, D. (2009) Photoselective netting: an emerging approach in protected agriculture. *Acta Horticulturae* 807, 79–84.

Sholberg, P.L. and Conway, W.S. (2003) Postharvest pathology. In: Gross, K.C., Yang, C.Y. and Saltveit, M. (eds) *The Commercial Storage of Fruits, Vegetables and Florist and Nursery Stocks.* USDA Agriculture Handbook no. 60. USDA, Washington, DC, pp. 100–113.

Short, T.H. and Shoh, S.A. (1981) A portable polystirene-pellet insulation system for greenhouses. *Transactions of ASAE* 24, 1291–1295.

Sica, C. and Picudo, P. (2008) Spectro-radiometrical characterization of plastic nets for protected cultivation. In: *Proceedings of the International Symposium on High Technology for Greenhouse System Management: Greensys-2007. Acta Horticulturae* 801. International Society for Horticultural Science, Leuven, Belgium, pp. 245–252.

Sierra, L.M.F. (2003) Producción controlada de cultivos protegidos: la certificación AENOR. In: Camacho, F. (ed.) *Técnicas de Producción en Cultivos Protegidos.* Cajamar, Almería, Spain, pp. 245–277.

Sigrimis, N., Arvanitis, K., Pasgianos, G. and Pitsilis, J. (2003) Manejo de sistemas hidropónicos. In: Fernández, M., Lorenzo, P. and Cuadrado, I.M. (eds) *Mejora de la Eficiencia del Uso del Agua en Cultivos Protegidos.* Dirección General de Investigación y Formación Agraria, Hortimed, FIAPA, Cajamar, Almería, Spain, pp. 259–283.

Sirjacobs, M. (1988) Agro-climatological criteria for selecting the most appropiate areas for protected cultivation in Egypt. In: *Protected Cultivation in the Mediterranean Climate. Greenhouses in Egypt.* FAO, Rome, Italy, pp. 5–12.

Slack, G. and Hand, D.W. (1983) The effect of day and night temperatures on growth, development and yield of glasshouse cucumbers. *The Journal of Horticultural Science* 58, 567–572.

Smith, D.L. (1987) *Rockwool in Horticulture.* Grower Books, London.

Sonneveld, C. (1988) The salt tolerance of greenhouse crops. *Netherlands Journal Agricultural Science* 36, 63–73.
Sonneveld, C. (2003) Efectos de la salinidad en los cultivos sin suelo. In: Fernández, M., Lorenzo, P. and Cuadrado, I.M. (eds) *Mejora de la Eficiencia del Uso del Agua en Cultivos Protegidos.* Dirección General de Investigación y Formación Agraria, Hortimed, FIAPA, Cajamar, Almería, Spain, pp. 149–168.
Soriano, T. (2002) Validación de un modelo de cálculo de transmisividad a radiación solar directa en invernadero mediante maquetas a escala y determinación del prototipo óptimo para la costa mediterránea. PhD thesis, Universidad de Almería, Almería, Spain.
Soriano, T., Hernández, J., Morales, M.I., Escobar, I. and Castilla, N. (2004a) Radiation transmission differences in east–west oriented plastic greenhouses. *Acta Horticulturae* 633, 91–97.
Soriano, T., Montero, J.I., Sánchez-Guerrero, M.C., Medrano, E., Antón, A., Hernández, J., Morales, M.I. and Castilla, N. (2004b) A study of direct radiation transmission in asymmetrical multi-span greenhouses using scale models and simulation models. *Biosystems Engineering* 88(2), 243–253.
Spanomitsios, G.K. (2001) Temperature control and energy conservation in a plastic greenhouse. *Journal of Agricultural Engineering Research* 80(3), 251–259.
Stanghellini, C. (1992) Evapotranspiration in greenhouses with special reference to Mediterranean conditions. *International Society for Horticultural Science (ISHS) International Symposium on Irrigation of Horticultural Crops.* Almería, Spain. *Acta Horticulturae* 335. ISHS, Leuven, Belgium, pp. 295–304.
Stanghellini, C. (2003) El agua de riego: su uso, eficiencia y economia. In: Fernández, M., Lorenzo, P. and Cuadrado, I.M. (eds) *Mejora de la Eficiencia del Uso del Agua en Cultivos Protegidos.* Dirección General de Investigación y Formación Agraria, Hortimed, FIAPA, Cajamar, Almería, Spain, pp. 25–36.
Stanhill, G. (1980) The energy cost of protected cropping: a comparison of six systems of tomato production. *Journal of Agriculture Engineering Research* 25 145–154.
Steduto, P., Hsiao, T.C. and Fereres, E. (2007) On the conservative behavior of biomass water productivity. *Irrigation Science* 25, 189–207.
Stewart, B.A. and Nielsen, D.R. (1990) *Irrigation of Agricultural Crops.* American Society of Agronomy, Crop Science Society of America, Soil Science Society of America, Madison, Wisconsin.
Stitt, M. (1991) Rising CO_2 levels and their potential significance for carbon flow in photosynthetic cells. *Plant, Cell and Environment* 14, 741–762.
Sutcliffe, J. (1977) *Las Plantas y el Agua.* Ediciones Omega, Barcelona.
Swinkels, G.L.A.M., Sonneveld, P.J. and Bot, G.P.A. (2001) Improvement of greenhouse insulation with restricted transmission loss through zig-zag covering material. *Journal of Agricultural Engineering Research* 79(1), 91–97.
Takakura, T. (1989) *Climate Under Cover.* Laboratory of Environmental Engineering, University of Tokyo, Japan.
Tanji, K.K. (1980) *Agricultural Salinity Assessment and Management.* American Society of Civil Engineers, New York.
Tanny, J., Cohen, S., Grava, A. (2006) Airflow and turbulence in a banana screenhouse. *Acta Horticulturae* 719, 623–630.
Teitel, M. (2006) The effect of screens on the microclimate of greenhouses and screenhouses: a review. *Acta Horticulturae* 719, 575–586.
Terés, V. (2000) El riego en sustratos de cultivo. *Horticultura* 147, 16–30.
Tesi, R. (2001) *Medios de Protección para la Hortofruticultura y el Viverismo.* Ediciones Mundi-Prensa, Madrid.
Thompson, R. and Gallardo, M. (2003) Programación de riego mediante sensores de humedad en suelo. In: Fernández, M., Lorenzo, P. and Cuadrado, I.M. (eds) *Mejora de la Eficiencia del Uso del Agua en Cultivos Protegidos.* Dirección General de Investigación y Formación Agraria, Hortimed, FIAPA, Cajamar, Almería, Spain, pp. 375–402.
Thompson, R.B., Gallardo, M. and Giménez, C. (2002) Assessing risk of nitrate leaching from the horticultural industry of Almería, Spain. *Acta Horticulturae* 571, 243–254.
Tognoni, F. and Serra, G. (1988) Biological aspects of energy saving in protected cultivation. *Acta Horticulturae* 229, 17–20.
Tognoni, F. and Serra, G. (1989) The greenhouse in horticulture: the contribution of biological research. *Acta Horticulturae* 245, 46–52.

Toki, T., Ogiwara, S. and Aoki, H. (1978) Effect of varying night temperature on the growth and yield of cucumber. *Acta Horticulturae* 87, 233–237.

Tompson, J.F. (2003) Pre-cooling and storage facilities. In: Gross, K.C., Yang, C.Y. and Saltveit, M. (eds) *The Commercial Storage of Fruits, Vegetables and Florist and Nursery Stocks.* USDA Agriculture Handbook no. 60. USDA, Washington, DC, pp. 22–31.

Tompson, J.F. and Kader, A.A. (2003) Wholesale distribution center storage. In: Gross, K.C., Yang, C.Y. and Saltveit, M. (eds) *The Commercial Storage of Fruits, Vegetables and Florist and Nursery Stocks.* USDA Agriculture Handbook no. 60. USDA, Washington, DC, pp. 55–58.

Torrente, R.G. (2000) Posibilidades de expansión hacia nuevos mercados. In: Fernández, M. and Cuadrado, I.M. (eds) *Comercialización de Productos Hortofrutícolas II.* Dirección General de Investigación y Formación Agraria de la Junta de Andalucía, FIAPA, Caja Rural, Almería, Spain, pp. 455–473.

Tsujita, M.J. (1977) Greenhouse rose spacing under high intensity supplemental lighting. *Canadian Journal Plant Science* 57, 101–105.

Urban, L. (1997a) *Introduction a la Production Sous Serre: la Gestion du Climat* (Tome 1). Tec-Doc, Paris.

Urban, L. (1997b) *Introduction a la Production Sous Serre: L'irrigation Fertilisante en Culture Hors Sol* (Tome 2). Tec-Doc, Paris.

Urban, L. and Langelez, I. (1992) Effect of high pressure mist on leaf water potential, leaf diffusive conductance, CO_2 fixation and production of Sonia rose plants grown in rockwool. *Scientia Horticulturae* 50, 229–244.

Valls, M. (2002) Formación y gestión del conocimiento en la agricultura Mediterránea. In: Alvarez-Coque, J.M.G. (ed.) *La Agricultura Mediterránea en el Siglo XXI.* Instituto Cajamar, Almería, Spain, pp. 245–261.

Van Berkel, N. and Verveer, J.B. (1984) Methods of CO_2 enrichment in the Netherlands. *Acta Horticulturae* 162, 227–231.

Van de Braak, N.J. (1995) Heating equipment. In: Bakker, J.C., Bot, G.P.A., Challa, H. and Van de Braak, N.J. (eds) *Greenhouse Climate Control: an Integrated Approach.* Wageningen Pers, Wageningen, The Netherlands, pp. 171–179.

Van der Velden, N.J.A., Jansen, J., Kaarsemaker, R.C. and Maaswinkel, R.H.M. (2004) Sustainability of greenhouse fruit vegetables: Spain versus the Netherlands; development of a monitoring system. *Acta Horticulturae* 655, 275–281.

Van Eimern, J., Karachon, R., Razumova, L.A. and Robertson, G.W. (1984) *Windbreaks and Shelterbelts.* Technical Note no. 59. World Meteorological Organization, Geneva.

Van Eysinga, J.P. and Snilde, K.W. (1981) *Nutritional Disorders in Glasshouse Tomatoes, Cucumbers and Lettuce.* Centre for Agricultural Publishing and Documentation, Wageningen, The Netherlands.

Van Meurs, W.T.M. (1995) Artificial light control. In: Bakker, J.C., Bot, G.P.A., Challa, H. and Van de Braak, N.J. (eds) *Greenhouse Climate Control: an Integrated Approach.* Wageningen Pers, Wageningen, The Netherlands, pp. 240–241.

Van Straten, G. and Challa, H. (1995) Greenhouse climate control systems. In: Bakker, J.C., Bot, G.P.A., Challa, H. and Van de Braak, N.J. (eds) *Greenhouse Climate Control: an Integrated Approach.* Wageningen Pers, Wageningen, The Netherlands, pp. 249–261.

Van Uffelen, R.L.M., Van der Mass, A.A., Vermeulen, P.C.M. and Ammerlaan, J.C.J. (2000) TQM applied to the Dutch glasshouse industry: state of the art in 2000. *Acta Horticulturae* 536, 679–686.

Varlet-Grancher, C., Bonhomme, R., Chartier, M. and Artis, P. (1982) Efficience de la conversión de l'energie solaire par un couvert vegetal. *Ecología Plantarum* 3, 3–26.

Verhaegh, A.P. (1988) *Kostprijzen Tomaat, Komkommer en Paprika in Nederland en Spanje.* Mededeling 611. LEI-DLO, Den Haag, The Netherlands.

Verhaegh, A.P. and de Groot, N.S.P. (2000) Chain production costs of fruits vegetables: a comparison between Spain and the Netherlands. *Acta Horticulturae* 524, 177–180.

Verhoeven, J.Th.J., Roenhorst, J.W., Lesemann, D.E., Velasco, L., Ruiz, L., Janssen, D. and Cuadrado, M.I. (2003) Southern bean mosaic virus the causal agent a new disease of *Phaseolus vulgaris* beans in Spain. *European Journal of Plant Pathology* 109, 935–941.

Verlodt, I. and Verchaeren, P. (2000) New interference film for climate control. *Acta Horticulturae* 514, 139–146.

Vermeiren, I. and Jobling, G.A. (1980) *Localized Irrigation: Design, Installation, Operation, Evaluation.* FAO Irrigation and Drainage Paper no. 36. FAO, Rome.

Veschambre, D. and Vaysse, P. (1980) *Momento Goutte a Goutte: Guide Pratique de la Micro-irrigation par Gouteur et Diffuseur.* Centre Technique Interprofessionnel des Fruits et Légumes (CTIFL), Paris.

Viaene, J., Verbeke, W. and Gellynck, X. (2000) Quality perception of vegetable by Belgian consumers. *Acta Horticulturae* 524, 89–96.

Villalobos, F.J., Mateos, L., Orgaz, F. and Fereres, E. (2002) *Fitotecnia: Bases y Tecnologías de la Producción Agrícola.* Ediciones Mundi-Prensa, Madrid.

Villele, O. (1983) La serre, agent de modification du climat. In: *L'INRA et les Cultures Sons Serre.* Institut National de la Recherche Agronomique (INRA), Paris, pp. 21–27.

Vince-Prue, D. (1986) The duration of light and photoperiodic responses. In: Kendrick, R.E. and Kronenberg, G.H.M. (eds) *Photomorphogenesis in Plants.* Martinus Nijhoff, Boston, Massachusetts, pp. 269–305.

Von Elsner, B., Briassoulis, P., Waaijenberg, D., Mistriotis, A., Zabeltitz, C.V., Gratano, J., Russo, G. and Suay-Cortes, R. (2000a) Review of structural and functional characteristics of greenhouses in European Union countries. Part I: Design requirements. *Journal of Agricultural Engineering Research* 75, 1–16.

Von Elsner, B., Briassoulis, P., Waaijenberg, D., Mistriotis, A., Zabeltitz, C.V., Gratano, J., Russo, G. and Suay-Cortes, R. (2000b) Review of structural and functional characteristies of greenhouses in European Union countries. Part II: Typical designs. *Journal of Agricultural Engineering Research* 75, 111–126.

Vonk-Noordegraaf, C. and Welles, G.W.H. (1995) Product quality. In: Bakker, J.C., Bot, G.P.A., Challa, H. and Van de Braak, N.J. (eds) *Greenhouse Climate Control: an Integrated Approach.* Wageningen Pers, Wageningen, The Netherlands, pp. 92–97.

Wacquant, C. (2000) *La Construction des Serres et Abris.* Centre Technique Interprofessionnel des Fruits et Légumes (CTIFL), Paris.

Wang, C.Y. (2003) Chilling and freezing injury. In: Gross, K.C., Yang, C.Y. and Saltveit, M. (eds) *The Commercial Storage of Fruits, Vegetables and Florist and Nursery Stocks.* USDA Agriculture Handbook no. 60. USDA, Washington, DC, pp. 61–65.

Wang, S. and Deltour, J. (1999) Studies on thermal performance of a new greenhouse cladding material. *Agronomie* 19, 467–475.

Ward, G.M. (1964) Observations of root growth in the greenhouse tomato. *Canadian Journal of Plant Science* 44, 492–494.

Wardlaw, I.F. (1968) The control and pattern of movements of carbohydrates in plants. *Botanical Review* 34, 79–105.

Warren-Wilson, J. (1972) Control of crop processes. In: Rees, A.R., Cockshull, K.E., Hand, D.W. and Hurd, R.G. (eds) *Crop Processes in Controlled Environments.* Academic Press, London, pp. 7–30.

Watson, D.J. (1958) The dependence of net assimilation rate on leaf area index. *Annals of Botany* 22, 37–55.

Welles, G.W.H. (1999) Fruit quality of glasshouse cucumber (*Cucumis sativus* L.) as influenced by cultural factors. *Acta Horticulturae* 492, 113–119.

Welles, G.W.H., Jause, J. and Peron, J.Y. (1992) L'influence des techniques modernes de production et des varietes sur la qualité analytique des legumes de serre. *PHM Revue Horticole* 324, 43–54.

Whatley, J.M. and Whatley, F.R. (1984) *Luz y Vida Vegetal.* Ediciones Omega, Barcelona.

Wijnands, J. (2003) The international competitiveness of fresh tomatoes, peppers and cucumbers. *Acta Horticulturae* 611, 79–90.

Winsor, G.W., Scharz, M. and Baudoin, W. (1990) *Soilless Culture for Horticultural Crop Production.* FAO Plant Production and Protection Paper no. 101. FAO, Rome.

Wittwer, S.H. (1969) Regulation of phosphorus nutrition of horticultural crops. *Hortscience* 4, 320–322.

Wittwer, S. and Castilla, N. (1995) Protected cultivation of horticultural crops, worldwide. *HortTechnology* 5(1), 6–23.

Woodward, F.I. (1987) Stomatal numbers are sensitive to increase in CO_2 from preindustrial levels. *Nature* 327, 617–618.

Yanagi, T., Okamoto, K. and Takita, S. (1996) Effects of blue and red light intensity on phoyosynthetic rate of strawberry leaves. *Acta Horticulturae* 440, 371–376.

Zabeltitz, C.V. (1990) Appropiate greenhouse constructions for mild climates. In: *Proceedings of the International Congress on Plastics in Agriculture.* New Delhi, India, pp. 117–126.

Zabeltitz, C.V. (1999) Greenhouse structures. In: Stanhill, G. and Enoch, H.Z. (eds) *Greenhouse Ecosystems*. Elsevier, Amsterdam, pp. 17–69.

Zabeltitz, C.V. and Baudoin, W. (1999) *Greenhouses and Shelter Constructions for Tropical Regions*. FAO, Rome.

Zacarías, L. and Alferez, F. (2000) Avances en el conocimiento de los mecanismos de regulación de la maduración del fruto. In: Marrero, A. and Lobo, H.G. (eds) *Actas V Simposio Nacional y II Ibérico de Post-Recolección de Frutos y Hortalizas*. Instituto Canario de Investigaciones Agrarias, La Laguna, Tenerife, pp. 11–19.

Zhang, Z.B. (2006) Shading net application in protected vegetable production in China. *Acta Horticulturae* 719, 479–482.

Zhibin, Z. (1999) Update development of protected cultivation in mainland China. *Chronica Horticulturae* 39(2), 11–15.

Zuang, H. (1982) *La Fertilisation des Cultures Legumieres*. Centre Technique Interprofessionnel des Fruits et Légumes (CTIFL), Paris.

Zuang, H. (1984) Les aléas climatiques liés á la production quantitative et qualitative de la tomate. In: *Agrométéorologie et Productions Légumiéres. Les Colloques de l'INRA* no. 33. Institut National de la Recherche Agronomique (INRA), Paris, pp. 77–79.

Zuang, H. and Musard, M. (1986) *Cultures Legumieres sur Substrats*. Centre Technique Interprofessionnel des Fruits et Légumes (CTIFL), Paris.

Index

- soilless cultivation 170–171, 178
  - characteristics 171–173
  - closed systems 175–176
  - joint selection 175
  - salinity 176–177
  - substrate types 174–175
  - systems 171